EMERALD CITY

The Lamar Series in Western History

The Lamar Series in Western History includes scholarly books of general public interest that enhance the understanding of human affairs in the American West and contribute to a wider understanding of the West's significance in the political, social, and cultural life of America. Comprising works of the highest quality, the series aims to increase the range and vitality of Western American history, focusing on frontier places and people, Indian and ethnic communities, the urban West and the environment, and the art and illustrated history of the American West.

RECENT TITLES

Vicious, *by Jon T. Coleman*
Frontiers, *by Robert V. Hine and John Mack Faragher*
Revolution in Texas, *by Benjamin Heber Johnson*
Murder in Tombstone, *by Steven Lubet*
Making Indian Law, *by Christian W. McMillen*
Fugitive Landscapes, *by Samuel Truett*
Bárbaros, *by David J. Weber*

FORTHCOMING TITLES

The War of a Thousand Deserts, *by Brian Delay*
The Bourgeois Frontier, *by Jay Gitlin*
Defying the Odds, *by Carole Goldberg and Gelya Frank*
The Comanche Empire, *by Pekka Hamalainen*
The Far West in the Twentieth Century, *by Earl Pomeroy*
César Chávez, *by Stephen J. Pitti*
Geronimo, *by Robert Utley*

EMERALD CITY

An Environmental History of Seattle

Matthew Klingle

To Beth— who is doing the work to remind us that all histories + all places matter.

S.F. 2013
OAH

Yale University Press New Haven & London

"Duwamish Head" and "The Towns We Know and Leave Behind, the Rivers We Carry with Us," from *Making Certain It Goes On: The Collected Poems of Richard Hugo,* by Richard Hugo. Copyright 1984 by the Estate of Richard Hugo. Used by permission of W. W. Norton & Company, Inc.

Set in Minion type by Tseng Information Systems, Inc. Durham, North Carolina.
Printed in the United States of America by Vail-Ballou Press.

The Library of Congress has cataloged the hardcover edition as follows:

Klingle, Matthew W.
Emerald city : an environmental history of Seattle / Matthew Klingle.
p. cm. — (The Lamar series in western history)
Includes bibliographical references and index.
ISBN 978-0-300-11641-0 (alk. paper)
1. Human ecology—Washington (State)—Seattle Metropolitan Area. 2. Urban ecology—Washington (State)—Seattle Metropolitan Area. 3. Environmental degradation—Washington (State)—Seattle Metropolitan Area. 4. Conservation of natural resources—Washington (State)—Seattle Metropolitan Area. 5. Seattle (Wash.)—History. 6. Seattle (Wash.)—Social conditions. 7. Seattle (Wash.)—Environmental conditions. I. Title.
GF504.W2K55 2007
304.2 09797'772—dc22 2007025159

ISBN 978-0-300-14319-5 (pbk. : alk. paper)

A catalogue record for this book is available from the British Library.

This paper meets the requirements of ANSI/NISO Z39.48-1992 (Permanence of Paper).
It contains 30 percent postconsumer waste (PCW) and is certified by the Forest Stewardship Council (FSC).

10 9 8 7 6 5 4 3 2 1

To Lloyd and Dorothy, the past;
Connie, the present;
And Benjamin, the future

"Just to amuse myself, and keep the good people busy, I ordered them to build this City, and my palace; and they did it all willingly and well. Then I thought, as the country was so green and beautiful, I would call it the Emerald City, and to make the name fit better I put green spectacles on all the people, so that everything they saw was green."

"But isn't everything here green?" asked Dorothy.

"No more than in any other city," replied Oz; "but when you wear green spectacles, why of course everything you see looks green to you."

—L. Frank Baum, The Wonderful Wizard of Oz (1900)

"Seattle, the Emerald City. Seattle is the jewel of the Northwest, the queen of the Evergreen State, the many-faceted city of space, elegance, magic and beauty."

—Promotional slogan for the Seattle–King County Convention and Visitors Bureau, Seattle Post-Intelligencer, May 11, 1982

Contents

Photograph galleries appear following pages 76 and 172

Preface

This is a book about a particular place, but it is also about ideas that make and sustain all places. It is also a history with its own history. The ending did not turn out as I expected.

I stumbled upon my topic more than ten years ago. My inspiration came while driving across the Ship Canal Bridge. It was the rarest of occasions in Seattle: a blue-sky winter day. As my car reached the high point of the span, I saw the Space Needle to my right, Mount Rainier directly ahead, snowcapped peaks and open water in all other directions. It was a familiar view, but as a historian in training, I saw everything anew. Here, I thought, was a city that exemplified Americans' long desire to harmonize city and countryside. Having recently read William Cronon's pioneering study of Chicago, I began to think of Seattle in similar terms—not exactly as another nature's metropolis, perhaps, but instead as a city that was urban by nature.

Or so I thought. Having found my topic literally at my doorstep, I rushed to the archives. I began with the accounts of Seattle's builders and leaders who had reshaped the city's physical landscapes. Seeing Seattle through the eyes of engineers, landscape architects, and urban planners, all of whom literally moved water and earth and people, seemed to confirm my hunch that city and nature were reconciled. As I delved deeper, this reading soon proved too tidy. Behind every victorious account of a tamed river, a scenic park, or a flattened hill was another story of levees breached by muskrats, squatters catching fish and shooting birds, or landslides wiping out homes. The poor and minorities, like the Native peoples of Puget Sound, fought those who annihilated their homes, fishing and hunting grounds, and workplaces. Wealthier residents rallied to stop the earthmoving, pollution, and highways that threatened their scenery or pastimes. Other sources of resistance were less familiar. Often, nature itself pushed back and imposed its own limits. Sometimes, combined energies of people and environment played out in predictable ways—poisoned rivers, vanishing wildlife, dispossessed peoples—but in other instances, nature did not disappear, and the seemingly powerless did sometimes prevail.

My findings corroborated what many historians were discovering: that transforming nature often meant controlling people as well. But this conclusion still seemed too neat because, for all of their differences, antagonistic Seattleites were fighting over the same things. All wanted to harness nature to sustain community, and all had fierce attachments to the places they had transformed. Stepping back, I began to ask what about my drive had originally stirred me. My answer took me to the common but neglected ground that unites our attachments to city and nature. The history of Seattle, like the history of America, was the evolution of an ethic of place.

The quest to find such an ethic is long-standing in both urban and environmental history. Urban historians have explored how the rise of the American city has been a dynamic process of inclusion and exclusion unfolding in union halls, residential neighborhoods, public parks, and shopping malls. But few historians consider the role that nature has played in forging the places urbanites call home, or how it has been an instrument to define and enforce the idea of community. For their part, environmental historians have expanded what counts as history, but the field remains trapped, at times, in an analytical device of its own making. By juxtaposing nature and culture as pure categories, environmental historians have demonstrated the independence of nature as well as the consequences of human actions. Yet this technique often yields stereotypical stories of decline, resulting in what some criticize as environmentalist history. This is a problem given the growing evidence of the unsavory, often hidden history underlying the conservation and environmental movements.

The historical scholarship on environmentalism's blinders is forcing us to reject the tendency to treat nature and culture so simplistically, and it may be fostering a more complex approach toward history. Unfortunately, this perspective is still wanting in environmental and urban policy, but thinking historically is different from finding a usable past. To think historically is to see the world as always contingent, as an impure and imperfect product of human actions and environmental processes through time. If historians are to contribute toward the pursuit of a just and sustainable society, we need to show how and why history is relevant.

That is why we need histories that see humans and nature as tangled together, but we need something more. We need a new ethic of place, one that has room for salmon and skyscrapers, suburbs and wilderness, Mount Rainier and the Space Needle, one grounded in history. We alone are responsible for splitting nature from culture, and for injuring it and ourselves as a result. "One of the penalties of an ecological education," Aldo Leopold wrote, "is that one lives alone in a world of wounds." Perhaps, but one of the potential gifts of a historical education is knowing that some wounds heal in time or can be endured, and that we do not have to go it alone. History is no panacea, but thinking historically can help us live with the consequences of being imperfect creatures in an uncertain world.

Maps

Acknowledgments

Those who love books know that acknowledgments are like family trees. These roots run deep, the branches spread wide.

This project began in graduate school, but three teachers at Berkeley planted the seed: Paul Starrs, Mark Eifler, and Gunther Barth. At the University of Washington, Keith Benson and James Gregory pushed me to tackle big questions while reminding me to keep academia in perspective. John Findlay served as my co-adviser and became a good friend in the process. His stern but sensitive comments made me a better scholar. I hope that I have met his high expectations. Richard White was a magnificent mentor. His insights into so many fields, his attention to lucid writing and rigorous analysis, make him one of the best scholars of this or any generation. But he is also extraordinarily kind. He fielded phone calls at home, wrote countless critiques, and tolerated my many false starts. You have to know Richard personally to appreciate the man behind the name. Anne Whiston Spirn and William Cronon were never part of my doctoral committee, but they deserve near-equal billing for their attention and encouragement over the years.

Many institutions cultivated my work. The University of Washington's History Department awarded me several fellowships, teaching assistantships, and lectureships. Grants and fellowships from the James J. Hill Reference Library, the Western History Association, the National Science Foundation (SBR 9810796), the U.S. Environmental Protection Agency (U-915597), the National Endowment for the Humanities, the American Council of Learned Societies, and Bowdoin College funded the research and gave me time to write. A

Fletcher Family Research Award from Bowdoin paid for the illustrations and permissions. An Environmental Leadership Program Fellowship helped me to imagine how my research might contribute to an environmentalism grounded in diversity and justice. I am grateful to Sally Loomis, Angela Park, and Paul Sabin, plus past and present ELP staff, fellows, and trustees for their support. Thanks also to Sheryl Shapiro, who invited me to join the Longfellow Creek Watershed Council for an all-too-brief time. If a workable ethic is possible, it will begin in a place like Longfellow Creek.

I relied on many people to help unearth the evidence I found, from Seattle to Cambridge. I am especially indebted to Carla Rickerson, Karyl Winn, Gary Lundell, Richard Engemann, Lisa Scharnhorst, Kris Kinsey, Janet Ness, Nan Cohen, Sandra Kroupa, and Nicolette Bromberg of the Special Collections Division, University of Washington Libraries; Carolyn Marr, Howard Giske, and Mary Montgomery at the Museum of History and Industry in Seattle; Scott Cline, Ernie Dornfeld, and Anne Frantilla at the Seattle Municipal Archives; Philippa Stairs and Greg Lange at the Washington State Archives, Puget Sound Regional Branch; Leslie Beck and Deborah Kennedy at the King County Archives and Records Center; Dave Hastings and Patricia Hopkins at the Washington State Archives in Olympia; Joyce Justice and Valoise Armstrong at the National Archives and Records Administration, Pacific Northwest Branch; W. Thomas White and Eileen McCormack at the James J. Hill Reference Library in St. Paul; Ruth Anderson at the Minnesota Historical Society; and Mary Daniels at Harvard University's Frances Loeb Library. Patrick Farrell entered my life as my research assistant in Seattle, and stayed on as my friend. Another undergraduate, Kathryn Ostrofsky, helped me as well at Bowdoin, whose indefatigable librarians proved that even at a small college library in Maine, you really can get there from here. Several people outside academia also helped me: Phil Katzen of Kanji and Katzen, PLLC, with legal research, Robert McClure and Lisa Stiffler of the *Seattle Post-Intelligencer*, plus Sue Joerger of Puget Soundkeeper Alliance and Nat Scholz of the Northwest Fisheries Science Center with pollution data, and David Neiwert with the Freeman family and Bellevue.

At Washington, advice from many senior colleagues across campus, from history to zoology, propelled my research: David Beauchamp, George Behlmer, Gail Dubrow, Susan Glenn, Thomas Hankins, Bruce Hevly, Richard Johnson, Richard Kirkendall, Suzanne Lebsock, Robert Paine, Bill Rorabaugh, Charles Simenstad, Matthew Sparke, Lynn Thomas, and Robert Wissmar. I am particularly obliged to my fellow "U-Dub" graduate students, notably Jennifer Alexander, David Biggs, Ned Blackhawk, Kate Brown, Bonnie Christensen, Andrea Geiger, David Louter, Michael Reese, Amir Sheikh, Susan Smith, Rob Smurr, and Michael Witgen. Beyond campus, others shared their ideas, too, including Carl Abbott, Mansel Blackford, Amy Brown, Michael Ebner, Sarah Elkind, Caroline Gallacci, Greg Hise, Andrew Hurley, Frank Leonard, Lorraine McConaghy, Martin Melosi, Gregg Mitman, Janet Ore, Adam Rome, Douglas Sackman, Ted Steinberg, Joel Tarr, Ann Vileisis, Richard Walker, and William Wilson. I have presented material from this book at meetings of the American Historical Association, the Organization of American Historians, the American Society for Environmental History, the Western History Association, the Canadian Historical Association, and the International Conference of Historical Geographers. I also gave talks at the Chicago Historical Society, Lake Forest College, Colby College, the University of Minnesota, Princeton University, and Bowdoin College. I am indebted to everyone who shared their ideas on those occasions.

As the book came together, many colleagues and friends pushed me to think through my conclusions, including Thomas Andrews, Kathy Brosnan, Kent Curtis, Jared Farmer, Mark Fiege, Drew Isenberg, Ari Kelman, Linda Nash, Jared Orsi, Sara Pritchard, Jeff Sanders, Robert Self, Bruce Stadfeld, Ellen Stroud, and Bob Wilson. Kathyrn Morse, Jay Taylor, and Kim Todd read the entire manuscript. Three other scholars kindly offered their expertise as well, saving me from many blunders: Alexandra Harmon, William Seaburg, and J. Ronald Engel. Ari Kelman and Carl Abbott reviewed the manuscript for publication, as did several anonymous readers, and their thoughtful comments made this a better book. Coll Thrush's work on urban Indians, which he shared over the years, dovetails with my story. He once said that he could not wait to see our books together on the shelf. Now you have your wish.

The professional and the personal become inseparable as one writes, and many close friends sustained my spirits during the roughest parts of my journey: Matthew Booker, Liz Escobedo, Kathy Morse, Monica Rico, Jennifer Seltz, Jim Snell and Tara Russell, Ellen Stroud, Cam Turner and Christine Tebben, Margaret Paton Walsh, and Bob Wilson. Two families in Seattle—Norm Fox and Tracy Schmitz, and Matt Segal and Corrie Greene—offered up their couches during return trips, and Norm and Matt kept me sane by dragging me into the pub or the backcountry. Finally, to return a long overdue tribute: Jay Taylor has been a true friend.

I was fortunate to find a home at Bowdoin College, an institution that prizes excellence in research and teaching, and supports each generously. I have fantastic colleagues, particularly those in history and environmental studies, and several helped me finish this book by providing time or counsel: Rosie Armstrong, Chuck Dorn, Munis Faruqui, Dorothea Herreiner, Guillermo "Ta" Herrera, DeWitt John, Eileen Johnson, Josephine Johnson, Mike Kolster, John Lichter, Craig McEwen, Sarah McMahon, Jill Pearlman, Eric Peterson, Sam Putnam, Patrick Rael, Larry Simon, Susan Tananbaum, Dharni Vasudevan, Allen Wells, and Lindsay Whitlow. My students merit special mention for their smart questions and ideas, some of which found their way into these pages. All teachers should be so lucky. By chance, Seth Ramus, my oldest friend from Berkeley and a gifted neuroscientist, also wound up here. He has heard earfuls about my work over countless beers for years. Thanks for listening. The next round is on me.

Writers are nothing without editors to make them shine. Nancy Scott Jackson and Julidta Tarver encouraged me to picture this book long before I earned my degree. As a first-time author, I had the good fortune to work with Jeannette Hopkins, who deserves her formidable reputation. She saw things in my writing that I ignored or was afraid to see, and then encouraged me to wrestle with them fearlessly. At Yale University Press, Chris Rogers was enthusiastic from the start, buoying my confidence while demanding the very best in writing. Laura Davulis cheerfully answered my many questions and skillfully steered the book into print. Phillip King pruned my often overgrown prose, making me think or laugh with every cut, and Bill Nelson made the superb maps.

Some material here appeared previously in *Journal of the West,* copyright 2005 by ABC-CLIO, Inc., and in *Journal of Urban History,* copyright 2006 by Sage Publications, Inc. I am grateful for permission to reprint this material here. Some material in Chapters 5 and 6 appeared previously in the anthology *The Nature of Cities: Culture, Landscape, and Urban Space,* edited by Andrew Isenberg (2006), used with permission of the University of Rochester Press. I also thank W. W. Norton & Company and the Estate of Richard Hugo for their permission to reprint selections from two poems: "Duwamish Head" and "The Towns We Know and Leave Behind, the Rivers We Carry with Us."

To complete the opening metaphor, my family is the trunk that holds me up. My younger brother, David Bowman, is more than blood relation; he is one of my dearest friends and toughest critics. Linda Bowman, my mother, gave love and support, but her bravery in the face of the most horrifying circumstances inspires me daily. Larry Frost joined us when we needed his humor and patience. John and Jollin Chiang welcomed me into their large circle of kin long before ritual made it official.

I dedicate this book to three generations of my family. First, to my grandparents, Lloyd and Dorothy Bowman, whose lives spanned the twentieth century, beginning on the plains of eastern Wyoming and ending in the suburbs of San Diego and Salt Lake City. They lived the past that I study and teach. I miss them terribly. Second, to my wife and fellow historian, Connie Chiang, whose creativity and brilliance dazzle me. She endured this project for almost a decade, giving generously of her time and intellect so that I might finish it to the highest of standards even as she labored to finish her own remarkable book. For me, Connie is my true love, my best friend, my closest colleague, and my home. Finally, to our son, Benjamin Lloyd Chiang Klingle, who loves reading books even if the words do not yet make sense. Whenever I faced the fears that can cripple every writer, doubts made darker by nights without sleep, watching Ben discover his world reminded me why place matters.

Like planting and tending a tree, writing is an act of faith and love. The faults in this book are mine alone, but the fruits that it might bear are the collective work of everyone listed here, plus many others I have neglected to mention. Thank you all.

PROLOGUE

The Fish that Might Save Seattle

Every year the salmon come home. From the deep waters of the North Pacific, millions turn toward their natal streams, driven to reproduce. All salmonids, as scientists label the fish, spawn in freshwater, but a few species undergo a miraculous physiological metamorphosis as juveniles, then leave for the sea. There, feeding off the ocean's bounty, they grow rapidly before changing again as they return to mate and, for most species, die in freshwater. No other member of the family *Salmonidae* has been as successful at this evolutionary adaptation, called anadromy, as the genus *Oncorhynchus*, which includes the seven species of Pacific salmon.

As genetically diverse as the waters that birthed them, Pacific salmon have inhabited a great hourglass-shaped arc, from Russian Siberia to Japan's Kyushu Island and from the Canadian Arctic to Baja California. Across this great arc, runs of fish have returned again and again so long as the waters remained cool, clean, and unobstructed. Some salmon battled the rapids of the Copper River in Alaska or the mighty Columbia; others returned to the small rivulets that trickled down the forested slopes of Sakhalin or Vancouver Island. Over time, cities came to encircle some runs, and a few salmon went home to the cities where they were born.

Seattle is such a city. Sometime in late July or early August, occasionally September, spectators flock to the Lake Washington Ship Canal in Ballard, a neighborhood of Seattle, to watch the salmon's homecoming. Schoolchildren and the elderly, tourists and longtime residents, plan their visits and vacations around those occasions even if the exact day and time are uncertain. They descend a flight of stairs, below the waterline and beneath the locks, the massive concrete and steel gates that separate freshwater Lake Washington from saltwater Puget Sound. There, in this submarine cave, they file into an observation room, press their noses to the thick glass, peer into the blue-green waters, and look for flashes of silver. Others stand on platforms above the labyrinthine fish ladders to watch jumping salmon struggling to complete their migration marathons.

As these salmon return, they seem to swim out of nature and into culture, yet the fish are never truly apart from humans. For millennia they have been, in part, the products of our desires. In the past 150 years, however, humans have transformed their physical nature. The salmon in the Ship Canal, for example, are accidental fish swimming in an artificial river, dug nearly a century ago to redirect water and lower Lake Washington in the name of real estate development and flood control. The canal itself was not intended for salmon. But some fish are unintended beneficiaries. In this mimetic river, maintained by the United States Army Corps of Engineers, fish culturists raised sockeye salmon in fish hatcheries and planted them here just after the canal's opening. The project eventually established a permanent population of sockeye, but the native chinook runs that probably roamed throughout the watershed began to vanish, a paradox that came to haunt this city of salmon.

In March 1999, the National Marine Fisheries Service, the federal agency that oversees the nation's stocks, demanded action to halt ebbing chinook salmon runs. As of that year, Seattle and its neighbor to the south, Portland, Oregon, shared the dubious distinction of becoming the first urban areas in the United States to face an Endangered Species Act listing. More than a century of commercial and sport fishing in local rivers, Puget Sound, and the ocean had diminished salmon numbers, especially the chinook or king salmon, which was prized for its fighting abilities and its sweet flesh. Reliance on artificial hatcheries only worsened the problem by diluting the genetic vitality of wild populations already exposed to unpredictable oceanic changes, like El Niño, known formally to climatologists as the El Niño–Southern Oscillation, a periodic warming of sea temperatures that depletes nutrients, breaks food chains, and robs salmon of the provisions needed to complete their epic runs. When salmon did return, they sometimes could not find their streams. Storm sewers or culverts entombed creeks. Loggers and bulldozers hauled away the forests that kept spawning streams cool and young fish hidden from predators. Lakes and rivers became sinks for household waste and industrial pollution. Dams that generated electrical power or impounded water to irrigate farmlands also chewed up juvenile salmon in their turbines and penstocks. The Ship Canal lowered the water level of Lake Washington by more than ten feet, drying up entire rivers in the lake's basin in less than two years. Shopping malls and subdivisions, floating on tides of concrete and asphalt, smothered estuaries beneath parking lots and pavement.[1]

Endangered salmon, in the words of then-mayor Paul Schell, had become "the fish that might save Seattle" from itself. Thrown by fishmongers at the Pike Place Market to the cheers of photo-snapping sightseers, raised by schoolchildren in classrooms for release into nearby creeks, and emblazoned on the sides of city buses—salmon had become Seattle's totem. They were to the Pacific Northwest and its largest city what lobsters were to New England, blue crabs to Chesapeake Bay, and walleye to the Great Lakes. Yet salmon were much more than a symbol to Mayor Schell. Rescuing the fish, he believed, would protect what made Seattle unique: its postcard perfect beauty. Critics wondered whether salmon were worth the price. Unnecessary restrictions on development, skeptics argued, could crash Seattle's skyrocketing economy, then at the apogee of the high-tech boom. Besides, after decades of artificial propagation, who could really tell which salmon were born wild or came from a hatchery tank? Rob McKenna, a King County council member, elaborated this point most forcefully when he claimed on national television that the human costs of protecting so-called "wild salmon . . . in the city" were too high. James Vesely, editorial editor for the *Seattle Times,* offered a more ominous prediction. Saving salmon would change the Pacific Northwest irrevocably by sparking confrontations over "property rights, tough limitations on recreation time and space, and residential development." Little wonder that John Findlay called the notion of a regional identity based on wild salmon "a fishy proposition" and "little more than an ill-founded conceit."[2]

Seattle's troubles were also America's troubles. Their shared plight stemmed from a long and complicated history that places humans and their environment at an impasse. The history of salmon and the city is tangled, and, when Seattleites tried to pry the two apart, they found themselves confronting a history embedded in the city's very foundations. The resulting questions spin heads: Are cities natural? Are salmon unnatural? Are salmon raised in captivity wild? Can a wild salmon live in a man-made river? But to ask these questions is to fall into the dualism that pervades so much environmental thinking. Nature must be pure, yet historically speaking that has turned out to be an elusive state. The way to move beyond these dualities is to think of history, both human and non-human, as a process grounded in time and space.

History is inseparable from place. It is a complicated word, but when defined plainly, place is the stuff of memories, or all the sensory delights and fears connected to particular locations in time. Ultimately, place is our sense of home, of belonging. Put more abstractly, it is a spot on a map, or perhaps a territory, demarcated formally or informally and imbued with perceptions, some subtle and intangible, others seemingly obvious. Places exist in and imply "space," yet place differs from space in our familiarity with it. As the eminent geographer Yi-Fu Tuan explains, place is "an organized world of meaning." We live, link, and work in a place. Many people have long feared that modern technologies, from the railroad to the automobile to the Internet, have erased our connections to place, but these concerns tend to ignore how travel and communication can create new places in turn, even if some are more virtual than real. To cast the human mind into matter and create places is to make geographic ideas real and allow human values to carry their full weight. Place is an inescapable product of change over time. Place *is* history.[3]

Blinded by the daily demands of our busy lives, we tend to lose sight of how the earth we stand upon contains and embodies history. We live on shifting ground. Places emerge from the physical world, reformed by our actions so that, in turn, we are continually constrained or enabled by our interactions with one another. As a result, place making is neither disinterested nor innocent. Sometimes, this architecture of domination is invisible or benign. Our world is full of such invisible places, like the ubiquitous marketplace, which once denoted a bazaar, a center, or town square but now is ostensibly everywhere and nowhere. In the abstract, the marketplace is not a real "place" at all, like the old town square or today's farmers' markets, even if stockbrokers and real estate agents act as if it is. However, to some people, like a farmer joined to the land by debt and familial obligation, place matters; it is a location known and made real through work. Physical changes to that place—by drought or blizzard or pestilence—can ruin the farmer and the distant commodities trader alike. Making places, then, is more than just building a home or making a living. Place is never incidental. Struggles over where and how people labor, where they live, how they and their creations move about, and the ideas they hold are matters of grave importance—and all unfold in time and space.[4]

In Seattle, the human practice of creating places began long before the

American conquest, when Native peoples believed wisdom and power rested in a magical topography populated by living spirits that, if respected, would deliver health and sustenance. Driven by the imperatives of modernity and the engines of industrial capitalism, European and American invaders tried to erase these places intimately tied to the natural world. Their vision was to improve upon nature, to surmount it, to turn it into property. The transcontinental railroads, seeking easy access to the deep waters of Puget Sound, helped drive this process as they and individual entrepreneurs turned water into land for terminals and shipping docks. These changes yielded great benefits for those with capital to invest or political muscle to flex, but they unleashed manifold changes to the region's waterways and coastlines that would later drive some salmon runs to the brink of extinction. They also had grave consequences for the working poor, immigrants, and disenfranchised American Indians who considered the waterfront commons their home.

Rapid growth turned Seattle's waterfront into a confusing snarl of trestles and wharves crisscrossing cesspools and garbage dumps, spurring a new and potent caste of experts to remake the despoiled urban landscape. City engineers employed awesome new technologies—electricity, hydraulic cannons, gasoline-powered dredges—to uplift the environment and its citizens by remaking the city's physical fabric in the name of efficiency, sanitation, and social order. Landscape architects, steeped in the philosophy of Frederick Law Olmsted, the great builder of America's parks, carved public greenswards out of Seattle's remaining forests in hopes of guaranteeing social peace through artistic loveliness. For all of their aesthetic differences, the engineers and the landscape architects shared one critical assumption: a sense that nature altered was nature perfected and society harmonized. Both classes of experts considered themselves enlightened arbiters of capricious markets and erratic nature. Through their labors, Seattle, like any city, had become what Thomas Parke Hughes calls an "ecotechnological system." This system was the manifestation of its builders, and in many ways it was remarkably successful, yet for all of their triumphs the engineers and park builders, together with railroad magnates and realty speculators, had created a city of poisoned waters, wounded lands, and social conflicts.[5]

Progress alienated as it uplifted. Lamenting what they took to be the death of nature in the city, some Seattleites lit out for the country to hike, fish, or camp. For them, the pursuit of unspoiled nature was possible only in the mountains and waters surrounding Seattle. Others moved out of the city altogether into new suburbs sold as alternatives to city life. The rest, Seattle's poor and dispossessed, those without the economic means or political power to move or play, shouldered the blame for disappearing salmon, littered parks, dirty beaches, and impure waters. A new urban ecology of inequality had emerged. Urban environmental injustice was more than the product of preexisting inequities, however, because the amalgam of city and nature became the source of new disparities and conflicts that magnified the power of some humans over others. Yet in the history of Seattle, as in the history of any city, or any *place,* those lacking power were not wholly helpless. People on the margins often had the unexpected aid of nature in their struggles. Landslides and floods that afflicted the poor also afflicted the powerful, compelling authorities to push back the earth and hold back the waters for the good of the entire community. Waning salmon populations empowered previously neglected groups, such as many Puget Sound Indians, to confront state and federal agencies to address the crisis. Human history in Seattle was thus never entirely human.

We need to rethink what an attention to history, human and non-human, can offer. To do this we must fully understand the notion of an *ethic of place.* At its core, such an ethic links the necessity for social justice to the importance of protecting the environment. According to legal scholar Charles Wilkinson, this ethic rests upon the recognition that humans rely on "the subtle, intangible, but soul-deep mix of landscape, smells, sounds, history . . . that constitute a place, a homeland." The ethical obligations flowing from this recognition encompass not only the people of a given place but also the animals, plants, and elemental forces that comprise the non-human nature surrounding us. We must go beyond the root of the word "city," *civitas,* a term originally used by the Romans to denote the leading town of a conquered region, to see how true citizenship depends on valuing human and non-human alike.[6]

Emerald City traces this evolving ethic of place through the history of one major North American metropolis that has long been associated with the stunning splendor of unblemished nature. Its title has two sources. One is a marketing slogan from 1982 that described Seattle, the "Emerald City," as a "many-faceted city of space, elegance, magic." That slogan does not lie. At approximately 47 degrees north latitude, Seattle is the most northerly major city in the continental United States. Summer days last almost sixteen hours at the solstice, and winter nights are dark for that long in December. What keeps Seattle from being as cold as Minneapolis–St. Paul, Toronto, or Boston is the Pacific Ocean. The vast inland sea of Puget Sound pulls moist clouds and temperate air over the city like a damp gray blanket. Less rain falls in Seattle than in New York City, but it is often overcast or misty. The clouds moderate the temperature and turn the sunshine into dappled, liquid beams. Local television meteorologists use the term "periodic sun breaks" in their forecasts, day after partly cloudy day.[7]

Seattle sometimes seems more fluid than solid. The city is perhaps best seen from the water at dusk and the prime vantage point is from the deck of an eastbound ferry departing Bainbridge Island, Bremerton, or Vashon. As Coke-bottle green waters churn beneath the bow, the landmark Space Needle and approaching towers of glass and steel seem to float on a carpet of green and gray. Sea and lakes encircle the city, which is built on a narrow and hilly isthmus bordered to the west by Elliott Bay, a deep inlet of Puget Sound, and to the east by freshwater Lake Washington. The glacier-tipped Olympic Mountains retreat astern. To starboard looms Mount Rainier, one of the highest peaks in the contiguous United States, home to more permanent glaciers than any single mountain below the Canadian border. To port, on clear days, rises Mount Baker, nearly as majestic as Rainier. In between are the snowcapped and forest-cloaked Cascade Mountains. Even when clouds and rain obscure the peaks, this is a view that rivals the vistas from the headlands above San Francisco's Golden Gate Bridge or aboard ships approaching New York City. If the West has been the symbolic compass point for hope in the American imagination, Seattle can make a strong claim as the capital of that state of mind.

The second source for this book's title is the imaginary metropolis of L. Frank Baum's *The Wizard of Oz,* and like that fictional city, Seattle's allure has been its natural splendor. To Dorothy, Toto, and their traveling companions—the Tin Woodsman, the Scarecrow, and the Cowardly Lion—the Emerald City of Oz looked beautiful from afar. They were awed by the towers of jewels and glass that rose above fields of flowers and grain. When they entered its green gates, though, they discovered a city built by conscript labor and ruled by a capricious Wizard. Similarly, from the deck of an approaching ferry, Seattle is striking and ready to be emblazoned on postcards or slick brochures from the city's Convention and Visitors Bureau. But the history of Seattle, like the history of the West, is not an enchanted romance. History offers a more complicated understanding of this Emerald City, and it forces us to consider several core questions. For example, why have many in Seattle, as in other places, not joined in the making of their city, or why have their contributions gone unheard? More important, why did the technically impressive and far-reaching

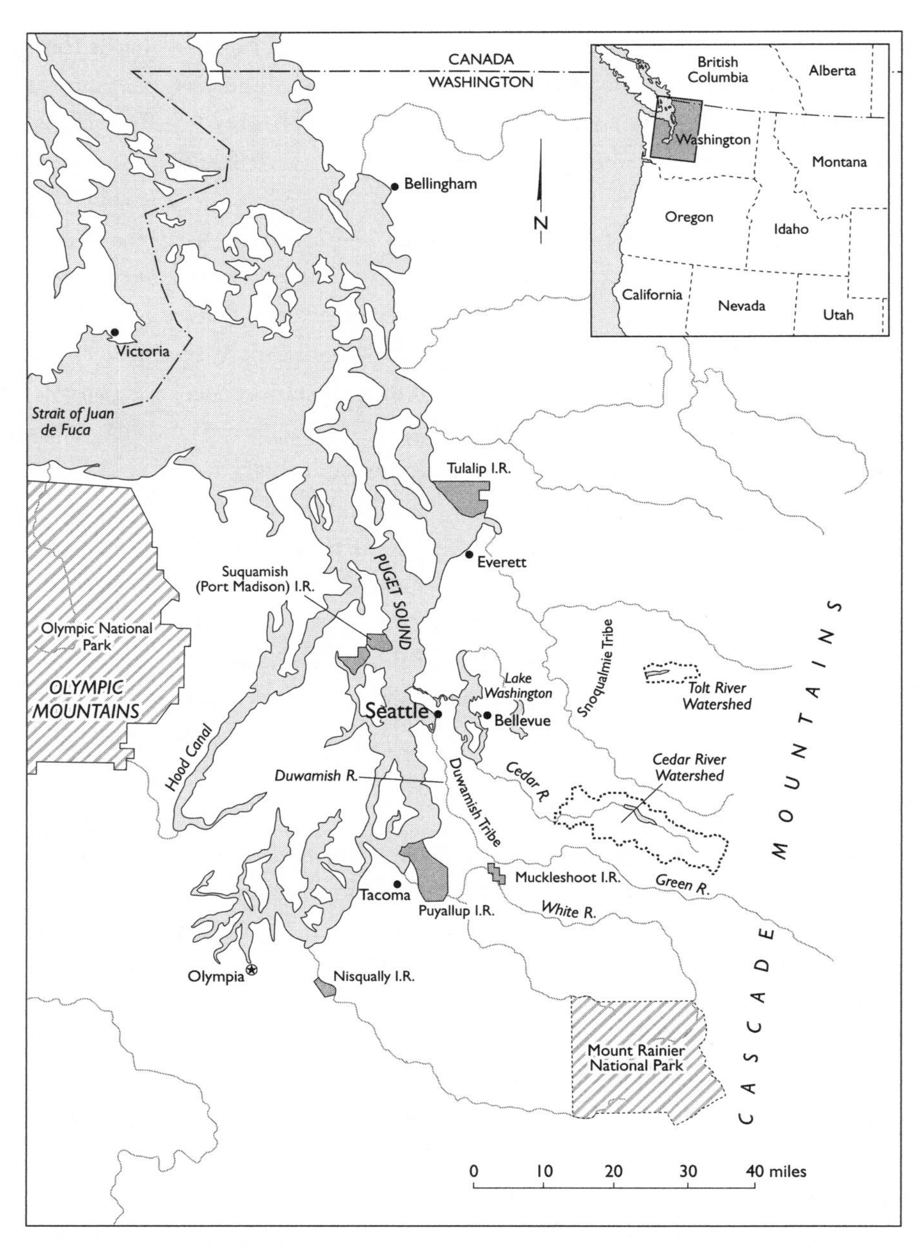

The Puget Sound region today, including Seattle municipal watersheds, selected Indian reservations (I.R.) and tribal territories, and major physical features.

changes to the physical fabric of Puget Sound generate an equally enduring landscape of injustice? And are there opportunities for a realistic kind of hope flowing from this past? To answer these questions demands expanding the scope of history to include nature. After all, history itself has been a chronicle of human achievements (and failings), and it is the singular pride of humans to place themselves at the center of the universe they did not alone create.[8]

Nature is not an agent in the way historians traditionally define the term. It is not a moral force, although some humans grant it such standing through religion or philosophy, but it is a physical agent. Nature is an integral part of the messy planet on which human action unfolds. It may not be an actor in the strictest sense, yet its actions force people to make moral choices—to dam or let flow running rivers, to protect or exterminate imperiled creatures, to clean or befoul poor neighborhoods. Nature, sometimes masked and disguised, constantly intrudes upon their actions, often because humans have underestimated their environments. Cities fall apart when earthquakes liquefy the filled shorelines on which they stand, or hurricanes weaken earthen levees holding back rivers or seas.[9]

Nature thus permeates the city, co-evolving with humans who aspire and struggle to control it, blurring any clear sense of where the biological ends and the cultural begins. This is especially the case with salmon. In the course of its journey, an individual fish, hatched and reared in a state-run hatchery or born in the gravelly bed of a stream emptying from a storm culvert into a city park, makes a remarkable journey. If the salmon survives gestation, it may grow in waters closely regulated by Seattle or, perhaps, King County, before traveling to sea and into international waters. If it eludes pods of orcas and packs of sea lions and roaming fishing nets, a returning adult may swim up the concrete-lined Ship Canal and into Lake Washington. Continuing on, it might charge upstream next door to a Microsoft executive's multimillion-dollar mansion in the high-priced suburb of Redmond, or, if it travels still farther, into the swift current of the Cedar River, the source of Seattle's drinking water.

Salmon trace a hopelessly knotty natural and human history. If the salmon turns into the Duwamish River, a once crooked waterway reengineered almost a century ago to flow in a straight line for industry and trade, it will pass some of the poorest neighborhoods in the urban Northwest and more than a half-dozen toxic Superfund sites. Where the fish returns matters to the people who try to catch it. Along the Duwamish, it may become entrapped in nets cast by a member of the Muckleshoot Indian Tribe, one of almost a dozen federally recognized and sovereign Puget Sound Native nations, or impaled on a fishhook cast by a Cambodian immigrant fishing illegally under the cover of twilight to feed her family. On the journey to or from Seattle, salmon, near the top of their food chain, ingest smaller creatures—aquatic insects, zooplankton, squid, herring—that contain the chemical residues of our modern computer-driven, automobile-dependent, vinyl and plastic-wrapped prosperity, including dioxins, polychlorinated biphenyls, or PCBs, fire retardants, and heavy metals. An Indian or a poor immigrant, standing even higher on the food chain, may ingest these pollutants after grilling the fish for dinner while a fly fisherman, in the name of sport, may let the fish go unharmed and thus escape poisoning. In their travels, salmon traverse Seattle's many geographies of inequality, where the legacies of the past have disconnected rich from poor, Indian from white, and immigrant from native born.

The majority of homecoming salmon never make it to the dinner plate. Seattleites, like most Americans, buy their salmon at the grocery store, packed in ice or in cans. It might be Pacific salmon netted in far-off Alaska and flown by jet to waiting customers, or captive Atlantic salmon raised like aquatic chickens in Puget Sound, coastal British Columbia, or the fjords of Norway or Chile. It is even hard to call Seattle's salmon wild anymore. Upon returning to Puget Sound, many salmon swim back to the hatcheries that birthed them, where state fish wardens will strip the eggs from females' bellies and the milt from males, and then inseminate and nurture the progeny in concrete ponds. In a city of salmon almost devoid of its original streams, humans now must run fish factories if they want to make certain that the salmon keep coming back.[10]

Seattle residents have betrayed both salmon and one another by failing to heed the historical consequences of their evolving ethic of place. In making choices, in trying to shape or control nature, Seattleites often tried to dominate one another through nature. Injustice was more than a political act or a state of being. It was manifest in the way Seattleites degraded (or tried to rescue) their environments. Injustice became specific to places over time, woven in the physical and social fabric of the city's neighborhoods. Salmon were more than the proverbial miner's canary, kept in a cage and carried deep into the earth to warn workers of poisonous gases. The fate of the salmon and the city were and remain linked in time, place, and the stories we tell. The history of Seattle, as of any city, is not a simple narrative of political and economic or social planning.

It is always a story of complex causes and complex effects constantly unfolding in place and through time.[11]

"In landscape," Anne Whiston Spirn writes, "ideas are tangible and values have consequences." Her claim speaks to the analytical heart of this book: place matters. The historian's challenge is to peel back the layers of time that envelop any place and ask which groups or individuals had the power to impose certain ideas and values before weighing the consequences. Ethics, like places, are not transcendent; they are immanent. This is one story of one place, Seattle, and the changing ethic of place in that great city, but in telling Seattle's story, this book rethinks the history and ethics of all American places, urban and otherwise. Ultimately, the dilemma of salmon in the city is a dilemma of our own making. Understanding that dilemma begins by finding the human nature in salmon and the non-human nature in the all-too-human city. It begins by coming back to place, as salmon return to their home.[12]

CHAPTER 1

All the Forces of Nature Are on Their Side

The Unraveling of the Mixed World

Salmon swam through the stories the first peoples of Puget Sound told themselves. They were more than sustenance. At one time, salmon, like other animals, were more akin to people because salmon in their current form did not exist. They were one of many kinds of Animal People, protean beings that were neither entirely person nor creature who lived in the Myth Age before the arrival of humans. For eons beyond measure, the Animal People called this epic world their home, a place in which their powers and attributes were unbounded by time or space. Some could change size and shape at will, while others could converse with one another or fashion tools with which to build their homes. Then everything changed, suddenly. In some accounts, the demiurge responsible was a being called Changer or Transformer, who, as the storytellers later said, turned the Animal Peoples into animals, plants, or physical features—a mountain peak, a bend in a river, or a shoal in the ocean—then imbued them with personalities and traits befitting each. Changer thus prepared the way for the After Peoples, that is, humans.[1]

In one story collected in 1916 by Arthur C. Ballard, a self-taught anthropologist, Snuqualmi Charlie told him the tale of Moon, the Transformer. While still a baby, the Dog Salmon people kidnapped Moon from his home in the Sky Country, brought him down to earth, and raised him as one of their own. When Moon came of age, he recognized his powers and began to prepare the world for a great change: "The new generation is coming now and you shall be food for the people, O Dog Salmon!" In the initial confusion during creation, Moon told the Dog Salmon people to go downstream, then, wondering if he had made a mistake, told them to come back upstream. Had he not made that error, the salmon would not have run down to the ocean and back again. Moon then set about transforming the rest of the world, frightening the many Animal Peoples who did not want to be changed. When Moon encountered Deer, he was sharpening a spear. Moon asked Deer what he was doing. "Making a weapon to kill the Transformer," Deer replied, whereupon Moon turned Deer into a deer, saying, "You shall be something good to eat." In this manner, Moon peopled the land with animals familiar to humans: a group of noisy fishermen became "sawbill ducks," or Merganser, a particularly greedy fisherman became Otter, and a thief who stole Moon's salmon dinner, roasting on a fire, became Wildcat, the soot of the flames turning into stripes on his face. After Moon had changed the world and before he created light, the Sun, he created the rivers, placing a man and a woman on each and telling them "fish shall run up these rivers; they shall belong to each people on its own river." "You shall make your own living from the fish, deer and other wild game," Moon commanded, and the couples had children "until many people were on these rivers." Without salmon, the After Peoples would not have endured and they knew this.[2]

Telling stories about Salmon's origin was more than mere allegory or a simple sign of respect toward the fish. The stories sustained the world and the world sustained the stories over time. Salmon had power in the deep past, when they lived in the Myth Age, and they still had power in the present, too. Animal Peoples were transformed, but they did not disappear. After Moon changed the Dog Salmon, they remained both animal and people, spending part of their lives in the Dog Salmon world, wherever that was located, before returning to the human world. Indeed, describing what the characters of the Myth Age looked like is problematic because there was no "nature" as far as Native peoples understood their world. The dualities we take for granted in the modern era did not apply. Instead, the world was a shifting continuum in which the living past and the living present connected. For many Indians today, this is still the case. To see the stories or any other part of Native spiritual life through Western eyes flattens the historical complexity of Native culture. Social subsistence and spiritual survival intertwined in Natives' daily lives, yielding a readily understood ethic grounded in place and practice. But this system would change as new peoples arrived, first from Great Britain, later from the United States, looking for land and other resources, motivated by a very different kind of ethic. They would insert themselves into this mixed world with unanticipated and far-reaching results.[3]

At first, Indians, as the interlopers called the inhabitants of the world Moon had transformed, accepted these new peoples, if somewhat warily. The interlopers also wanted to cooperate. After all, Indians knew where to hunt game, catch fish, and trap furs. They knew the lands and waters in detail and that intimacy translated into leverage in dealings with the newcomers who now lived among them. They knew where power dwelled. Natives incorporated these strangers into what became another sort of mixed world, one where Indian and non-Indian cultures and customs blurred and blended. It would be a short-lived accommodation for, as the tales told, change was constant and by the time the Americans came to the inland sea they called Puget Sound, bent on acquiring land, cutting trees, and building a city of their own, the mixed worlds Moon had created were beginning to unravel.

The Geography of the Transformer

Geologists say that Puget Sound began in fire and ice. Starting in the late Eocene, the slow process of continental drift pushed the Pacific Plate beneath North America, supplying the tectonic and volcanic energy that uplifted the Cascade Mountains and builds them still. In the early Pleistocene, the first wave of ice sheets, more than three thousand feet thick, began to carve out the deep fjords that became Puget Sound. Smaller grooves became basins for freshwater lakes, flat valleys for streams: Green Lake, Lake Union, Lake Washington, the Duwamish River, and the scores of small creeks scattered across Seattle today. The last ice sheets, the scientists say, began to retreat around fifteen thousand years ago, one mile every twenty-five years, leaving behind huge mounds of debris, glacial moraines that became islands in the Sound and ridges flanking its coasts. Soon after the ice receded, Mount Rainier erupted in the Osceola Mudflow. The paroxysm of ash and lava and mud swept north toward the sea almost six thousand years ago, inundating the foothills and coastline in a wall of rubble.

Eventually, the mountain waters pushed their way back to the ocean to create a tangled watershed encompassing nearly 4,100 square miles from the high Cascades, down the White River and the Green River, and into the flat lowlands. Here, the gradient mellowed, the current slowed, and the rivers either joined the Stuck River, which elbowed south into the Puyallup River near present-day Tacoma, or the Green and White Rivers, which flowed north to join the Duwamish and empty into Elliott Bay. The Green and White channels were often only a few hundred feet apart. Periodic seasonal flooding also sent the waters of the Stuck into the White River any time logjams plugged the path. North of this snarled intersection, the Black River, a short and slow watercourse, more swamp than stream, emptied Lake Washington into the Duwamish River. In most years, the waters reached Elliott Bay. In other years, when the tides were high, the Black and Duwamish sometimes flowed backward into Lake Washington.[4]

The source of all this water was the confluence of even larger rivers of air overhead that delivered the Northwest's now-infamous rain. Air currents above the North Pacific Ocean collided with one another, generating moisture-laden storms, especially during the autumn and winter, that smashed into the Olympic and Cascade Mountains, rising steeply to elevations above nine thousand feet only a few miles from sea level. As the clouds struggled to surmount the peaks, they dumped snow and rain, as much as one hundred inches or more annually, nurturing the thick coniferous forests that spread across the land in the glaciers' wake. The resulting runoff swelled watercourses with silt and debris, engorging already swollen rivers to jump their banks and plow through anything in their path. The sea was just as unpredictable. The channeling effects of Puget Sound created extreme tides that pushed against freshwater tributaries. Tidal bores driven by fast-moving squalls raced up the Duwamish River all the way into Lake Washington, some twenty miles distant, overwhelmed the small Black River, and elevated the lake level by as much as ten to fifteen feet overnight.[5]

The land that became Seattle was just as unstable and intricate as the waters around the narrow isthmus, only seven miles across at its widest point, measured west to east. American surveyors, sent by the federal General Land Office to map the interior reaches of Puget Sound in the 1850s, found that running survey chains over the rocks and through the mud was nearly impossible. Entry after entry in their field books included the stipulation "not on a true line between sections." It was an understatement that turned geology into a wry aside. As the last glaciers receded toward the Canadian Arctic, the irregular topography characteristic of present-day Seattle took shape, a crazy-quilt of ravines and hills, the tallest almost five hundred feet high, interlaced with skeins of creeks, rivulets, and springs. The ridges and hills tended to run north to south, trailing the glaciers' advances and withdrawals, while gravity pulled the water oozing from the gravelly soil downhill. Some waters poured into Puget Sound, Lake Washington, Lake Union, or the many small ponds nestled between the heights. Some of it sank into hollows carved out by the great ice sheets, turning dry land into marshes or, over time, bogs filled with sphagnum moss and mats of thick peat. On the high prominences and headlands fronting the ocean, like Alki Point, winds and weather pounded the soil. Here, water was scarce or brackish. Hardy prairies and grasslands dominated instead of the widespread forests with thick undergrowth of ferns and salal, a native tough-leafed shrub, common in the less exposed lowlands.[6]

The verdure was deceiving because the retreating glaciers had pulled much of the nutrient-rich soil away with them, making growing conditions difficult for all but the toughest plants despite the frequent rain. For all of the jokes about Seattle being perpetually gray and rainy, precipitation has never been distributed equally across the city or the region. The complex maze of hills and valleys generates equally complex microclimates. Even today, one high-lying neighborhood, like Capitol Hill, can receive up to a foot more rain than the University of Washington campus, downhill and only two miles to the north. The meteorological properties of Puget Sound can magnify these effects. Sandwiched between the Olympic Mountains to the west and the Cascade Mountains to the east, clouds and precipitation can become trapped in the Sound's central and southern sections. Labeled the Puget Sound Convergence Zone, this phenomenon, most common from April to June but frequent during the rest of the year, ensures that the region is often cloudy. The city's yearly average precipitation is thirty-eight inches, less than New York or Boston, but annual variations, now or in the past, are not uncommon. Whenever ocean temperatures in the tropical eastern Pacific rise by more than half of one degree Celsius, an occurrence commonly known as El Niño, the resulting effects on sea and atmospheric currents can push seasonal weather systems south and bring widespread drought or destructive winter storms to the Pacific Northwest.[7]

Contrary to earlier theories of ecology, there was never a climax or steady state of affairs for the living creatures around Puget Sound. The constant instability of the physical environment ensured that. Indeed, the region's immense trees could not have grown to their spectacular heights without this turmoil. At the end of the last glacial period, the forests were dominated by lodgepole pine, a tree that loathed shade and reproduced quickly. As the climate warmed and grew wetter, Douglas fir and western hemlock began to predominate along with red cedar, which clustered in damp places. Douglas fir, in particular, thrived on chaos. The needles and dead wood at the foot of the giant firs often ignited huge forest fires that leveled tens of thousands of acres. The hot flames forced the firs to expel their seeds. Afterward, the nourishing ashes and abundant water made for fast recovery. To be sure, many firs and cedars lived for hundreds of years, later astounding European and American invaders unaccustomed to such huge trees. Captain George Vancouver, traveling south along the southern shore of the Strait of Juan de Fuca, just north of Puget Sound, in 1792 said the whole country "had the appearance of a continued forest extending as far north as the eye could reach." Later, as he journeyed south into Puget Sound, exploring the region for the Royal Navy of Britain, he described an "impenetrable wilderness of lofty trees, rendered nearly impassable by the underwood, which uniformly encumbers the surface." Yet Vancouver's notion of a static, unchanging wilderness was inaccurate. It compressed the deep scales of evolutionary time into human dimensions, ignoring how the forests were always changing in response to a volatile environment.[8]

The great trees that commanded Vancouver's attention were only one part of this complex environment. Pacific salmon, the region's signature creature, co-evolved with the changing forests and mountains to develop their unique reproductive cycles over time. Cold and clear waters shaded and protected by dense forests made for perfect habitat—save one crucial deficit. Well before human times, sometime back in the early Pleistocene, the various species of salmon had become anadromous. Over time, they began to spawn and die in freshwater lakes and streams after spending their adult lives at sea. Ancestral salmon were forced to acclimate to rising and falling ocean levels each glacial period, which cut the fish off from freshwater in one cycle, salt water in the next. Most biologists agree that Pacific salmon, members of the genus *Oncorhynchus* (from the Greek words *onkos,* "hook," and *rynchos,* "nose"), evolved from a more primitive form of the genus *Salmo.* When the last great ice sheets receded, sometime between twenty-five and fifteen thousand years ago, salmon began to colonize the rivers around the North Pacific. Today, there are seven principal *Oncorhynchus* species in western North America—pink or humpback salmon, sockeye salmon, coho or silver salmon, chum or dog salmon, chinook or king salmon—plus two additional anadromous species: steelhead trout, a saltwater rainbow, and cutthroat trout, which has a freshwater variant like the steelhead. (Sockeye also have a strictly freshwater variant commonly known as kokanee.) Unlike the Atlantic salmon, a species that also migrates to and from fresh and salt water, most Pacific salmon spawn once, then die. It is, perhaps, romantic to think that salmon die so their young might live, and there is some truth to this. After the final glaciers withdrew and as the climate began to moderate, the rain and snow typical of the Pacific Northwest leached vital minerals from the porous, nutrient-poor volcanic soils and swept them into the sea. These trace minerals, essential to the proper growth of young salmon, were the biochemical triggers that enabled the fish to switch from freshwater to salt water and back again. So several Pacific salmon species evolved over time to replenish what their environment could not provide. After growing fat on the riches of the cold North Pacific and returning to spawn, dying salmon, energies spent, fertilize their nurseries with the sea's mineral wealth.[9]

Out of evolutionary necessity, salmon have become creatures so critical to particular ecosystems that, according to some ecologists, if they decline or disappear, their absence may have widespread effects. As salmon spawn and expire in lakes and rivers around the North Pacific, their carcasses release phosphorous and nitrogen, plus other trace elements, to feed the nutrient-starved forests and other species along the food chain. Along the banks of the most fecund salmon streams, some scientists have found evidence that dead salmon give life to larger and more luxuriant trees, denser populations of insects and the fish that feed upon them, and more numerous predators, from eagles to bears, which, in turn, help to propel the cycle of life by leaving behind the remains of killed fish to rot. But like the forests that dying fish may help to sustain, the salmon populations of the Pacific Northwest were never static. Anadromy was an elegant evolutionary strategy, but it was often not enough to ensure reproductive success. Shifts in weather patterns, like periodic El Niño events, could, depending on their intensity, scope, and duration, disrupt salmon runs along the entire northwestern Pacific Coast. Elevated water temperatures might suppress colder upwelling currents transporting nutrients from the depths to the surface and disrupt the region's oceanic food web. Adult salmon, dependent on this web directly or indirectly, could suffer as a result. Other fish species, like hake and mackerel, favoring warmer waters, come north during El Niños and feed on juvenile salmon as they arrive in the ocean. This dynamic shapes the size and vigor of salmon runs to this day.[10]

Into this fusion of the biological and the geological came the first people, so the archaeologists say, arriving around the time of the last glacial stade, a period of glacial advance and retreat, about fifteen thousand years ago. Over time, they became the people anthropologists labeled as speakers of Whulshootseed. These people, considered members of the Puget Sound Salish by anthropologists, were the first tellers of the Changer or Transformer stories. Glaciers, volcanoes, and floods were not inanimate forces to them, but active and numinous beings. According to one oral history, the valleys between modern-day Seattle and Tacoma once lay beneath a shallow arm of the sea until the ocean receded, trapping several great whales in a deep landlocked lake. The whales grew tired of their cramped quarters and bored subterranean channels to the sea, releasing the lake's waters and creating the White, Green, Stuck, and Duwamish Rivers. Another tale describes a great battle between North Wind, a spirit force of cold and treachery who wanted to shroud the world in ice, and Storm Wind, who triumphed and brought moderate temperatures and seasons with his victory. The landscape today chronicles their epic struggles. A stone ridge on the bed of the Duwamish River is the remnant of North Wind's massive ice weir, built to deny returning salmon to Storm Wind's people living upstream. Three hills standing above the Duwamish River's valley floor, flanking North Wind's abandoned weir and his former village, mark his final and fateful clash with Storm Wind.[11]

Natives acknowledged that their world was in constant flux because landscapes themselves had powers and the physical environment possessed a kind of sentience. This was the legacy of Transformer, and Natives celebrated or feared the most unstable places as imbued with spiritual significance. Nearly any place (or anything) in the world could be a source of power. The challenge was to find where power resided and learn how to retrieve it, because some places had more power than others. On the western shore of Lake Washington lived a supernatural flying snake with retractable horns called Changes-Its-Face that could see in all directions at once. Shamans visited its home, near present-day Leschi Park in Seattle, to draw on Changes-Its-Face's fearsome strength, often at their own peril because the spirit could compel weak-willed humans to do wicked things. Along the Duwamish River, Natives called one place Hand Causing Ill Will, because a giant hand, missing several fingers, punched up from the muddy water to obstruct passing canoes. Indians believed they could tap into the power these places generated because the original inhabitants of the mythic era, like the Animal Peoples, still resided in the lands and waters. Humans could encounter these beings or spirits anywhere and in almost any form, from rough seas to vicious animals. Some spirits could enhance human abilities by increasing strength or bravery in battle, others had the power to kill or cure, but all were respected. Shamans invoked them in stories and ritual and song to harness their force. Individual families and entire bands placated particular spirits in the hope of protecting themselves from disease or hunger.[12]

Natives thus organized their lives to match the rhythms of the rivers rising and falling, the tides ebbing and flowing. The ethnographer Thomas Talbot

Waterman, who interviewed Indian elders around Puget Sound to collect place names, was astonished by their complex "geographical psychology" of space. The various Puget Sound Salish bands, later called "tribes" by whites, also gave names to places as a way of indicating how they used or shaped the land to their own purposes. Indeed, in many cases referring to specific locations, the names were the land and the land became the names. A sand spit jutting into Elliott Bay through the muddy tide flats, near where the stadium for the Seattle Mariners baseball team stands today, was called Little Crossing-Over Place. A bend in the Duwamish River lined with Oregon ash trees was named Much Paddle-Wood, and another curve just downstream, called Lots of Douglas Fir Bark, provided ample supplies of this vital material for starting and stoking fires. Names described the turn of the seasons and the flow of the waters, times to fish and to burn forests to cultivate salal, berries, or camas bulbs, or to stake nets to catch flying waterfowl.[13]

Natives saw themselves integrated into and dependent on the material world, but this did not mean that nature dictated their actions. Social standing mattered because not everyone had equal access to all places. Later European and American invaders mistakenly thought Indians had no sense of property. To the contrary, proximity to valuable resources shaped their property regimes, and the greatest concentration of natural wealth lay where the water met the land. Native bands tended to form around specific watersheds as told in the story of Moon, the Transformer. Kith and kin ties deeply intertwined with particular rivers, turning them into highways for transportation and conduits for food. The lower reach of Duwamish River, uncoiling into Elliott Bay, was home to the people later called the Duwamish; on the river's upper stretch, at the split from the Green River, lived the people called the Muckleshoot, a name later applied to the bands who moved to the federal Indian reservation on the Muckleshoot Prairie, north of Tacoma. All along the waterways flowing through their homes was a plethora of resources. Dense clumps of eelgrass, hidden beneath the tides at the Duwamish River's mouth, covered the mudflats, protecting salmon, rockfish, and crabs. Beneath the slick green mats, buried in the mud or clinging to exposed rocks, lived an astounding variety of mollusks: horse clams, cockles, mussels, native oysters, and the enormous geoduck clam. Salmon spawned on the gravelly beds of the Duwamish and Green Rivers, plus smaller tributaries. Wooded banks housed deer, game birds, and other quarry. Lowland forests, tall Douglas fir or red cedar trees, became canoes and homes. Natives around Puget Sound traded with one another and with more distant bands on the far side of the Cascades for things they could not acquire or make for themselves. Access to this wealth, shared with family and friends through exchange ceremonies, determined both survival and social standing. Ownership was not used competitively to force out one's rivals, as in the case of some European and American property arrangements, but rather to make known who had provided food and protection.[14]

Productive labor thus marked patterns of access enforced through customary use. Granting access yielded social power. Coming together to dig shellfish, net salmon, or enjoy amusements like the "bone game," a highly competitive sport in which individuals from two teams placed bets and then faced off with unmarked and marked bones, each trying to guess first in which hand the other held the unmarked bone, strengthened existing ties and provided opportunities for new alliances and friendships. It was an adjustable property regime, designed to fit changing physical and social conditions. To Native peoples, their own lives also deeply intertwined with the Animal People that had preceded them. Because animals possessed specific powers and granted gifts, the Puget Sound Indians honored their relations with these creatures through elaborate rituals and rites. This is not to say that Native peoples did not use animals to their advantage: in contests over specific fishing sites between rival kin groups, for example, particular Indian bands sometimes clashed, occasionally violently, for access and control. Natives did overexploit their resources, often to hurt rivals or enforce territorial boundaries. But such contests were usually short-lived because cooperation and magnanimity were truer and more lasting forms of social power.[15]

No creature was as central to this system among the original people of Puget Sound as the Pacific salmon. Deer, elk, waterfowl, shellfish, game birds, camas bulbs, berries, and other fresh and saltwater fish—these were important but, unlike salmon and other migratory fish, like herring, they did not appear in abundance at expected times of the year. Salmon was also the most versatile of all. Freshly caught and fire-roasted on wood planks, the fatty fish was a seasonal delicacy. Smoked or dried, it could be stored for lean times. With their very survival dependent on reliable runs, Natives developed an intimate knowledge of fish ecology and physiology. They knew that each of the various kinds of salmon had specific spawning requirements. Sockeye were the fattest because they spent their earliest years in lakes before returning to the ocean. Chum salmon were called dog salmon because they had big canine-like teeth or because their lean flesh was best suited for dogs; they spawned in late fall in streams near the ocean along with the less desirable pink or humpback salmon. Coho or silver salmon reproduced in small, fast-flowing streams every autumn. Chinook or king salmon, prized for its firm, sweet flesh, needed larger rivers far from the sea for its spring or fall runs. Sockeye also required big rivers for spawning as well as the cold lakes in which their smaller juveniles lived for one or two years, gaining size and strength before migrating out to the ocean.[16]

Stories told of the consequences humans faced if they abused the fish who were both food and putative kin. In one tale, Hado, the humpback salmon, worried at his first coming upstream, did not want to be caught and tossed aside on the bank for the dogs. He wanted to be dried for food, so he threw himself on the bank instead. It was his annual gift. If the people did not respect his offering, Hado could become angry and might bring sickness, like smallpox. Salmon appeared to die at the end of their spawning season, but Natives believed their souls went back to the ocean to return in the next cycle. In another account, a rash young man speared a chinook salmon and cast it ashore without proper reverence. When he repeated the mistake the next season, he fainted and died. After the shaman said that the salmon had taken the young man's soul in place of his own as punishment, the fishermen retired their bows and arrows during the chinook run. Unforeseen forces could sometimes prevent salmon from returning to spawn, a disastrous event, compelling humans to seek assistance from a spirit helper. One time, the Green River turned dry. The people sang to find relief, and one man said that Grandmother Bullhead (a common name for sculpin, a bony fish that preys on salmon fry and eggs) might have a solution. "We will give her half of the salmon eggs if she will help us," the people replied, and the man appealed to Grandmother Bullhead. She agreed to cooperate, donned her finest clothes, and went to the people in their village and began to sing; the rains returned, and so did the salmon. The people, grateful and surprised at her powers, paid her in salmon eggs as promised. These stories and others illustrate that the three guiding principles of Native fisheries before contact with Europeans and Americans were dependence, moderation, and respect—the foundations of what could be called a Native ethic of place.[17]

Although the stories scientists tell today and the stories Natives have passed down through the centuries are, in many ways, irreconcilable, both speak of a place in constant change where humans and nature intertwined. Where scientists saw an unstable earth and an ever-changing but evolving assortment of life as the arc uniting past to present, Natives saw the will of intelligent, immortal beings exerting their influence upon the world. Reciprocity was at the core of this relationship. Human beings had obligations to fulfill if they were to enjoy the world's abundance. In the words of anthropologist Jay Miller, the first peoples of Puget Sound lived in "an anchored radiance," a culture or way of life rooted in the land—an ethic of place learned through physical and emotional experience, rehearsed through ritual, and enshrined in storytelling.[18]

To Be Enriched by the Industry of Man

When, in May 1792, Captain George Vancouver of the Royal Navy nosed HMS *Discovery* into what he would later name Puget Sound, he was exhausted and depressed, rattled by an earlier accident in Hawaii that almost cost him his life. Vancouver's mood only worsened the farther north he traveled along the western North American coast. He bestowed his melancholy on the maps he made of territories he sailed past: Cape Disappointment, Destruction Island. But when he turned south to chart a vast inland sea to the east of Cape Disappointment, the terrain brightened his morale. The concurrence of forests and meadows and mountains and water evoked "certain delightful and beloved situations in Old England." Sailing past the eventual location of Seattle, Vancouver wrote:

> To describe the beauties of this region, will, on some future occasion, be a very grateful task to the pen of a skilful panegyrist. The serenity of the climate, the innumerable pleasing landscapes, and the abundant fertility that unassisted nature puts forth, require only to be enriched by the industry of man with villages, mansions, cottages, and other buildings, to render it the most lovely country that can be imagined; whilst the labor of the inhabitants would be amply rewarded, in the bounties which nature seems ready to bestow on cultivation.[19]

Vancouver was on a mission of empire. He had been sent to lay claim to the rich sea otter fur trade around Nootka Sound and establish a basis for British rule in the eastern Pacific. His journal and maps were instruments of conquest; he was writing lines on the land by naming and charting the region's physical features. He often imagined himself traveling in a vast and beautiful wilderness populated by ignorant brutes. Unlike the Indian peoples, who he viewed with detachment or contempt, Vancouver and the explorers who followed him did not see places. Instead, they saw unknown spaces to fill with meaning and make into locations their countrymen would understand. As he sailed along the coast of the deep inland sea, the names Vancouver conferred invoked Old England—Restoration Point, in honor of the British Crown; Mount Rainier, to recognize his friend and patron, Rear Admiral Peter Rainier; and Puget Sound, named after his faithful Lieutenant Peter Puget, who surveyed most of its shores and depths. He showed little interest in what the Indians had named their homes, names fashioned from deep knowledge and use of the land; he did not even recognize it as a home to anyone before him. In his testaments to beauty and utility, in the names he gave to the region, Vancouver used mapping to chart the imperial rule he assumed would follow.

Yet Vancouver also saw ugliness and misery. European explorers had already inflicted their arrival on Native bodies. When Vancouver exchanged pleasantries with local Indians, he observed the "indelible marks" of a "baneful disorder." It was smallpox. Like so many Indian peoples, those of Puget Sound had no acquired immunity to smallpox, measles, influenza, or a host of other infectious diseases Europeans and Americans brought with them. The Native peoples were the victims of what epidemiologists call "virgin soil epidemics." The first cycle of infection hit in the late 1700s, unleashed by Spanish traders in the Strait of Georgia, by Russian surveyors in the Kuril Islands, or by other Indians trading with whites on the Northern Plains who then carried the disease downstream along the Columbia River, then northward over land into Puget Sound. Natives in this region lived in some of the most densely populated communities north of Mexico, so the misery spread quickly. The exact origins of the first smallpox outbreak remain unclear, but for the next hundred-plus years, cycles of disease reduced the Native peoples of Puget Sound by as much as two-thirds, perhaps even more. Estimates of the pre-contact Indian population in the Pacific Northwest remain in dispute, but most scholars agree that the total number (including present-day Alaska, British Columbia, Washington, and Oregon) fell by an average of at least 80 percent across the entire region by the 1850s. Smallpox was far from the only killer. Influenza and malaria, the latter more common to the warmer southern part of the region, were also notorious. These epidemic diseases ripped families apart and undercut the ability of bands to fend off attackers or acquire food. Yet the diversity of Puget Sound Salish life, where the boundaries between bands shifted according to marriage and trading alliances, made Natives around the sound resilient. Families and bands often merged and reconstituted themselves when disaster struck, their separate identities part of a continuum.[20]

The pockmarks of the *Variola* virus, which scarred its victims with deep and painful boils, were the harbingers of another mixed world. By the second half of the eighteenth century, along a north-south axis, the Pacific Northwest sat at the crossroads of invasion. Tlingit, Kwakwaka'wakw, and other Native slavers periodically swooped down from the northern coast of what is now British Columbia in giant war canoes to seize captives from the villages of Puget Sound. Maritime incursions of Spanish, Russian, British, and finally American explorers looking for resources to exploit and lands to claim came after them. By the early nineteenth century, another east-west axis of invasion emerged as fur traders from the North West Company and the Hudson's Bay Company crossed the mountains looking to establish commercial connections between interior British North America and markets in Russian Alaska, Hawaii, and California. Earlier historians called it settlement by Europeans and Americans. It was more accurately called resettlement because the intruders, at first, moved in to live among Native peoples. Contrary to the descriptions of explorers like Vancouver, this was not empty land even if alien disease, a forerunner of colonization, had dislodged Natives from many of their ancestral places.[21]

At first, the influx of Europeans was limited and localized. The potential for trade in furs that had compelled Vancouver to visit the Pacific Northwest served as the basis for Great Britain's claim to the region, leading to the formation of the North West and the Hudson's Bay Companies. The North West Company, with less capital and fewer white employees, relied on Indian labor, both local bands and workers imported from Quebec and Upper Canada, to seek out fur-bearing animals and help run its operations on the far side of the Rocky Mountains. When the Hudson's Bay Company swallowed its smaller rival in 1821, it adopted the same practices for its outposts at Fort Vancouver on the Columbia River, and at Fort Nisqually and Fort Langley on Puget Sound. A slump in the fur trade in the 1830s prompted the Hudson's Bay Company's governor George Simpson and its chief factor John McLoughlin to diversify the operations beyond fur trapping into farming, fishing, and logging. This reorganization plan, driven largely by Native labor, boosted the company's fortunes. Native workers received pay in trade goods such as guns, worked metal, tobacco, and textiles, plus protection from rival bands and slavers from the north. By the early 1840s, it seemed as if everything in this new order was for sale: furs for export, provisions for trappers, and Native women for sex.[22]

Through the middle of the nineteenth century, the new trade networks around Puget Sound yielded a fluid and mixed culture where whites and

Indians could cooperate around common rules that governed intercultural relations. Just as corporate fur traders exploited Indian resources and labor, Natives exploited the "King George's Men," their name for Hudson's Bay employees. They borrowed and elaborated upon an existing pidgin or trade language from the Columbia River and Vancouver Island, called Chinook jargon, to facilitate communication amid the linguistic complexity of the Northwest coast. Many King George's Men also intermarried with Natives, since blood relations determined access to trade and resources; some Britons and their native wives communicated with each other only in Chinook jargon. Native peoples who participated in trade and intermarriage made occasional concessions to their new neighbors, such as working at the Hudson's Bay Company forts in exchange for food and protection from enemy bands. In doing so, they did not descend into dependency but instead folded wage labor and trade into their preexisting seasonal rounds.[23]

As of 1840, despite the desolation of disease, Indians still outnumbered Britons and their non-Indian employees in the Pacific Northwest. By then, Americans began pouring into the Oregon Country. At first they were a trickle, with the earliest coming by sea. Puget Sound Natives called these first Americans "Bostons" or "Boston men," in reference to the fur traders who had traveled to the Northwest, beginning in the 1810s, on clipper ships chartered by New England and New York merchants. In the 1830s, missionaries seeking to convert heathen Indians came west, most of them traveling overland across the continent. Reports from fur traders and missionaries, touting the region's climate and soil, publicized the Oregon Country to the rest of the nation. By the 1840s, homesteaders from New England and the Ohio River valley came west, eager for what they considered free land. They moved into the Willamette Valley, south of Fort Vancouver, and the Palouse Country of the upper Columbia River. The Oregon Country was still jointly occupied by the United States and Great Britain, an agreement negotiated in 1818, but the Americans proceeded to ignore diplomatic protocol. They cited manifest destiny, the implacable faith in limitless expansion, on behalf of their own national interest. In 1838, eager to assert its entitlement to the Oregon Country, the federal government dispatched the U.S. Exploring Expedition, or ExEx, under the command of Captain Charles Wilkes, to reconnoiter the jointly claimed lands north of the Columbia River, and then to circumnavigate the globe in a parade of naval power.

In 1841, the USS *Vincennes,* the ExEx flagship, dropped anchor in Puget

Sound and Natives paddled out to the ship, greeting the sailors in canoes flush with furs, fish, and venison to trade. Visiting ships were nothing new to Puget Sound Natives, and trade was an accustomed ritual for welcoming newcomers. Wilkes, a notorious martinet whose obsessive mistreatment of his sailors later earned him a court-martial, forbade his men to purchase the pelts, instructing them to acquire food in exchange for fishhooks, tobacco, cloth, and gunpowder. Wilkes wrote that the Indians were "greatly surprised that so large a ship should want no furs, and it is difficult to make them understand the use of a Man of War." Like Vancouver, Wilkes considered Indians "a lazy and vicious set and exceedingly dirty," but he, too, extolled Puget Sound's scenery even as he calculated and measured the landscape for its resources. Echoing Vancouver's earlier eulogy, Wilkes said that the smell of the pines, ferns, and flowers "savored of civilization." There was, perhaps, "no country in the world that possesses waters equal to these," where a "seventy-four gun ship could sail unimpeded and in safety." The tall trees, some more than thirty feet in circumference and one hundred feet high, were perfect for naval stores or lumber mills and, best of all, the Hudson's Bay Company had already mollified the Indians, or so Wilkes believed, making future "peaceful occupation an easy and cheap task."[24]

Wilkes's contempt for Indians surpassed Vancouver's in tone and volume. He repeatedly called Natives "beyond measure the most provoking fellows to bargain with that I have ever met," as "lazy lounging & filthy," prone to thievery and beggary. He saw Indians as a debased race that would move aside for civilization. His choleric temperament did not blind the naval officer to the need for diplomacy, and in early July 1841, stopping for provisions at Fort Nisqually, a Hudson's Bay Company outpost near present-day Tacoma, the ExEx crew celebrated Independence Day with a parade led by a drum and horn corps. Wilkes may have held the ceremony to impress upon both the British fur traders and their Indian neighbors the power of the United States. Afterward, at a banquet, the crew feasted on salmon and game while the Indians watched. According to a Native present at the celebration, Slugamus Koquilton, everyone "had a splendid good time" and he and others "went away saying that Wilkes and the Boston men were good." A change was at hand. By 1846, Great Britain realized that demography was stacked against it and signed a treaty with the United States to set the international boundary at the 49th parallel. By order of the British Government, the Hudson's Bay Company shut down its outposts and decamped north of the border to Canada. All of Puget Sound came under

American possession, though most western emigrants still chose to go south to the rich agricultural lands of the Willamette Valley, leaving the rugged and forested northern end of the new Oregon Territory for lumbermen and the determined homesteader. Some came north merely because they had come to Oregon too late.[25]

One such migrant was Arthur A. Denny, a surveyor from Knox County, Illinois, who turned his attention to Puget Sound when he learned that earlier emigrants had claimed all of the best land along the Willamette River. On November 13, 1851, Denny, sailing on the schooner *Exact* from Portland, Oregon, led a party of twenty-four men, women, and children to a rainswept beach in what is now West Seattle. After exchanging greetings with a band of curious but friendly Indians, Denny's wife, Mary Ann Boren, wept at the prospect of a log cabin home in the wilderness. Arthur Denny named the new settlement New York "Alki," meaning "by-and-by" in the Chinook jargon, to reflect the party's modest hopes. Contrary to popular belief, he was not the first to arrive. Others had already preceded him, lured north because of dwindling opportunities for homesteading to the south. In the summer of 1850, John Holgate, an Iowan living in Oregon, explored the Duwamish River by canoe, but did not claim any land. The following September, in 1851, a party of homesteaders led by Luther M. Collins, an Illinois farmer who had joined forces with a Dutch immigrant, Henry Van Asselt, and a German-American father and son, Jacob and Samuel Mapel, coming north from the now-crowded California gold fields, staked their claims in the Duwamish River valley, at the southern end of Elliott Bay, intending to farm.[26]

The next spring, Denny and his party paddled their canoes from their camps across Elliott Bay to found the town they christened in honor of an Indian leader who had greeted them on the beach. To his unfamiliar name, which sounded to them like "Sealth," the Americans added an extra syllable plus a hard vowel sound to create "Seattle." Arthur Denny and his companions would later imagine themselves as masters of their own fate, bravely hacking a city out of forested landscape that, as Denny remembered it, was so "rough and broken as to render it almost uninhabitable." But settlers' reminiscences tell a different story, too. August V. Kautz, a U.S. Army surveyor, wrote in his journal in 1853 that "when the tide is out the table is spread." That table was the acres and acres of mollusks living on the stinky, briny mudflats, exposed twice daily, that edged Puget Sound. The new arrivals from the Midwest had to learn that the clams and oysters Kautz referred to were delicacies. It was a hard lesson to learn. Denny and other settlers would have starved had they not relied on Natives to provide them with recognizable provisions—venison, berries, and potatoes—and expose them to new food, namely salmon and shellfish. During the first winter of 1852–53, Denny made "a canoe voyage" to an Indian village on the Black River, some fifteen miles to the south, a round-trip journey of several days over muddy tide flats and through marshy waters, "to get a stock of potatoes" and replace depleted stores of pork, flour, and bread. In the first years of the new settlement, the Americans traded not for profit but for their continued existence. Sometimes, Indians gave the Americans gifts of provisions. Other times, they struck deals in exchange for cloth or wrought metal, usually on the Indians' terms. One American settler, A. B. Rabbison, later told the historian Hubert Howe Bancroft, "we were guided entirely in dealing with the Indians by their own laws and not ours."[27]

Denny, Collins, and other Americans may have imagined themselves stepping ashore onto a new land, but to the Indians they met on the beach or along the Duwamish River, the Americans were just the latest in a long line of visitors. They expected the Bostons to behave as the King George's Men had. Many of the first Bostons had accommodated themselves to the Natives to gain access to resources and services. The entrepreneur Henry Yesler was one such Boston. An Ohioan of German descent, he was by turns ambitious and footloose, drifting from job to job, mining gold near Sacramento, cutting logs in Portland, before coming to Seattle to build a sawmill in October 1852. Discovering that all of the lands bordering Elliott Bay were taken, he persuaded two prominent landowners, David S. "Doc" Maynard and Carson Boren, to give him an easement to the waterfront. Boren and Carson, eager to bring investors and more residents to Seattle, willingly complied and Yesler ordered equipment for the mill shipped from San Francisco. Many of Yesler's original employees were Indians who lived intermittently in his bunkhouse between visits to their homes on the Duwamish River or along Elliott Bay, sawing logs and running the machinery that was helping Seattle to grow.[28]

Yesler adopted the customs of the country in other ways as well: he married a Native woman named Susan, the daughter of "Chief Curley," a Duwamish leader. Susan and Henry Yesler later had a daughter named Julia Benson Intermela. Yesler's legal wife, Sarah, joining him in Seattle in 1858, supported the bigamous arrangement. One possible reason was profit. Yesler received a twofold benefit, because under American law at the time, white men who married Indians were entitled to a double share according to the Donation Land

Claim Act of 1850. Sometime in the late spring of 1853, thanks to Indian labor and Indian relatives, a thin plume of smoke and vapor rose from Yesler's mill, spiraling above the rough-hewn log homes clustered along the forested shores of Elliott Bay. Yesler's mill and adjacent cookhouse, built of cedar shakes and clapboard siding, provided the first salaried jobs in the young town. The editors of the *Columbian,* a newspaper in Olympia, the territorial capital, predicted Yesler's enterprise would eventually attract "thither the farmer, the laborer, and the capitalist." As one local historian later wrote, the mill's "huffing, buzzing, and blowing steam made the music of the bay, and the hum of its saws was the undertone of every household."[29]

Yesler's mill did attract Indians to work and live nearby, something that George Gibbs, the federal Indian agent assigned to the Pacific Railroad Survey, noted in his 1854 report. "Whenever a settler's house is erected," he wrote, "a nest of Indian rookeries is pretty sure to follow if permitted." Proximity provided protection from rival bands and gave nearby Indians a sense of power. A dam and sawmill erected at the outlet of Lake Washington had unexpectedly improved the fishery there: the Indians boasted, Gibbs said, "as if it had been of their own construction." Gibbs misunderstood the Indians' actions. To them, mills and dams were proof of the Bostons' commitment to provide for their neighbors, but as more and more came to Puget Sound, the Americans' entrenched notions of racial superiority and Indian inferiority subverted preexisting and fragile diplomatic arrangements. It was a time-worn pattern in the history of American expansion westward, and the results in Puget Sound followed a similar trajectory. A Methodist minister, David E. Blaine, who had emigrated from Seneca Falls, New York, to Seattle, complained in a letter to his family back East that "moral and upright" white men, like Yesler, went native when they came to the Pacific Northwest. "They live with savages and as savages," because "when they left the states their only aim was to get rich." Blaine, never a popular minister, tended to dismiss his congregants, white and Indian, as ignorant and uncouth. His wife, Catherine Blaine, an avowed suffragist and one of the signatories to the Declaration of Sentiments and Resolutions of women's franchise passed at Seneca Falls in 1848, was equally dismissive of their new home and of the Indians as "at best a poor degraded race."[30]

In early 1853, the Blaines' worst assumptions came true when an Indian known as Mesatchie Jim, or "Bad Jim," as translated from Chinook jargon, murdered his own wife. A trio of white vigilantes, led by the Duwamish valley farmer Luther Collins, hunted down Mesatchie Jim and lynched him in downtown Seattle. In a town where the rule of law was weak and federal power still nonexistent, armed vigilantism was not uncommon, so local and territorial authorities tried to reestablish order by putting the three men on trial. After a short hearing, the court acquitted two and dismissed the charges against Collins. Some white Seattleites, like David "Doc" Maynard, who employed Indians in his many business enterprises, were dismayed with the outcome. They had reason to be concerned. The lynching unleashed a cycle of retribution said to have led to the killing of an emigrant, James McCormick, in July and, the following spring, the lynching of two Snohomish Indians by a mob of angry whites in downtown Seattle, again egged on by Collins. Only the forceful intercession of Carson Boren, now the local sheriff, who arrived after the first two hangings, saved the life of a third Indian. Collins, indicted and tried yet again, was never convicted, perhaps exonerated by the local jury because it saw the lynching as retaliation for both the death of McCormick and the murders of two other emigrants by the Snohomish the year before. "Whenever there was trouble," Henry Yesler recounted afterward, "it was the fault of some worthless white men" who, "instead of leaving the Indians to settle matters in their own way," incited the Natives to "revenge."[31]

More and more Bostons came to see Indian mores and livelihoods as improper, even destructive. They saw the profligacy of the Indians, who had never developed the resources surrounding them into thriving enterprises, as further proof of their inferiority, not understanding the complex relationship with the natural environment the Native Americans had achieved. One early American settler, Charles Prosch, reminisced that shellfish, "gathered and eaten at all seasons of the year . . . relished by Indians and whites alike," as well as the profusion of berries, game, and fish, had made "a paradise for lazy people who were content to live like Indians." Steeped in a Jeffersonian tradition of improving the natural world to secure economic progress and political independence, white Americans were repelled by what they took as lack of initiative, ignoring the centuries of Indians' shaping their lands through fire and cultivation to promote more abundant game and food plants. Unless they showed evidence of improvement themselves—putting up a home, clearing the forest, draining swamps—settlers could not file patents under the terms of the Donation Land Claim Act, which was intended to encourage rapid settlement in the contested Oregon Country and cement American rule there. Emily Inez Denny, Arthur Denny's niece, told of men in early Seattle who spent the day "clearing, slashing and burning log heaps, cutting timber, hunting for game to supply the larder," while the women, "with only primitive and rude appliances," upheld domestic order.[32]

Settlers maintained their entitlement to landed property, what local historian Clarence Bagley later called their "foothold of civilization upon the remote frontier," through daily toil and viewed the Indians' less frantic labor as a sign of moral inferiority. In contrast to their Indian neighbors, white Americans by and large enforced clear distinctions between ownership and access; private property was meant to be private. The first coastal surveys of Puget Sound in the 1840s, General Land Office surveyors in the 1850s, and individual settlers all inscribed property lines on the land. Maps and land claims divided the terrain into discrete categories: timbered plots, navigable rivers, shellfish beds, and homesteads. Surveyors relied on the technology of the town plat—the grid system created by Thomas Jefferson and made into law by the Northwest Ordinance of 1787—that layered the land into discrete blocks to ease surveying, settlement, taxation, and eventually social order. Jefferson's idealized arrangement, designed to encourage settlement in the trans-Appalachian West, unfolded like an imaginary net over the rugged Puget Sound country to turn rough-hewn nature into workable property. By 1876, sixty-nine separate plats covered the isthmus, all radiating outward from Yesler's mill and wharf.[33]

Land was one pillar of Seattle's economic foundation, and lumber was the other, but the lumber trade and the real estate market were often at odds. Lumber mills needed capital to finance and expand their operations and labor to cut and ship the tall trees. Both were in short supply in the isolated Pacific Northwest, and in Seattle, affordable real estate was becoming scarce, too. And Yesler soon had competition, because the massive trees augured lucrative profits for ambitious lumbermen. The Puget Mill Company at Port Gamble opened on the western edge of Puget Sound the same year Yesler's mill began operations, and another mill opened at Port Blakeley. The high cost and backbreaking labor of conveying logs over rugged land restricted Washington's burgeoning timber industry to the coastline, which, in Seattle, was quickly becoming the prime location to build.

Yesler's sawmill was now the social and industrial center of the town, but as development hemmed in his operations, he relied heavily on his Indian employees to gain the advantage over his competitors, as did almost every other Seattle business. Without Indian labor, the young town's economy would have collapsed. Indians packed hundreds of barrels of salmon for Doc Maynard's business; they pressed dogfish carcasses into oil to lubricate the region's sawmills and illumine lamps; and they helped the Bostons traverse the inlets and waterways of Puget Sound in their canoes, the "Siwash buggies," as one early resident later remembered. Natives helped build the town and made it run. Yet, to prospective real estate investors and homesteaders, the presence of Indians reduced the value and integrity of their property and propriety. The Blaines, the Methodist minister and his wife, were even more fearful after the cycle of murder and retaliation following Mesatchie Jim's lynching, and worried that consorting with Indians would be the end of their town. "We feel considerably alarmed for ourselves," Catherine Blaine wrote to her relatives. Charles Prosch wrote that cohabitating with Indians, as Yesler and others had done, gave "birth to a class of vagabonds who promised to become the most vicious and troublesome element in the population."[34]

The old world was slipping away, and Indians would change from being family and neighbors to strangers and trespassers, even enemies.

Indians, Retarding the Wheels of All True Progress

In the winter of 1851, cannon fire echoed across the cold waters of Puget Sound. The United States sloop-of-war *Vincennes,* the former flagship of the U.S. Exploring Expedition, had returned minus its former commander Captain Wilkes, who was later court-martialed for mistreating his crew during the voyage around the globe. One specific charge was his reckless order to cross the dangerous bar at the mouth of the Columbia River, a decision that had cost the expedition a ship, which sank in the bar's treacherous currents. A military tribunal later acquitted Wilkes of the charges but removed him from command. As the *Vincennes* nosed into Elliott Bay, he was in Washington, D.C., preparing the scientific reports from the ExEx voyage. Thanks in part to his efforts, all of the Oregon Country was under nominally American rule by this time, and Wilkes's former warship saluted the Denny party, encamped on the shores of Alki Point. The Americans were not the only witnesses to the show of firepower. According to Thomas Prosch, an early settler, the cannonade invoked "a strong and respectful impression upon the hundreds of Indians." The Denny party, "noticing the effect upon the Indians," pronounced the noise "music of a delightful character."[35]

In 1853, when Seattle formally incorporated as a town, fewer than two thousand American colonists lived among twelve thousand Native people, maybe more, across Puget Sound. That same year, after almost eight years of rancorous debate, Congress authorized the U.S. Army Corps of Topographical Engineers to launch four expeditions to survey possible routes for the first transcontinental railroad. Isaac Ingalls Stevens, an army officer who had resigned his commission to become the first territorial governor of Washington in 1853, rejoined the military to lead the northernmost expedition through the region between the 47th and 49th parallels from St. Paul, Minnesota, to Puget Sound and the Columbia River. One biographer called Stevens "a young man in a hurry," another "a little Napoleon." A graduate of Phillips Andover and West Point, where he was first in his class, and a wounded veteran of the Mexican-American War, Stevens threw himself into the railroad survey with characteristic energy. His engineering experience surveying and building coastal defenses in New England prepared him technically for the job he faced: how and where to bind the nation with tracks of steel.[36]

On Stevens's expedition were civilian scientists, such as the naturalist George Suckley, and military personnel, including Captain George B. McClellan, later the Union Army commander in the early years of the Civil War. As Vancouver and Wilkes had before them, they measured and recorded to make the locations they passed through landscapes useful for the railroad builders and homesteaders who would follow. When the survey reached Puget Sound in the late spring of 1853, they were astounded, as their predecessors had been, by the fecundity of its waters and mountains, with Suckley especially impressed by the fantastic salmon runs in the "cunningly-contrived aqueous labyrinth." Stevens praised the "capacious harbors" and "commodious" spaces for railroad construction and shipping terminals, perfect for future commerce with East Asia. George Gibbs, another civilian who described Indian populations for Stevens, also noted what previous settlers had already found: abundant coal deposits at the south end of Lake Washington that appeared "to burn well with strong flame, leaving no slag," perfect for locomotive boilers. Stevens seemed poised to achieve the fame he craved. Suckley complained that Stevens's future was "wrapped up in the success of the railroad making its Pacific terminus in his own territory," yet the Pacific Railroad Surveys faced problems from the start. They lacked competent railroad engineers; the estimates of grade and altitude and track mileage were inaccurate; measurements of mountain snowfall and rain were wildly optimistic; and sectional politics intervened. Secretary of War Jefferson Davis (the future Confederate president) backed the southern route along the 32nd parallel to San Diego, and infuriated northern congressmen refused to approve it. The transcontinental railroad remained trapped in congressional debates until the Civil War freed the Union to begin building the first route in 1862, a compromise, from Omaha to Sacramento.[37]

Stevens's greatest influence, however, was not surveying but treaty making with the Northwest's diverse and far-flung Native peoples. The American invasion of Puget Sound had preceded treaty negotiations, unfolding in relative peace and in slow motion, upending the prescribed system of westward American expansion in which treaty negotiations opened Indian land to surveyors, land agents, and claimants. The United States government had allowed emigrants to seize land it did not own, and under federal law, it was difficult to make such land titles valid. Treaties turned "unclaimed" land into the public domain, which could then become private property, thereby avoiding a legal mess for the federal government. Stevens also had to reach agreements with Indian groups and prevent further legal confusion since American law, from the beginning of the republic, had treated tribes as dependent sovereign nations. But Puget Sound's Natives were not discrete groups ruled by one or even several leaders; instead, numerous and hard-to-define bands moved seasonally out of villages and camps in the lowlands to the mountains and rivers and seaside to obtain game and fish. These diverse bands were often connected by blood and trade yet remained independent. Leaders who resembled chiefs often had limited authority over waging war, or over their own immediate relations. In his survey of the Indians of Washington Territory, Gibbs urged Stevens to achieve "the union of small bands under a single head" to prevent the inevitable "mischief" when "scattered and beyond control." Gibbs played to the young officer-engineer's sense for order, advising him to invent chiefs and create tribes where none existed.[38]

Beginning in December 1854, and working at a breakneck pace, Stevens negotiated three Puget Sound accords—the Treaty of Medicine Creek for the south, the Point Elliott Treaty for the northern and eastern reaches (including Seattle), and the Point No Point Treaty covering the Hood Canal to the Strait of Juan de Fuca. Heeding instructions from Washington, D.C., Stevens agreed to establish reservations for recognized tribes signing the treaties, but, in a nod to Indian insistence and businessmen like Yesler who depended on Indian labor, he agreed to locate such reservations close to industries. Yesler called on acting territorial governor Charles H. Mason, who served while Stevens was in the field, to let the local band along the lower Duwamish River remain. Stevens assumed that Indians who moved to reservations would eventually assimilate into white American society as hardworking farmers and homesteaders, yet he listened to Gibbs, who warned that moving coastal tribes away from their traditional fisheries would breed violence. Stevens thus promised the treaty tribes payments for their ancestral lands, and left unrestricted the right to fish, hunt game, pick berries, and harvest shellfish in their "usual and accustomed places" on any "open and unclaimed lands." In pomp-filled ceremonies, Stevens planned an entirely new Indian social landscape.[39]

Under the Point Elliott Treaty, people that Stevens knew as Duwamish and Snoqualmie could live on any of the reservations created by that treaty, and some of them did. The treaty as written specified only a small number of reservations: the Tulalip Reservation, north of Mukilteo and site of the Point Elliott treaty ceremony; the Port Madison Reservation, located west from Seattle on the Kitsap Peninsula; the Swinomish Reservation on Fidalgo Island; and the Lummi Reservation near present-day Bellingham at the far northern end of Puget Sound. Several of the Native bands with villages in the east-central Puget Sound region, particularly those near the town of Seattle who lived along the Duwamish River, on Elliott Bay, or in the Cascade Mountain lowlands fronting Lake Washington, proved unwilling to move to the designated reserves. They probably hoped or expected they would get their own reservations but never did. Several prominent Americans felt differently. The Seattle minister David Blaine had hoped before the treaty councils to see the Indians "removed from our midst." Now he despaired that their continuing presence was "retarding the wheels of all true progress." "Nature has contributed bountifully to the healthfulness, prosperity, and happiness of those who may inhabit this new country," he said, but Indians and alcohol, which consumed the lives of whites and Natives alike in the mind of the censorious preacher, prevented Seattle from being "one of the most delightful regions of country in the world."[40]

Many whites considered Indian beggary and thievery an affront to their security and morality, and many Indians in turn blamed whites for not feeding and clothing them in times of need. The new wage economy enabled some Indians to try their hand at farming, lumbering, or raising livestock for sale in addition to fishing. Being compelled to move to the reservations often meant giving up prosperity to live alongside rival bands on alien lands. Rancor over resettlement erupted into warfare on the eastern side of the Cascades in 1855, spreading to the western slope. After a failed attempt by territorial militia to capture Leschay (also known as Leschi), a Nisqually Indian leader critical of the reservation policy, Natives attacked settlers' farms along the White River and killed nine homesteaders. The survivors fled for Seattle, but Seattle itself was attacked the following January, and frightened citizens crowded into a blockhouse under the protective cannon fire of the U.S. Navy sloop *Decatur* anchored in Elliott Bay. In August 1856, Stevens officially concluded hostilities. Arthur Denny later recalled that homesteaders who had considered all Indians welcoming now declared "that all were hostile, and must be treated as enemies."[41]

Stevens accused Indians like Leschay, who had once been good neighbors, of turning into murderers, and he sent Leschay to trial. After a lengthy legal contest, he was sentenced to death by hanging for allegedly killing an American militiaman. In the wake of his execution, the federal government merged the Indian bands along the Green and White Rivers and told them to settle in the Muckleshoot Reservation south of Lake Washington, created in 1857 after the war. Those bands along the lower reaches of the Duwamish River became part of the Port Madison Reservation on the western shore of Puget Sound. Stevens and the leaders of Seattle considered them a beaten people, but many whites left Seattle in fear for their lives and property, and the region's farms and lumber mills slipped into a decade of economic depression that many blamed on the Indian troubles. A newspaper in the territorial capital Olympia proclaimed that Seattle, in 1856, was "literally used up and rubbed out."[42]

Isaac Stevens's son and biographer, Hazard Stevens, later wrote that his father considered the treaties "the most important, beneficial, and successful services he rendered" to whites and Indians alike. For their part, Indians generally accepted the terms of their supposed removal because they understood their treaty rights literally—they were still entitled to fish and hunt in the usual and accustomed places. The reservations in Puget Sound were designed less to contain Indians than to keep them accessible for labor and to permit them to subsist for themselves rather than depend wholly upon government subsidies. Indians continued to fish, hunt, and forage, peddling their catch to white Seattleites. Emily Inez Denny remembered that Indians filled her family's larder with "fish of many excellent kinds . . . brought fresh and flapping to our doors." Even the raiding bands of Tlingit and Kwakwaka'wakw, who had been a historic threat to Puget Sound aborigines, now substituted wage labor for slaving raids and worked side-by-side with their former victims picking hops or cutting trees. The census of 1880 listed 3,553 people in the city, with only 47 identified as Indians, but census workers likely overlooked the many more Indians passing in and out of town daily, en route to fishing grounds, agricultural jobs, or logging camps.[43]

A perceptive visitor to Seattle in the 1860s might have discovered a city that was at once both white and Indian, even if the invaders held the upper hand over the Natives. A closer look would have further revealed the evolution of a new, perhaps hybrid ethic of place, one that mingled Indian and non-Indian lives and attitudes toward the landscape. For their part, Indians continued to see the lands and waters that were becoming the city of Seattle in terms of their seasonal subsistence rounds and social activities, regardless of the changes to the physical environment. Charles Kinnear, an emigrant who arrived as a ten-year-old with his family, remembered how Indians encamped along the shoreline, occupying more than three city blocks, in the midst of wharves and warehouses, selling salmon to passersby and playing the bone game before a crowd of spectators. Other Indians sold baskets on Seattle street corners, worked as domestic servants for whites, or joined the burgeoning prostitution trade that grew up next to Henry Yesler's mill. As one Methodist minister recalled before 1876, these "Indian bawdy houses or squaw brothels" were of the "lowest and vilest character," filled with the "frantic cries of those whose sin had brought them to the verge of madness and despair." Prostitution was only one response to the ongoing effects of Indian displacement, which when combined with the continuing ravages of diseases to which Indians had no immunity, forced many others to beg from white residents and federal Indian agents. The commissioner of Indian affairs, ostensibly exasperated by those under his supervision, recommended disbanding the Muckleshoot Reservation to put an end to their wandering so close to the small city of Seattle. What the commissioner did not realize or want to admit was how thoroughly intertwined Natives had become with the building and workings of the young city.[44]

For their part, the white invaders increasingly began to see Indians as the true impostors and as an unwanted presence in the city. Their ambitions, geared toward commercial enterprise and social order, had no place for Indians who did not stay put, physically or socially. Fears of violence and immorality, coupled with worries that roving Indians brought disease with them into the city or threatened to spark a major conflagration with their cooking fires, had led local papers to call for their expulsion. In 1865, town officials passed an ordinance to restrict Indian encampments to the most outlying regions on the southern side of town, next to the swampy mudflats at the mouth of the Duwamish River. Almost ten years later, after reports of a smallpox outbreak afflicting "a couple of beastly squaws" in one of the Indian enclaves along the waterfront, the *Seattle Daily Intelligencer* inveighed in an editorial that "as yet nothing has been done in the way of preventing these filthy animals from visiting our city at will." Over the next two decades, city officials passed still more laws restricting Indian movement and reinforcing new anti-miscegenation laws passed by the territorial legislature in 1866.[45]

Henry Yesler, the sawmill owner who had married a Duwamish woman, had opposed the new reservation policy and helped to wage peace after the 1855–56 war by serving as an Indian agent for Seattle. He brokered an arrangement for almost 150 Natives, likely Duwamish, to move either to Bainbridge Island, across Elliott Bay from Seattle, or to the new Port Madison Reservation. Yesler, however, was not entirely consistent in his role as Indian protector. In 1866, he signed a petition, along with 165 other white residents, asking Arthur Denny, now the congressional territorial delegate, to block a reservation for the Duwamish Indians on the Black River, about ten miles south of Seattle. The proposed reservation, according to the petitioners, threatened "the quiet and flourishing settlements" that had always "justly and kindly protected" the Indians living there. Natives, however, did not stay still. They continued to move about following the seasonal wage economy up and down the coast, up into the forested Cascades and back down into the farmed lowlands. Meanwhile, Yesler, the crude lumberman, went on to become one of the largest individual property owners in Seattle and a two-term mayor. He died a prosperous man in 1892, living up to his earlier billing by a local newspaper as that "indefatigable" entrepreneur who "contributed more largely to the building of business structures in this city than any other property holder."[46]

Left unstated in the obituaries that followed his death was that Yesler's mill had given birth to another local industry: vice. The mill and cookhouse stood at the terminus of a "skid road," a name applied to the wood and mud ramps used in lumber camps throughout the Pacific Northwest to slide logs downhill and into the teeth of buzzing saws. Soon the term became a proper noun, shorthand for Seattle's red-light district. On Skid Road, later known by the more infamous term Skid Row (the two were used interchangeably in the city's early years), were the brothels of Indian and white women, opium dens, card houses, and saloons intended to relieve gullible young, often white laborers of their wages in a town full of footloose bachelors. Seattle was not a city entirely without law, but the higher law that prevailed was profit and fast growth, so local authorities ignored or tolerated the gambling and sex trade despite the protests of the city's self-styled moral guardians. This new geography of commercial vice put Seattle on the map. It soon became a destination for loggers from even more distant lumber camps, fishermen and sailors back from long trips at sea, and miners laying over en route to the next big strike. A popular folk song of miners during the Fraser River gold rush in British Columbia, in 1858, set in Chinook jargon, praised Skid Row's temptations:

> There'll be mowitch [venison]
> And Klootchman [Indian women] by the way
> When we 'rive at Seattle Illahee [Seattle country] . . .
> Row, boys, row!
> Let's travel to the place they call Seattle
> (That's the place to have a spree!)
> Seattle Illahee

An early Seattle settler, Edith Redfield, described her young city's lumber mills as "little kingdoms, a law unto themselves." "Here white men, Indians, Chinamen, and Kanakas [native Hawaiians]," she wrote, "worked side by side and boarded at the Company's cook-house." But these polyglot "little kingdoms" were only shadows of the older integrated world. Seattle Illahee had become a white man's country.[47]

Helen Hunt Jackson, author of *A Century of Dishonor,* a caustic critique of Indian policy published in 1881, traveled west for the *Atlantic Monthly* two years later to find a country being invaded by an army more "irresistible than warriors—men of the axe, the plow, the steam-engine." "The siege they lay," Jackson observed, "is a siege which cannot be broken for all the forces of nature are on their side." Emigrants to the Northwest had aroused the attention of the railroads that were "straining muscles of men and sinews of money" to carry "this great tide." Sawmills were chewing up the forests and spitting out a hundred thousand board feet a day with "no sense of responsibility of the future," producing so much lumber that the sawdust was used to feed the boilers in the mills. "Such waste of tons of fuel makes one's heart ache," Jackson wrote, when "thinking of the cities of poor, shivering, and freezing every winter."

The wilderness was doomed in Jackson's mind, but what saddened her more than the decadence of petty frontier capitalists was the robbery of original Indian names. Mount Rainier now stood where Tacoma used to rise. She preferred the native name, Tacoma, but the white residents of Seattle wanted the name bestowed on the peak by George Vancouver. Indian names that had once given shape and form to the world were now appropriated for commercial or political ends, or rubbed out completely. "There seems a perverse injustice in substituting the names of wandering foreigners, however worthy," she concluded, "for the old names born of love . . . the only mementos which, soon, will be left of a race that has died at our hands." What Jackson did not realize was that although the Indian names might eventually fade from the historical record, the descendants of the first peoples would keep them alive. White Seattleites might disregard what came before, but the Natives would neither disappear nor forget.[48]

Jackson's romantic gloss over Seattle's Indian past, not to mention its history after contact, was typical of white Americans of her era. The aim of her essay was to evoke nostalgia and loss as much as anger and action. The effect was to erase the very real role that Native peoples had played in making the new city. The same sleight-of-hand continues to this day whenever someone invokes the city's namesake, Seeathl, also known as Sealth. A leader in what were later known as the Duwamish and Suquamish tribes, Seeathl was present at the negotiations for the Point Elliott Treaty, and supposedly had met Isaac Stevens, the lead negotiator for the United States government. The brash army officer first visited Seattle in January 1854 after his initial appointment as territorial governor the previous autumn. A large crowd turned out to see the new governor, who probably came to tell the Indians they should expect a treaty conference at a later date. In Stevens's honor, Seeathl gave a speech at the gathering, first purportedly transcribed by a white settler and since revised and reprinted many times.

The problem is that the provenance of Chief Seattle's speech is uncertain. There is no existing written transcript. As was typical of many Native cultures, the Whulshootseed speakers of Puget Sound originally lived in a world defined by oral tradition and not written words. The first written version comes from Henry A. Smith, who published it in the *Seattle Star* of October 29, 1887. Smith, an Ohio physician who moved to Puget Sound in 1852, working for a time on the Tulalip Indian Reservation before moving his practice to the city, was supposedly present at the address. The shores of Elliott Bay, he remembered, were "lined with a living mass of swaying, writhing, dusky humanity." "Old Chief Seattle," Smith wrote in his article, "broad-shouldered, deep-chested, and finely proportioned," silenced the crowd with his "trumpet-toned voice," as "deep-toned, sonorous, and eloquent sentences rolled from his lips like the ceaseless thunders of cataracts flowing from exhaustless fountains." In the speech as published in 1887, Seattle said that the dust beneath Stevens's feet "responds more lovingly to our footsteps than to yours, because it is the ashes of our ancestors." He also issued a warning: "the white man will never be alone" because "the dead are not altogether powerless."[49]

Soon after Smith published his account, Chief Seattle's speech took on a momentum of its own. Four years later, it was published in a history of Washington State, only to be reprinted again and again. Its most famous phrase, "the earth is our mother," quoted by environmentalists as holy writ, was in fact a fabrication made in 1970 by screenwriter Ted Perry, who used a version in an ecological film. A local journalist and historian in the 1990s called the old Indian chief Seattle's King Arthur. The speech, if it was such, thus joined a long tradition in American letters of invoking Indians as spurious oracles to soothe white anxieties about their own privilege. All historians really know is that an Indian named Seeathl spoke to whites on several occasions, and that the city of Seattle is named for him. The rest borders on what historian Jill Lepore has called "a shabby sort of historical ventriloquism."[50]

Nonetheless, Seattle's speech contained two historical truths of a sort: the first, that the lands Seeathl and his people inhabited and lost were places they knew and cared for deeply, consecrated by memory and labor. The second was a truth by omission. The Americans had seized the lands and waters that Natives had called their home; they even appropriated the words of the Indians themselves to legitimize their invasion. And in the Native world, the words were the places, too. What the speech could not hide was that the Indians did not vanish, nor did the lands and waters they saw as the geography of the Transformer who had preceded everyone and turned the chaos of creation into homes for mortal beings. The Bostons could not wipe away the Native names, let alone the people who had conferred them. The names gave voice, in part, to an older ethic that once bound people and place together in time. Indeed, if there was a new fusion of Indian and non-Indian ethics of place in early Seattle, it was a short-lived union.

In 1916, the anthropologist Arthur Ballard, after transcribing the Snoqualmie legend of Moon, the Transformer, asked his informant, Snuqualmi Charlie, to offer his own assessment of the story. "I am an Indian today," he told Ballard, and he continued:

> Moon has given us fish and game. The white people have come and overwhelmed us. We may not kill a deer nor catch a fish forbidden by the white men to be taken. I should like any of these lawmakers to tell me if Moon or Sun has set him here to forbid our people to kill game given to us by Moon and Sun. Though white people overwhelm us, it is Moon that placed us here, and the laws we are bound to obey are those established by Moon in the ancient time.

Snuqualmi Charlie, likely born in 1850, alive in 1916, was no relic of a dying race. His story was not a simple memento. It was a protest and a challenge.[51]

CHAPTER 2

The Work Which Nature Had Left Undone

Making Private Property on the Waterfront Commons

In a booster publication from 1914, *The City That Made Itself,* Welford Beaton, a Seattle Chamber of Commerce official, wrote: "Nature apparently grew tired before she finished Seattle. She made a wonderful harbor, produced an empire of timber-hung pictures on the horizon, spread three lakes among the hills, and then left the town site to itself like a tousled, unmade bed." Only when "Man" completed "the work which Nature had left undone" could commerce "pour unhampered [into] its natural channels." This real work could begin after Man intervened to give seminal shape to an unfinished and feminine Nature. Seattle was to be the triumphant product of Nature endowed with human ingenuity.[1]

To shape the tousled town site, Seattle's earliest business and political leaders had to find the labor supply: first Indians, then native-born white and immigrant workers. But the city also needed capital and a medium to convey that capital. It was not the preordained task as told by Beaton in his masculine hyperbole. By 1914, Seattle was on its way to becoming the premier city of the Pacific Northwest, home to a growing and ever more diversified economy of shipyards and foundries, retailers and slaughterhouses, printers and flour mills. Yet originally, Seattle had little to offer prospective investors. Until the end of the nineteenth century, its primary enterprises were exporting the natural wealth of Puget Sound and importing manufactured goods for local consumption and distribution to its nascent hinterland: the far-flung logging camps, mines, and fishing stations across western Washington. The rainy, muddy, logged-over hills, smoking with the fires of burning slash, the debris left over after the cutting, looming above the small town offered little to potential investors except one thing: inexpensive and available land. Local civic and business leaders tried to entice risk-takers or patrons eager to invest in the dream of a greater Seattle to supply the money. When that money was mixed with the land, it became real estate.[2]

Cities in nineteenth-century America were ravenous creatures, hungry for capital to help them capture trade, sell real estate, and solidify their positions as commercial and jobbing centers. Cities in the Far West were especially voracious. Some, like San Francisco, though blessed in 1848 by the discovery of gold along the American River, still had to have merchants and political leaders working relentlessly, even brutally, to insure its position as the gateway to the California gold fields. Seattle's collateral was more modest but provident—landed property, capital fixed in time and space, adjacent to a deep ocean port ready to ship the area's rich natural resources of timber, coal, and fisheries to world markets. Real estate developed in the right places and in the right amounts was the potential key to its prosperity. Still, many prospective buyers in the Midwest and the East hesitated to assume the risks of acquiring western property. Homesteaders who labored hard to earn a few hundred acres were a poor advertisement for investors. It fell to the federal government, as patron of internal improvements, and to the transcontinental railroads, all but bribed by the government to build their lines, to bring catalytic power to this capital-poor but resource-rich region.[3]

In addition, Seattle of the 1880s had many faults to counter its many virtues, principally steep hills and mountains covered by thick forests bounded by acres of mudflats, flood-prone rivers, and extreme tides. Making property in Seattle had to begin with constructing a permanent and distinct boundary between water and land. Unless land remained solid and water remained navigable, investors would never sink their money into such an impermanent and chancy enterprise. Civic leaders and private citizens set out to renovate Seattle's original watershed; by changing geography, they believed they would conjure up real estate. They would simply remove excess water and earth, and advertise the transformed landscape as an opportunity ready for the right buyer.

Of course, behind this plan was an assumption: that everyone agreed what property was and to what ends it would be used. Urban Americans, in Seattle as elsewhere, were deeply conflicted over the role private property and capital should play in building cities. Those private investors who wanted the most favors in creating new property or the transportation networks necessary to make real estate accessible cited the public interest, even if their definition of public interest was eventually exclusive. Opponents countered that unbridled development, concentrated in the hands of the few, would close off once-public space and derange social order. In this lay the second assumption, that real estate was a solid term for a solid thing. In reality, it was fluid.

In turn-of-the-century America, a tension expanded between public authority and private power in the arena of municipal governance and the shaping of the metropolitan landscape. Urban historians have tended to consider public and private power as separate, framing the battle as civic politics versus the commercial marketplace. Less well understood has been the dependency of property as an institution on the transformation of physical and social environments, a process that rested on lowering the boundaries between political and commercial enterprise. It would take another century—and more—for the descendants of Seattle's original builders to understand and face the consequences of creating property. What Seattleites then and later faced was, in a sense, a collision of two opposed codes that linked making property to social behavior. One defined property as a public, common good, while the other saw property in terms of private gain. Yet neither code accounted for the particulars of place. Real estate, by its very definition, implies permanence. The problem is that real estate is a socially constructed, politically contested, and physically unstable product that often eludes the most ingenious of its owners or creators. Since all of the physical world had the potential to become private goods, the social inequities of making property indelibly became part of the city's new environment.[4]

Railroads, the Alchemy of the Age

A dream had seized Marshall Moore, the governor of Washington Territory, the same vision that had compelled his predecessor, Isaac Stevens, to resign his post as territorial governor and lead the Pacific Railroad Survey from the Mississippi River to Puget Sound. It was the dream of a transcontinental railroad, a transportation system impervious to the seasons and geography, operating at great speed and greater efficiency, generating wealth wherever it went. Nineteenth-century Americans tended to valorize technology, and the railroad was a favorite subject. To be the terminus of a transcontinental railroad was to capture something magical, and, in the 1870s, the Great Northern Pacific Railroad was the magician that the region needed. If the Northern Pacific, aided by a generous land grant from Congress in 1864, reached Puget Sound, Moore predicted wonderful things for his constituents. "Nature has made this body of water the key to more than fifteen hundred miles of the north Pacific ocean," he proclaimed in an 1867 address. "Railroads are not a mere convenience to local populations, but a vast machinery for the building up of empires. . . . They are the true alchemy of the age, which transmutes the otherwise worthless resources of a country into gold." To Moore, one worthless resource was his territory's extensive coastline. The thousands of square miles of Puget Sound tidelands, exposed twice every day, were stinking mats of muck. With the railroad's arrival, they might house terminals and wharves that would free Washington Territory from California banks and Oregon shippers.[5]

By 1870, with approximately ten thousand Americans living around the Puget Sound basin, the region had become a de facto colony of San Francisco, which was the principal buyer of its coal and timber as well as its primary lender. Portland, with its larger population almost equal to that of the entire Puget Sound region, had become the commercial hub of the Pacific Northwest. Shippers portaging past dangerous rapids on the Columbia River, crossing the treacherous bar at the river's mouth, and contending with periodic flooding, exacerbated by persistent logging and mining, would arrive at Portland only to face the powerful Oregon Steam Navigation Company. Owned by local capitalists, it monopolized the trade from the rich farmlands of eastern Washington and Oregon. As a result, Seattle businesses tended to see their city as frozen in a state of suspended animation, waiting for meteoric growth. The small town had its vast inland sea but no great watercourse equal to the Columbia penetrating its mountainous and forested interior, only skeins of rivers and mudflats fed by fast-flowing rivers descending steeply from the high peaks. Once the rapid-filled rivers reached the coastal valleys, they turned so sluggish and sinuous and slow that easy passage was impossible.

One local merchant, James Edwin Whitworth, who carried supplies to and from the coal mines at the south end of Lake Washington, complained in 1869 that it took at least a week to travel almost eight linear miles by way of the Duwamish and Black Rivers. Both were jammed with snags and sandbars. He often had to pay local Indians to guide him through this navigational nightmare because flooding changed the channels every winter and every spring. The rivers played havoc with those living on their banks as well. Homesteaders along the lower Duwamish of midcentury tried to erect dikes to reclaim wetlands and protect their fields from floods and tidal surges. By the end of the century, tired

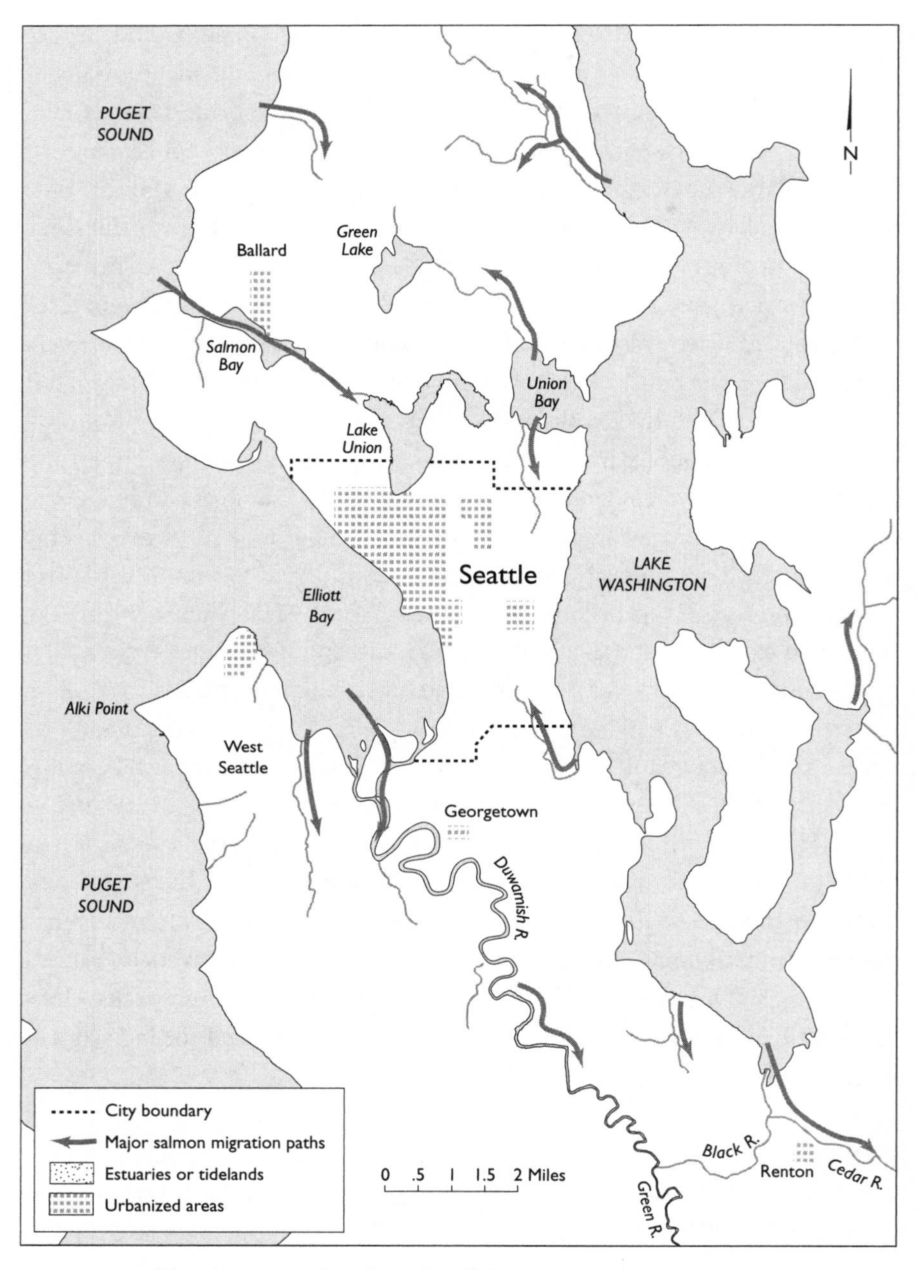

Seattle and its immediate hinterland, showing the original drainage of Lake Washington and the Duwamish River, c. 1890.

of being drowned twice a year, the farmers petitioned King County to create drainage districts and empower citizens to repel the waters. Farmers in the regions at the southern end of Lake Washington even tried turning the Cedar River to flow directly into the lake instead of the debris-choked Black River, only to find that large floods, every four to six years, washed away the dikes and jetties. The Muckleshoot Indian Reservation, in the Green River valley farther upstream, was particularly hard hit whenever the rivers surged. T. J. McKenney, the territorial superintendent for Indian affairs, wrote after one round of floods in the winter of 1867 that the "disasters" were "very serious upon their property" and sent many of the Muckleshoot into "shock."[6]

Little wonder that Marshall Moore longed to see the railroad liberate this waterlogged region from perennial isolation, and, in August 1870, the Northern Pacific sent a survey party to scout for potential terminal sites. Every major Puget Sound town offered the railroad cash subsidies and waterfront lots. Seattle extended one of the most generous packages: a quarter million dollars in cash and bonds, plus three thousand acres of land and seventy-five hundred town lots. Still, Seattle lost out, as an investment banker, Jay Cooke, took control of the Northern Pacific, then acquired the Oregon Steam Navigation Company. He persuaded the Northern Pacific to designate Portland as its western end, and Tacoma as the tip of the branch line running north. Cooke cared nothing for Puget Sound, only for land open to development. Consequently, he lobbied to amend the original congressional charter that had required the railroad to reach Puget Sound first. Left in the cold by Cooke's gambit, many Puget Sound towns now began to construct short local lines of their own, many only on paper, designed to obtain tideland and waterfront land grants. The Seattle and Walla Walla Railroad, organized in 1873 by a consortium of leading businessmen to reach coal mines at the southern end of Lake Washington, received such a grant, even though they were intent not on building a railroad empire but on luring a railroad empire to come to them.[7]

The Seattle and Walla Walla Railroad, which received the first tideland grant, also received the choicest point of entry into the city, from the north around the "Ram's Horn," a knobby protrusion, on the harbor line. The only consistently level land was on the shores of Elliott Bay and Lake Washington, or down the Duwamish River, and railroads needed reliable grades for smooth construction and operation. Controlling the Ram's Horn thus meant controlling the Seattle waterfront. Once construction began on the Seattle and Walla Walla line in 1874, the value of the tidelands quickly became apparent. A visitor from Rochester, New York, said that anyone in Seattle with capital "from $5,000 to $10,000, if he understands the 'dynamics of cities,' can in a . . . few years make a millionaire of himself."[8]

Thomas Burke, a lawyer from upstate New York who had come to Seattle in 1875 at the urging of family friends, understood this dynamic. He soon became one of the city's most powerful real estate owners. Forming a partnership with John J. McGilvra, the former federal attorney for Washington Territory, he was later elected county probate judge and further added to his prestige by marrying McGilvra's daughter, Caroline. "The Judge," as he became known, was a popular Seattle attorney and an advocate for social order. He faced an early test as white laborers throughout Washington Territory, worried over job security and fair wages, turned their anxieties against the Chinese, accusing them of working for lower pay and crowding out native-born and European immigrant workers. Such fears were not without merit. In the labor-poor West, the Chinese were an integral part of the dual-labor system, relegated to the most demanding jobs for the lowest pay. Despite territorial laws barring Chinese from voting or testifying against whites in court, employers continued to hire Chinese to cut timber, harvest hops, or mine coal at the lowest wages possible. The Chinese Exclusion Act of 1882, passed by Congress to limit further immigration, only fueled the bigotry. Across the West, from Wyoming to California, white mobs taunted, beat, and often murdered Chinese laborers. A similar cycle of cruelty unfolded in the Puget Sound country, and in November 1885, a mob in Tacoma forced almost 350 Chinese residents onto Northern Pacific trains bound for Portland. The following February, in 1886, a crowd in Seattle bullied nearly 200 Chinese onto boats for San Francisco but were thwarted from expelling more by Judge Thomas Burke. "It is to our interest to see them go," he told the *Seattle Post-Intelligencer,* "but it is not to our interest, but just the opposite, to see one drop of innocent blood spilled or a single breach of the law." Burke's defense of the Chinese, a calculated business decision to keep a pliable, inexpensive labor force nearby, emboldened Governor Watson Squire to dispatch the territorial guard and quell the unrest.[9]

Judge Burke had another business reason for keeping order. In the land boom incited by the railroad fever, Burke strategically concentrated his purchases in lots adjacent to the city's waterfront and in its new suburbs around Lake Union. He confessed in 1881 to a friend that he was "in speculations here up to my eyes." He made a risky investment by buying tidelands in the West where, unlike in the East, their status was in a judicial and regulatory twilight zone. Eastern cities, like Boston and New York, had already filled large sections of waterfront in the late eighteenth and early nineteenth centuries, but by the late nineteenth century, jurists considered tidelands common public property, open to all. In Washington, as in the rest of the Far West, the federal government held in trust all submerged lands beneath the ordinary high tide line until a territory matured into a state. No settlers could seize these lands. Upland owners whose property abutted submerged lands could neither build structures that obstructed navigation nor claim adjacent tide flats as their own. The theory was that tidelands should remain public because making them private would choke off maritime commerce and fishing.[10]

But in Seattle, the actual configuration of tidelands made the theory of common property often impractical. Elliott Bay provided an excellent harbor, but the steep slopes surrounding the bay restricted the size and length of dock construction. The local lumber magnate Henry Yesler had filled the tide flats at the base of his mill with sawdust, wood chips, and ballast disgorged from ships over the years, a process technically known as "adverse possession," though it was an illegal seizure of common property. A small island even emerged at the foot of his wharf, aptly named Ballast Island. Yesler's illicit confiscation soon became common practice. Thanks to faulty General Land Office records, tidelands were inadvertently included in many original Donation Land Claim Act patents filed by Seattle's first white arrivals. A number of landowners whose property abutted tidewater, hearing of the mistakes, tried to build over submerged lands they sneaked into their claims. Once land speculation began to gather strength in anticipation of a transcontinental railroad, Seattle officials had to fend off hundreds of these claimants.[11]

The schemes of land speculators who hoped to make money once the locomotives reached Seattle were soon dashed. Competition had dried up available capital for railroads, whose high fixed costs and large debt load relied on heavy infusions of capital or high traffic volume to survive. A spasmodic national economy added further burdens to railroad operators. The Panic of 1873, the first serious economic downturn after the Civil War, was the direct result of the railroads and their abnormally fast growth. Nationally, railroads laid more than thirty-five thousand miles of new track between 1866 and 1873, spurred on by a loose money supply and overeager financiers. Jay Cooke and Company was one of the most enthusiastic lenders, but in September 1873, its books were bleeding red and the firm declared bankruptcy. The decision helped throw the nation's economy into turmoil, forcing numerous banks and brokerage houses to fail, from New York City to Memphis. Jay Cooke's fortune evaporated and so, too, did his leadership of the Northern Pacific.[12]

As the Panic waned, the largest railroad lines, in an effort to survive, organized pools to standardize shipping rates and divide their territory to limit competition. Frederick Billings, a railroad investor, reorganized the Northern Pacific and announced plans to restart construction stalled in the Dakotas. A German immigrant and master of the railroad pool, Henry Villard, soon upended Billings's plans by acquiring the Oregon Steam Navigation Company, turning it into the Oregon Railway and Navigation Company, and using the new firm to take over the Northern Pacific in 1881 by blocking its operations in Oregon. With the help of a subsidiary firm, the Oregon Improvement Company, or OIC, Villard acquired the Seattle and Walla Walla, reorganized it as the Columbia and Puget Sound Railway, and folded it into his holdings. Newspaper stories stoked the superheated real estate market, much of it along the waterfront where the Columbia and Puget Sound would lay its tracks, urging those with cash to get in the game. Judge Burke and others in a clique of local businessmen, with the support of the new Seattle Chamber of Commerce, tried to organize their own railroad, in the image of the old Seattle and Walla Walla, whereupon Villard countered by agreeing that his Columbia and Puget Sound Railway would be the Northern Pacific's direct line to Seattle. His boldness came at a cost: stockholders, incensed by Villard's aggressive tactics, removed him as head of the Northern Pacific and his Oregon companies in 1884. The Tacoma–Seattle line, no longer a priority for the Northern Pacific, fell into near constant disrepair. Disgruntled local residents later called it the "Orphan Road."[13]

This rail line to nowhere epitomized Villard's sense of place. In his mind, as well as in the minds of other railway executives and the investors who backed their enterprises, Seattle was but a spot on a map, nothing more than a potential terminal for his locomotives and a market for the goods they carried. The city was not so much a place as a line in his ledger, so when his financing failed, Villard closed the books on Seattle. He later rebounded from his misfortune, creating a corporate conglomerate, Edison General Electric Companies, the forerunner to General Electric. But his personal collapse had local consequences, initiating a brief depression in Puget Sound that deflated real estate sales. Others had come to depend on Villard's boldness. Many also began to emulate him.

Thomas Burke, unable to convince legislators to punish the Northern

Pacific at congressional hearings investigating railroad forfeitures, joined forces with a honey-tongued attorney from Maine, Daniel H. Gilman, to build their own railroad line. Gilman had studied law at Columbia College in New York while working in the city's mercantile houses. Moving to Seattle in 1883, he joined with Burke, now one of the area's largest property owners, to make his fortune. The judge's opposite in personality, brash and confident, Gilman was prone to playing fast with money and gossip. A rainmaker, he traveled back east to raise funds for the venture while Burke stayed behind, drafting the articles of incorporation and soliciting local support. In 1885, Burke and Gilman announced the formation of the Seattle, Lake Shore, and Eastern Railroad, designed to run around Lake Washington and over Snoqualmie Pass to Walla Walla, with branch lines to the coal mines in the Green River valley south of Seattle. A New York banking firm agreed to capitalize the new line at $15 million if local investors contributed a mere $10,000 each of their own money.[14]

Anticipating another boom, Burke began to acquire still more properties and options on future purchases, notably on the tidelands at Smith's Cove north of downtown and along tide flats just south of the central business district, an ideal site for a switching yard. He cleverly disguised the new railroad and his real estate transactions as local enterprises undertaken in the public interest. With the city's grant to the old Seattle and Walla Walla now in the hands of the Oregon Improvement Company, which refused to share the thoroughfare with the upstart railroad, Burke floated a possible solution: to build yet another waterfront right-of-way beyond the high tide mark on trestles and pilings. If he could not command the shoreline, he would extend it into the ocean. Burke touted his plan as the best thing for Seattle, whereupon the city council let Burke acquire the necessary property. Privately, he persuaded the local trustees of the Columbia and Puget Sound Railway to deceive the OIC about his intentions. The ploy worked, and in 1887 the city granted the new right-of-way, later called Railroad Avenue.[15]

The Oregon Improvement Company, learning of Burke's deception, launched a counteroffensive by proxy. In 1888, lumberman Henry Yesler began construction of a new wharf on tidelands claimed by the OIC. When the company refused to challenge Yesler's invasion of their property, speculators poured onto the tide flats and a frenzied land rush was on. The move backfired. Some squatters sent pile drivers to stake their claims; others floated prefabricated shacks to the choicest properties. Competitors who could not afford piling or lumber claimed to be oyster cultivators, who, under territorial and common law, had access to tidelands for shellfish farming. Their claims were especially specious since, by the 1880s, Elliott Bay's waters were polluted and unfit for oyster beds, choked with sawdust and effluvia from sawmills and city sewers. Elliott Bay soon resembled a floating mining camp, complete with fistfights, gunplay, and drunken brawls.[16]

"The craze for salt water," Burke complained to a friend, had "broken out again with greater violence than before," with swarms of "lunatics of high and low degree" roosting "like so many cawing crows on the mud flats." To Burke, it seemed as if the whole city was helping itself to his investments. "Now this salt water crew plant themselves on my property," he said, "and tell me that inasmuch as it is below the line of ordinary high tide, they have as good a right to it as I have, and even better, because of presuming to buy it and pay for it, I was flying in the face of the law." His only consolation was that "the ever-charitable sea" might "drown such rascals" and "thus rid both sea and land of them."[17]

In 1880, Seattle's population had numbered around thirty-five hundred; by the decade's end, when the latest tideland boom was at full throttle, it approached thirty-one thousand. Seattle had become a roughneck Venice, built on trestles and pilings, stretching over tide flats and water in search of lucrative real estate, its epicenter the waterfront next to Yesler's sawmill, adjacent to Skid Road, a neighborhood also known as the Lava Beds for the smoldering piles of wood chips and sawdust beside the mill. Young Seattle, like most nineteenth-century cities, was a transmuted forest and often burned like one. Fires in 1875 and 1878 prompted the *Seattle Daily Intelligencer* to call the Lava Beds "a mere tinder box." After a major fire on July 26, 1879, destroyed twenty buildings, another editorial blamed vagrant Indians for increasing the fire risk: "The 'siwash' camps on the sand reef just across from Main Street at the edge of the sawdust flat should be broken up immediately." Indians encamped on the city's southern tide flats were not to blame. It was Judge Burke's "salt water crew" instead. Seattle residents, from the mighty judge to the lowliest speculator, had gone wild in their quest for property. The allure of cheap real estate propelled the population surge, leading to more and more buildings and creating the perfect environment for a significant conflagration—timber-frame structures opening onto wooden-planked streets and wharfs atop creosote-soaked pilings crisscrossing sawdust-filled mudflats.[18]

On the afternoon of June 6, 1889, a woodworker in a cabinet shop on the corner of Front and Madison Streets spilled hot glue into a pile of wood shavings, and within an hour all of downtown Seattle was ablaze. Mayor Robert Moran ordered an entire city block dynamited to form a firebreak, but the flames jumped the breach. Firefighters ran from standpipe to standpipe only to find the water mains had failed. When the fire burned itself out the next day, Skid Road was gone and downtown Seattle had been wiped out. Hundreds of residents were forced to live or conduct business in canvas tents on the edge of the burn zone. Rudyard Kipling, touring North America that summer and visiting Seattle soon after the fire, declared, "in the heart of the business quarters there was a horrible black smudge, as though a Hand had come down and rubbed the place clean." With nearly fifty blocks, or about sixty-four acres, destroyed, damages were estimated at nearly $10 million. Remarkably, no one was killed, and within a year most of the downtown had been rebuilt, at a cost of almost $13 million, in brick or stone and with improved water mains.[19]

Seattle's rapid recovery had less to do with the fire than the opportunities that the fire presented to the city in the form of misfortune. With the trestles carrying tracks around the Ram's Horn gutted by the flames and much of the city's docks and wharves in smoking ruins, the waterfront was the first place to be rebuilt. As workers drove piles into the mud and hauled away charred timbers, questions over the legal status of Puget Sound's tidelands reopened. Washington's statehood was pending and, at the 1889 constitutional convention in Olympia, delegates split over whether tidelands were public or private. Representatives from inland eastern counties, fearful that the railroads might monopolize Puget Sound ports and dictate unfavorable shipping rates to farmers, insisted that tidelands remain under permanent state control. Delegates from Washington's coastal counties, many tideland investors among them, responded that, without tideland grants, industrial and railway development would be impossible. In the wake of the great fire, such investments were imperative to Puget Sound's economic survival, so the coastal delegates maneuvered to protect the rights of those who held existing grants to improved tidelands or parcels slated for future development.[20]

After protracted negotiations, convention representatives agreed to recognize state ownership of beds and shores in navigable waters up to the line of ordinary high water. Improved land above that line would be available as private property. It was an imperfect compromise. The *Seattle Post-Intelligencer* editorialized that, by leaving state authority open to legal challenge, the compromise favored "tideland grabbers"—those who wanted to turn the mudflats into exclusively private property. The "tideland grabbers," acting to solidify their gains, then pushed through legislation after statehood to establish a Harbor Line Commission, its duties to survey and fix established harbor lines that would divide navigable water from improvable submerged lands. Afterward, individual county boards of tideland appraisers would dispose of surveyed parcels. The commission guaranteed the prior right of shore owners to purchase adjacent tidelands, even properties previously claimed but left unimproved. It also upheld the right of others to sell, lease, and purchase tidelands. Such actions enshrined the view that mud and water were private property. This sense of place, however, required that disputes over defining where the public interest ended and private property began could be adjudicated efficiently. Jacob Furth, a prominent Seattle financier and real estate speculator, explained all this in a letter to Harbor Line Commission head William F. Prosser, saying that it was "of vital importance for our state to have the cases settled without subjecting the people to long delays and vexatious lawsuits." His concerns rested on the premise that what was good for developers was good for the community. Whether tideland development was good for the city's physical environment was a question left unasked. But even before the Harbor Line Commission issued its findings in 1891 and 1893, residents were throwing up injunctions and petitions in its path. The legal reconstruction of the tideland environment, designed to favor private development, did not vanquish earlier notions of common property.[21]

Furth's appeal to the legal system to resolve property disputes eventually proved misguided. The courts were inconsistent arbiters of commonly shared definitions of place. When the state began to divide the tidelands in favor of the railroads and speculators, the courts became the primary arena for addressing citizen concerns about public control of the state's oceanfront lands. One group of tideland cases challenged the authority of the Harbor Line Commission to assess and divide the tidelands, another set addressed the ability of claimants to sell or lease their holdings. The state constitution dictated that prior claimants and upland owners adjacent to submerged lands retained their right to develop tidelands, but with new statutes in effect, the courts held that upland owners, in certain instances, had embarked on tideland improvement as a pretext to invade state property or expel other claimants, a clear violation of the law's intent. In other cases, the courts found that driving piles and erecting structures over submerged lands impinged upon the rights of other landowners or obstructed navigable waters. The mill owner Henry Yesler, who claimed that the Harbor Line Commission usurped his constitutional rights when it told him to stop extending his wharf into Elliott Bay, lost his appeal to the U.S. Supreme Court on these grounds.[22]

In response to the rising tide of lawsuits, the new public lands commissioner, W. T. Forrest, told appraisers to follow "prominent natural features" and established contours to the line of "low ordinary tide." Forrest fell back on what had become common practice in the distribution of coastal lands on the eastern seaboard, but his instructions did not make things easier for appraisers. The King County Board of Tideland Appraisers found their task nearly impossible. Tides and silt covered surveying markers, squatters confounded appraisers by moving stakes and transit lines, bogus oyster claims demanded attention, and new businesses, such as sawmills, changed existing harbor lines and preexisting property by building docks or filling in lands with debris. Appraisers also faced opposition from the individual buyers and corporate interests who wanted to develop their tideland properties in gigantic blocks with few streets to make room for potential railway and shipping terminals. Surveying larger blocks, in turn, served the interests of more capitalized investors by limiting the number of available properties. Estimates for tideland property varied dramatically and owners manipulated assessments to inflate values. H. H. Pease testified at an appraisal hearing that plots closest to the proposed railroad switching yards south of Marion Street were worth $13,000 per acre, and Amasa Miller said that lands south of Madison Street were worth at least $25,000 per acre. The old settler Arthur Denny questioned whether anyone knew what these lands were "really worth."[23]

The transcontinental railroads had a great deal at stake in this charade because they, unlike local land speculators, were playing a far bigger game on a much larger board. The difference was that in this contest the players could more easily deceive one another. Northern Pacific agents testified before the King County Board of Tideland Appraisers that "as public servants and common carriers" the railroad was "located here a long time before the State had required title to these tide lands." As a result, the Northern Pacific, "in the matter of appraisement," asked that lands "now occupied by the Railroads ought to be segregated from those to be acquired by individuals, and a lower value placed thereon." The railroad expected to spend still more money to develop its properties because its business served everyone, not just a few landowners. Its actions suggested otherwise. Fearful of opposing railway lines buying up choice waterfront lots, the Northern Pacific tried to drive down the price of existing parcels in order to corner the market.[24]

The competitor the Northern Pacific feared most was the Great Northern, the last of the transcontinental railroads and the only one built without significant government assistance. It was also the most profitable western railroad thanks to the business acumen of its founder, James J. Hill, a Scot of legendary ambition. Hill, who had built his line to carry freight, not simply to secure land or manipulate securities, was a capitalist who thought like an engineer. He was exceedingly careful about building sturdy bridges and track, finding the most level grades in order to save fuel, and extending branch lines to mines, farms, and lumber mills to generate traffic. Hill wanted a deep ocean seaport to expand his empire of rails to the Pacific, and in 1890 he proposed Seattle as the terminus for the Great Northern. He demanded easy access into the city and found the best man to get it for him: Thomas Burke.[25]

Burke's biographer Robert Nesbit wrote that the judge became "a satrap to the Empire Builder." This is a simplistic reading of their relationship. The judge was far more than a simple subordinate. He saw in Hill an opportunity to take revenge on the Northern Pacific for thwarting his own real estate and railroad ventures, but he also realized that a growing Seattle must have its own transcontinental railroad line. Hill mirrored his own temper, "a direct, positive, straight forward businessman" who shared Burke's own sense of propriety. Hill saw in Burke someone he could both trust and manipulate—an ideal combination in the railroad business, one that made corruption an effective business practice. Burke and Hill used each other willingly, though Hill let it be known that he was the more powerful of the two. In 1890, the Empire Builder, as Hill was known, retained Burke as his legal representative, asking that the judge draw up a franchise for his local branch line, the Seattle and Montana Railroad. He then loosed Burke against the increasingly aggressive Harbor Line Commission to prevent further public seizure of tidelands property, and Burke took to the task with relish, filing an injunction against implementing the commission's report and calling the proceedings "of a secret and star chamber character." By restricting tideland development along lines it deemed best, Burke charged, the commission had "succeeded in pillaging and bottling up the city of Seattle and in plundering her citizens." Burke had personal motives, his own real estate holdings, for resisting the commission, but he was also concerned about Hill. Great Northern land agents had acquired properties around Smith's Cove, north of downtown, for switching yards and now needed access to Seattle's industrial southern end. Burke had an elegant solution: he proposed selling to Hill the Seattle, Lake Shore, and Eastern franchise along Railroad

Avenue, which ran from Smith's Cove to the downtown waterfront, to bottle up the Northern Pacific. When the move became public, Oregon Improvement Company officials reminded the city that it had no right to grant the 1887 franchise to Burke's company because the tidelands at that time were territorial property.[26]

All of the principal railroads realized, in the words of W. P. Clough, vice president for the Great Northern, that the struggle was to "get through this town in excellent shape and with ample terminals located in the best part of town." That "best part" was at its southern end, along the tide flats fronting Elliott Bay, south of Skid Row among the shanties and saloons of Whitechapel, a salacious outpost named after a disreputable district of London. The Oregon Improvement Company had the superior legal position of owning the first easement along the waterfront. Yet the Great Northern had political capital and Burke spent it lavishly, portraying Hill as Seattle's savior, playing off deep resentments against both the Northern Pacific and the Oregon Improvement Company. He also quietly purchased tidelands north and south of the Northern Pacific and Oregon Improvement Company properties in south Seattle for Hill, forcing the OIC to propose a joint-terminal system in response. The Oregon Improvement Company wanted to bargain because it was in dire straits. Caught in the downdraft of the Panic of 1893, an economic emergency hastened, in part, by a run on the nation's gold supply that compelled bankers to call in their loans, the company's credit was overextended. Hill knew this and refused the offer, opting for a court battle instead.[27]

The "Railroad Crossings Case" that followed was all about private property but it was fought in the name of public access to Seattle's waterfront, with each combatant accusing its rival of a plan that would ruin the city's future. In their battle to sew up access, the railroads ignored what effect their digging, dredging, and building might have on the coastline. The superior court in Seattle found in favor of the Great Northern, thanks to heavy lobbying by Burke and the city's major newspapers, all pro-Hill, but the state Supreme Court, reversing the decision, forced the Great Northern to pay for the right-of-way across the Oregon Improvement Company's franchise. Although Hill lost in court, the Great Northern seemed to have succeeded where all other railroads had failed. It had secured a clear and uncontested claim to Seattle's harbor. The eventual outcome proved more uncertain. The new city engineer, R. H. (Reginald Heber) Thomson, who released a report in 1893 in the middle of Hill's legal battle, was critical of the rail baron's strategy to pick his terminal site without considering the needs of the city or other railroads. Thomson worried that laying more tracks would obstruct sewer drainage, increase train traffic, and constrain port facilities. Railroads had monopolized the waterfronts of other West Coast cities, and Thomson wanted Seattle to avoid the same fate. He told the Great Northern to build a tunnel under downtown to reach its proposed depot. Ever the thrifty executive, Hill, who had warned Burke that Thomson was "no dammed fool," promised to consider the idea.[28]

Seattle's business and political power bosses had captured their own transcontinental line. Now, they had to pay the costs. The railroads that made some local landowners wealthy had also made a mess of the city's waterfront, unleashing legally sanctioned mayhem masquerading as urban growth. Almost two decades of speculation had left a jumble of docks and flophouses, saloons and wharves, next to Railroad Avenue. The social results of this reckless growth were undesirable; the environmental effects were perhaps even worse. Locomotives on the elevated tracks above threw embers and oil onto businesses and passersby below. Another devastating fire was a constant threat. No provisions were made for adequate grade crossings, so rival railways laid their tracks wherever there was room, while beneath the wood-planked streets, saltwater snails called teredoes, the bane of wooden ships and piers, were eating away at the foundation on which the whole waterfront now stood.

The Opportunity Which Nature Has Prepared

It is easy today for most Americans to ignore "grade separation," the term used by engineers and urban planners to build systematically for transportation routes, utilities, and roads, an invisible and mundane function of cities. For urbanites at the turn of the century, separating trains from pedestrians, automobiles from sidewalks, and waterways from wharves was more than a matter of boring logistics. It was an issue of economic necessity and public safety, but grade separation was only the first of many challenges demanding systemic solutions. The arrival of the transcontinentals generated still more fury among businesses and city politicians over the best way to develop Seattle's waterfront and, by extension, the region's abundant rivers and waterways. Erecting dams or digging canals to alleviate chronic flooding problems and improve navigation would affect the scope and direction of development on the downtown waterfront by dictating where railroad tracks and switching yards could be located safely and permanently. All of these changes would ultimately influence the environment and ecology of Elliott Bay and the Duwamish River. At first, the costs went unnoticed. The individual speculators, investment syndicates, and railway companies that owned tidelands filled their plots in piecemeal fashion, close to shore, and in shallow water, to make landed property for their own purposes without concern for other users. Captain Thomas Symons, the regional chief of the Army Corps of Engineers, remarked in 1894 that resolving all of these problems simultaneously would "require a high order of engineering skill." It was a task unfit for realtors or corporate chieftains.[29]

Territorial governor Marshall Moore's metaphor of the railroad as the alchemy of empires in his 1867 address was inaccurate. It was engineers, backed by entrepreneurial financing or the federal government, not locomotives, who were the true alchemists of the age. The railroad was merely their vehicle. And if railway engineers were the true alchemists, a territorial governor turned real estate speculator was their wayward apprentice. Eugene Semple, Washington Territory's penultimate governor, was an advertisement for the excesses of the Gilded Age. A huckster with a title who was fascinated by waterways, Semple, like many gamblers, had an inflated sense of his skills and luck. Before embarking on his political career, Semple had been a newspaper editor in Oregon, inveighing against the Oregon Steam Navigation Company's cornering traffic on the Columbia River. Semple took this lesson from his battle: monopolies were always bad for business unless you owned the monopoly. After a two-year term as territorial governor in 1887–89, he lost a bid to become the new state's first governor, and then used his connections to secure a post on the Harbor Line Commission, the public agency charged with resolving the contentious tidelands dispute. In public, Semple was an aggressive proponent of citizen control of the tidelands; in private, he used his new position to determine which tidelands were best for his exploitation. He chose the marshy flats at the mouth of the Duwamish River, adjacent to Seattle's promising port.[30]

After helping to broker the 1890 tidelands act, Semple left the commission and incorporated the Seattle and Lake Washington Waterway Company. Former state governor Elisha Ferry served as titular company president, with Semple holding the real power. His plan, on paper at least, was deceptively simple: to excavate a sea-level canal through Beacon Hill, south of downtown Seattle, thereby connecting saltwater Elliott Bay with freshwater Lake Washington. Once the canal opened, it would lower the lake, cut off its original drainage through the Black River, and curb the disastrous flooding. But the real aim of what came to be called the South Canal was to make landed property. In an abstract sense, Semple's canal was no different from a railroad. Both were advertised foremost as systems of public transportation, yet both were ultimately used to acquire and sell off real estate for private gain. The ideals motivating Semple and the railway builders, from Villard to Burke, were essentially the same. Since the state government had defined tidelands as public property available for private development, Semple assumed he would receive the right-of-way grant to build in the name of the public good. Dredges would cut the canal through the tidelands at the mouth of the Duwamish River to the base of Beacon Hill. There, gigantic hydraulic cannons would sluice dirt and gravel from the hillside into enormous flumes to fill the tidelands, turning them into salable property. The land sales would help to fund operations, and Semple anticipated that the city and the state would meet any shortfall in light of additional future tax revenues from industries locating on his newly made lands.[31]

Semple's idea was unoriginal but his timing seemed impeccable. Seattleites had been talking about canals between Lake Washington and Puget Sound since the 1850s, with at least six separate routes already surveyed or proposed by the time Semple hatched his own scheme. In 1861, Harvey L. Pike dug a ditch between Lake Washington and Lake Union to transport coal from the mines in southern King County, but he found the task too overwhelming. The Lake Washington Improvement Company, formed in 1881 by several Seattle businessmen, promised to widen Pike's ditch and complete the canal, and, with the help of Chinese contract laborers, it linked Lake Union to Salmon Bay in 1884, building another extension, called the Portage Canal, to Lake Union and Lake Washington the following year. Meanwhile, farmers in the Duwamish and Green River valleys, besieged by ceaseless flooding, petitioned the state to build an even larger canal to lower Lake Washington. The railroads, too, had considered canals and waterways to aid their operations. The Northern Pacific's initial survey of 1863 had outlined the problems construction crews would face in Puget Sound: a high water table, flood-prone rivers, steep grades, and extreme tides. Reconnaissance by Henry Gorringe of the Oregon Improvement Company, in 1881, confirmed the earlier survey's findings. In both cases, the railroad surveyors proposed digging a sea-level canal between Lake Washington and Elliott Bay to lower the lake, reduce flooding, and help in the filling of the Duwamish River delta.[32]

Like many Americans in the nineteenth century, the citizens of Seattle had turned technology into a fetish, endowing it with miraculous powers to transform the physical environment, reorder society, and deliver endless material abundance. Canals and railroads garnered special attention, and both had shaped how Americans understood their connection to location as a result. After the completion of the Erie Canal in 1825, merchants in New York City came to consider Buffalo residents their neighbors, while in Chicago, every time railroads pushed farther into the pine forests of Wisconsin or the farmed prairies of the Dakotas, the mental maps of its citizens expanded accordingly. New technologies ushered in a new sense of place in both locales, and the same was true in Seattle. Yet in their adulation of railroads or canals, Americans tended to forget that such technologies were ultimately human, their powers often limited by politics and financing. Neither the railroads nor local citizens wanted to pay for the proposed canals, nor did the Army Corps of Engineers, which balked at taking on flood control because Congress had refused to expand its mandate to include river reclamation. Repeated entreaties from farmers and Seattle businesses, beginning in 1867, beset by flooding and needing canal transport, failed to move corps officials to do more than map possible canal routes. After concerted lobbying by Seattle's congressmen, the agency, in 1891, bowed to pressure and assigned a Board of Engineers to study the proposal the following year. After reviewing the plan, corps engineers dismissed a canal as unnecessary given Seattle's uncertain economy.[33]

Meanwhile, the building free-for-all along the city's waterfront had made an even worse mess of the wharves and docks, trestles and bridges, provoking R. H. Thomson, the new city engineer, to ask the Board of Tideland Appraisers for an immediate solution. Thomson recommended his colleague and fellow engineer, Virgil G. Bogue, architect of the Northern Pacific's line over Stampede Pass in the Cascades, an engineering marvel given the steep grades and massive snowfall that hampered construction, to devise a master plan for development. Bogue's report, released in January 1895, both praised Seattle's harbor, comparing it to San Francisco and Antwerp, and criticized the lack of a common carrier terminal site to handle trade between oceangoing ships and transcontinental lines. "Seattle has, in her tide flats," Bogue held forth, "the opportunity which Nature has prepared," but that ideal harbor was possible only if city and county officials planned systematically. The Board of Tideland Appraisers outlined the problem in its letter thanking Bogue for his "providential" report on the "contradictory and confusing" tidelands problem: a landscape "cut up by waterways and dotted here, there, and everywhere with improvements." "The difficulties connected with laying out a great addition to a city under such circumstances were many and soon became apparent," the appraisers lamented. "The more we struggled with them, the more we became involved in doubt and uncertainty."[34]

Semple's audacious canal and real estate scheme only added to the confusion because he had unveiled his idea to a land-hungry public. He had already divulged his plans to Captain Thomas Symons of the Army Corps, a close personal associate, and to Philip Eastwick, a civil engineer also working for the corps in Portland. Indeed, he may have known of the corps's decision not to build the proposed Lake Washington canal in advance, waiting to submit his proposed contract to the state land commissioner until after the federal government announced its position. In any case, he dispatched his nephews, Edgar and Henry Ames, to strong-arm Governor John H. McGraw into approving the contract. The governor easily assented. With loans from the Mississippi Valley Trust Company of St. Louis, Semple began work in July 1895. By the following summer, his dredges had carved nearly two thousand feet of waterway and filled almost seventy acres of tidelands.[35]

Jacob Furth, president of Seattle's Puget Sound National Bank and a powerful financier, had scoffed at Semple's boasting that he alone could stop the flooding and fill the tidelands. From the beginning, financing for Semple's venture was shaky. The motto of Semple's Seattle and Lake Washington Waterway Company—"we fly with our own wings and make the meat we feed on"—was at odds with his constant pleading for funds from bankers like Furth. "I have carefully considered the financial propositions connected with the establishment of the water ways," Furth explained to Semple, "and have come to the conclusion that I cannot do anything with it." The banker's prudence had basis in political fact. Several industries located near the canal route had asked the state government to stop Semple. "I am thoroughly convinced," wrote the president of the Seattle Malting and Brewing Company, one of the largest enterprises in the area, "that the proposed system would greatly retard the improvement of Seattle harbor."[36]

Like the paper railroads that crisscrossed Puget Sound, routes planned and licensed but never built, Semple's project seemed to be another corrupt creature of politics. His vision frightened many because it was a kind of uncontrolled growth machine. Rival tideland developers who had preceded Semple decried what they saw as the monopolization of public lands and sued Commissioner Forrest to stop Semple's dredges. Under state law, once tidelands were filled, patented, and sold, upland owners could extend their rights farther into the submerged lands. Each time tidelands became dry land, the water beyond their edge might be reclassified as tidelands and thus open to improvement as real estate. The cycle could continue until either the water became too deep or the limits of technology halted the forced march of the earth into the sea. Given the state laws favoring private tideland development, the courts could not compel Semple to quit.[37]

Uniting under the banner of the common good, opponents of Semple's canal now pushed for federal intervention by the Army Corps of Engineers. Advocates for a northern canal route, to follow the natural drainage basin of Lake Union into Puget Sound, cited the 1891 corps report, which focused on the preexisting Portage Canal. But their rhetoric, like Semple's own propaganda, was a stalking horse for personal interests. The attorney Daniel Gilman, owner of several large parcels fronting Salmon Bay and Lake Union, wrote to Judge Burke, who owned even more land in the same area, that a northern canal would spawn "a boom that will lift us out of want." Gilman and Burke's North Canal scheme was a mirror image of Semple's South Canal plan except that, instead of raising private funds and hoping to attract public support, the North Canal plotters proposed to enlist the help of the federal government first, and then entice investors to follow. It was a dangerous tactic, and Gilman cautioned Burke to be careful in lobbying so that public enthusiasm for the North Canal was not "let loose in too many directions."[38]

After repeated pleas from city businesses, the Army Corps, in 1895, appointed yet another Board of Engineers led by Symons, moonlighting as a private consultant for Semple's company, to determine the best canal route. The board reported that the South Canal posed significant risks. Digging through Beacon Hill, a ridge of glacial debris some three hundred feet high riddled with large boulders, was a major engineering obstacle. The northerly route, only twenty-five feet above sea level, on the other hand, was comparatively simple and the easiest to engineer. After the corps announced its findings, North Canal advocates, upon finding that Symons had colluded with Semple, complained to his superiors; Symons later withdrew from future deliberations over Seattle's canal. Not until 1898 did the corps even agree to consider building a canal along the route favored by Gilman and Burke through Shilshole Bay to Salmon Bay and Lake Washington. In anticipation, the city council condemned the lands, transferred the right-of-way to the United States two years later, and waited for the Army Corps of Engineers to start construction.[39]

What had finally convinced the corps to consider building a canal at all was the Klondike Gold Rush of 1897–98, Seattle's first sustained economic boom and its chance to escape California's economic clutches. When the news of the discovery in Canada's Yukon Territory reached Seattle in July 1897, gold fever gripped the town. The city mayor, William D. Wood, on business in San Francisco, wired his resignation and headed north. But the real riches lay in provisioning the thousands of would-be prospectors streaming north en route to the Klondike, because Canadian authorities demanded that they be prepared for a year's stay in the backcountry. Local merchants, such as Nordstrom, the now famous clothier, capitalized on the need for sturdy shoes. Other purveyors sold winter clothing, nonperishable food, guidebooks, and transportation to Alaska. Seattle businesses alone outfitted over 25,000 miners, more than any other city, and returns were staggering. Wholesale grocers reported sales of $6.5 million, and the clearings at eight banks increased from $33.3 million to $67.3 million in a single year. There was a demographic echo as well: in 1890, Seattle's population stood at 42,837, but two years after the Klondike strike the population had nearly doubled, to 80,671. Seattle merchants plowed the profits they made off the miners and new arrivals back into building more business. The city's newly energetic Chamber of Commerce employed Erastus Brainerd, a local journalist and entrepreneur, to spearhead a promotional campaign to sell Seattle as the gateway to Alaska and the Northwest. Brainerd wrote for the *Seattle Argus,* a local paper, invoking natural law and historical inevitability to explain Seattle's good fortune: "The current had set strongly toward Seattle, for the people of that city had been working for years and dug their channels wide and deep, foreseeing that a great stream might flow through them." Now, a "mighty torrent" was passing through Seattle and trade, "like water," would always follow "the line of least resistance."[40]

By comparing Seattle's burgeoning trade empire to a river, Brainerd employed a metaphor common to the language of urban boosters. In this formulation, trade followed the same God-given commandments as water or other physical phenomena with a few substitutions, like replacing gravity with capital. Getting the metaphor right was important because it made a city's location and consequent economic achievements seem as natural as a river flowing to the sea. Yet the image was incomplete without acknowledging that humans had worked hard to turn the rivers of trade to flow through their city. North Canal partisans recognized Brainerd's rhetorical talents and hired him to join their cause. The battle to build a canal for Seattle would now swing on which side best wielded their figures of speech.[41]

Brainerd tried to deny Semple the metaphor of nature's laws, so Semple had to fight all the harder against the shifting political currents to convince investors that his own project was still viable. Investors were backing out and the corps had cast doubt on his plans. In desperation, he asked the major railroads to form an alliance with him while trying to persuade the federal government to purchase his company's contract. Since construction was under way, he argued, the city and the corps would save money and time. But Semple had underestimated his opponents, and the resulting public outcry prompted congressional hearings in 1902 to settle the canal controversy for good. South Canal partisans at the hearing said their option, the shortest connection from Elliott Bay to Lake Washington, prescribed "a new regimen" for the flood-prone Duwamish River. Filling tidelands had also accrued visible benefits, making "solid and substantial land" and promoting growth "along legitimate and natural lines." North Canal backers sent Brainerd to lead the fight, and the busy journalist turned propagandist churned out broadside after broadside, asserting in one pamphlet that a chasm through Beacon Hill would split the city in two and quickly fill up with silt from the Duwamish River. The northerly route via Lake Union, in contrast, would "carry out nature's outline but uncompleted purpose." Brainerd cited geologists and other experts to demonstrate that the North Canal would reconnect what was really a "single hydrographic system." He ended his plea with a challenge: "Nature and man have done wonders in this region of the Aladdin touch. . . . Shall Congress alone block the road or will it smite the rock and cause the waters to flow?"[42]

Following the hearings, the corps remained unmoved and refused to proceed on the North Canal despite the Klondike boom. And Semple's troubles only grew. The clay-filled soils along Beacon Hill resisted the hydraulic cannons, slowing the excavation and tideland filling, and silt clogged the dredges tearing up the Duwamish River. Local residents also complained that the sluicing and dredging were wrecking their homes and businesses. With the canal sluicing operations consuming water needed for the city's reservoirs as well as public works projects, like the massive hydraulic cannons used to level Seattle's downtown hills, in 1903 city engineer R. H. Thomson forced the company to close down. Semple resigned soon thereafter as company president. He wrote his old confidant, Thomas Symons, comparing himself to the slave in William Shakespeare's *The Tempest,* bound by contract to keep working for nothing: "'The wand of Prospero' has been taken from us, and we will now 'toil like Caliban' in the common place business of filling tide flats." In his efforts to save his company, Semple had agreed to fill tidelands for the various railroads, and dredges slogged along for almost a decade until the contracts expired and the company disbanded. All that remained of Semple's dream afterward was a gash in the side of Beacon Hill.[43]

Semple's literary allusion contained something more than the bitterness of dashed hopes. Like the railroad, his dredge operators and engineers had alchemized Seattle's waterfront, turning what they saw as valueless mud and water into solid, salable land. His luckless adventure was thus more than a footnote of failure. The South Canal fracas laid bare an important difference of opinion over how best to develop Seattle's tidelands and commercial waterfront. In the minds of his supporters, Semple's scheme was the apotheosis of free enterprise harnessed to provide for the common good. Semple would dig the canal, transmute seawater into land, and share the profits with a grateful city. To his detractors, Semple was venality incarnate, a former public official who used his position for personal gain. If the South Canal succeeded, his opponents maintained, his company would lock up the waterfront as private property. In the end, Semple's failure did not matter. The alchemic reactions continued and even accelerated. In 1890, an apron of salt-encrusted mud and rocks hemmed in the city twice daily. Ten years later, warehouses and wharves sat where the tides had once flowed. Semple did not reap the rewards, but his idea of place had transmogrified Seattle.

A Demoralized and Dissipated Landscape

On a rainy night in January 1900, a group of men conspired to bomb a river into submission. At a point where the White and the Stuck Rivers flowed less than a thousand feet apart, about twenty-five miles south of downtown Seattle, farmers schemed to plant dynamite, breach the two channels, and steer the White northward. Heavy rains that autumn and heavier snow that winter had sent floodwaters of the White River into the Stuck River, and southward onto lands of rural Pierce County, near Tacoma. Neighboring farmers in King County, thankful to be spared, had refused to help. Their ill will was warranted. Two years earlier, the Pierce County farmers had successfully pulled off the same plan with unwanted results; the explosion threw the river back into their laps. Jubilant farmers in King County quickly built a dam to make the diversion permanent. Now, the Pierce County farmers wanted revenge. A watchman hired by King County to defend the river banks found the dynamite and squelched the plot. It was only the latest episode in an ongoing melodrama that had played out for almost a decade whenever logjams, the result of flooding and lumbering upstream, blocked the channels. Years of dynamiting and damming had permanently changed the course of the two rivers, making it all but impossible to determine who was at fault, if anyone, for the recurrent flooding.[44]

Flooding had played havoc with the new property regime that urban boosters and corporate landowners had tried to impress on Seattle and its adjacent watershed. Small farmers took the law into their own hands because, for all of its power, the law proved an inadequate instrument to safeguard cherished places. The long dispute over how best to develop Puget Sound's tidelands showed that the law tended to side with those who had money and power. Even then, the law could offer only limited protection against unpredictable nature. Entrepreneurs, like Eugene Semple, and speculators, like Thomas Burke, could fill and buy all the property they wanted, but their purchases were valuable only if spared from inundation. The same was true of the best land for laying level railroad track; it was almost always in the region's floodplains, and Northern Pacific Railroad engineers had complained for years that flooding endangered railway operations. Now, farmers and railroad engineers were at odds. The Northern Pacific tried to dredge certain sections of the Duwamish and White Rivers in 1901 and 1903 to protect its tracks. Nearby farmers, worried that the dredging would worsen flooding, threatened to sue and the engineers pulled back. With the arrival, in 1906, of the corps's new head engineer for the Seattle district, Hiram Chittenden, help for everyone seemed to be on the way. For if Semple was an advertisement for Gilded Age greed, Chittenden carried himself like a paragon of Progressive Era virtue, a man who claimed to weigh the facts in acting for the greater good.[45]

Before his assignment to Seattle, Chittenden, a West Point graduate, had designed the original road system in Yellowstone National Park and served as a consulting engineer on the Ohio River and in California's Central Valley. He originally opposed federal efforts to control flooding, but his experience on the Ohio River and the Sacramento River in California convinced him of his error, and in 1896 he wrote an influential congressional report that encouraged the corps to construct reservoirs and clear away river debris to protect property and promote irrigated farming. Chittenden's report revolutionized the agency and helped to transform the corps into a flood control advocate. In his spare time, Chittenden became an eminent historian of the fur trade, his scholarship steeped in the ideology of a progressively moving frontier advancing westward across the continent, his theory of history paralleling that of his contemporary Frederick Jackson Turner. But whereas Turner often lamented the passing of the frontier, which he considered the mainspring of American exceptionalism, Chittenden believed in manifest destiny as a renewable resource, justifying continued expansion into the Pacific basin. His own studies of the fur trade, which were infused by both Turner's idealistic nostalgia for an agrarian democracy and an aggressive sense of nationalism inspired by Theodore Roosevelt, presented a version of the frontier thesis in which imperialism was the fount of American regeneration. After arriving in Seattle, his final post, he came to see Seattle's languishing canal as another step in America's inexorable journey westward.[46]

An astute politician, Chittenden manipulated popular sentiment to push for the North Canal route as the most practical solution to Seattle's transportation and flooding woes. He found his opportunity when James A. Moore, a real estate speculator who owned property along the route, put in a private bid to finish the project. Moore's application won approval from Seattle voters and, in early 1906, gained congressional backing. His project was little more than yet another scam by local investors to alarm the corps into action. This time, however, the investors had the support of a corps engineer. Chittenden persuaded Moore to transfer rights to another group, the Lake Washington Canal Association, and prepared to renew the fight for federal assistance. North Canal boosters, buoyed by Chittenden's campaign, coined a new term for their city's determination to overcome adversity at any cost: the "Seattle Spirit." Local newspapers appealed to the Seattle Spirit in weekly articles and editorials; advertisers for homes, warehouses, and real estate parcels summoned prospective buyers with similar entreaties.[47]

Chittenden's biggest boost came in the late autumn of 1906, when a Chinook wind, a warm burst of air laden with moisture, melted the Cascade snows and turned rivers and creeks into battering rams. Rising waters tore through the misshapen channels separating the White and the Stuck, throwing up a logjam at the confluence. People living in small towns from Kent to Pacific awoke on the morning of November 14 to find the rivers racing down their streets. They fled for their lives. After the floodwaters ebbed, Chittenden convened a board of engineers, who reported that flooding had snapped railway bridges, wrapped farms in blankets of silt, and nearly obliterated whole towns. In vivid anthropomorphic language, redolent of Progressive Era principles, he said the flooding had "demoralized" the channels and left the rivers "dissipated." The disobedient rivers needed "discipline" and reform, and the corps would provide the supervision and therapy: new dams and dikes, dredging and straightening the Duwamish to channel floodwaters to the sea, diverting the White River into the Stuck so it would flow toward Tacoma, and completing the North Canal to lower Lake Washington. The ambitious design would require "a complete change in the route by which a large river [the Duwamish] is carried to the sea," but "the exceptional character of the work" justified "exceptional measures for carrying it into execution."[48]

After a flurry of meetings and hearings, state legislators approved a local improvement district, proposed by Chittenden's board of engineers, to raise $1 million for the Lake Washington canal. They ignored warnings from a few sportsmen that building dams and dikes to control episodic flooding would have far-reaching effects: closing off spawning habitats and eliminating wetlands would ruin salmon fishing and waterfowl hunting. "If you didn't know the west coast country and read newspapers alone," the editors of the *Pacific Coast Sportsman* wrote in November 1906, "you would think that the mountains were tumbling down and the whole country being swept into the Pacific Ocean." "The damage amounts to very little," they concluded, and the flood itself left behind a gift for farmers and wildlife alike: river mud, "which is the best fertilizer for land that the world holds." The majority of King County residents, waterlogged and fearful, disagreed and three years later, in 1909, voted to create a commercial waterway district to remove the bends in the Duwamish River and shorten its length by nearly ten miles. The following year, in 1910, the Army Corps of Engineers agreed to finish the canal between Lake Washington and Puget Sound. Representatives from King and Pierce Counties and the Northern Pacific Railroad settled on where the White River would go: a series of dams and levees would send it permanently south into the Puyallup River, near Tacoma. King County would pay the bulk of the cost, and its farmers cheered.[49]

By late 1914, with the start of World War I in Europe, dredges and steam shovels were slinging dirt and mud along almost every major river and lake in urban Puget Sound. In the span of almost five years, engineers rerouted the plumbing of an entire drainage basin. It was as if someone pulled a plug and a giant sink emptied. When corps contractors completed the Montlake Cut and the locks at Ballard in the summer of 1916, Lake Washington poured into Lake Union, dropping the water level around the lake by almost ten feet in three months. With the new Lake Washington Ship Canal, sloughs along the Sammamish River on the eastern shore dried up and marshes emerged from open water in Union Bay, near the Montlake Cut. As the waters receded, houseboats and businesses on Lake Washington and Lake Union found their sewer outfalls dumping onto exposed mudflats. At the southern end of Lake Washington, engineers had turned the Cedar River, which had flowed into the Black River, into Lake Washington to provide sufficient water for the canal. When the locks were opened, the Black River began to disappear. More than one hundred acres of new land emerged. Joseph Moses, a Duwamish Indian, remembered how the receding waters left canoes stranded on high ground and trapped fish in the remaining pools: "People came from miles around, laughing and hollering and stuffing fish into gunny sacks."[50]

Chittenden had imagined a plan resonant of Progressive Era ideals: public-minded engineers allied with private interests to produce solutions to protect the common good. Yet waterways that were public on paper were often private in practice, leaving them vulnerable to continual tinkering. With each subsequent "improvement"—a filled estuary here, a new waterway there—the economy and the physical environment both grew more erratic. Taking rivers and lakes apart to reassemble them meant taking apart and reassembling preexisting property as well, and Chittenden was unprepared for this and also unprepared to face the fact that the physical environment eluded easy human control. Almost every year after canal construction began, the corps and its private contractors battled muskrats that burrowed into the banks along the right-of-way for the North Canal and the small earthen dam at the outlet of Lake Union designed to regulate drainage into Salmon Bay. The corps built stronger barricades; the muskrats kept undermining the dam. On March 13, 1914, the muskrats won, the dam broke, and the level of Lake Union fell by ten feet in a few hours, ripping almost two hundred houseboats from their moorings. The gush of water swept away two city bridges and the pier to a major lumber mill.[51]

Noxious rodents were only the start of Chittenden's headaches. Many others were the Army Corps's own fault. To cut expenses, Chittenden had recommended a single set of locks for the canal at the head of Salmon Bay to equalize the elevation of Lake Union and Lake Washington by a water-level trench. But what made sense from an engineering blueprint threw the previous arrangements made by Burke, Gilman, and other landowners into disarray.

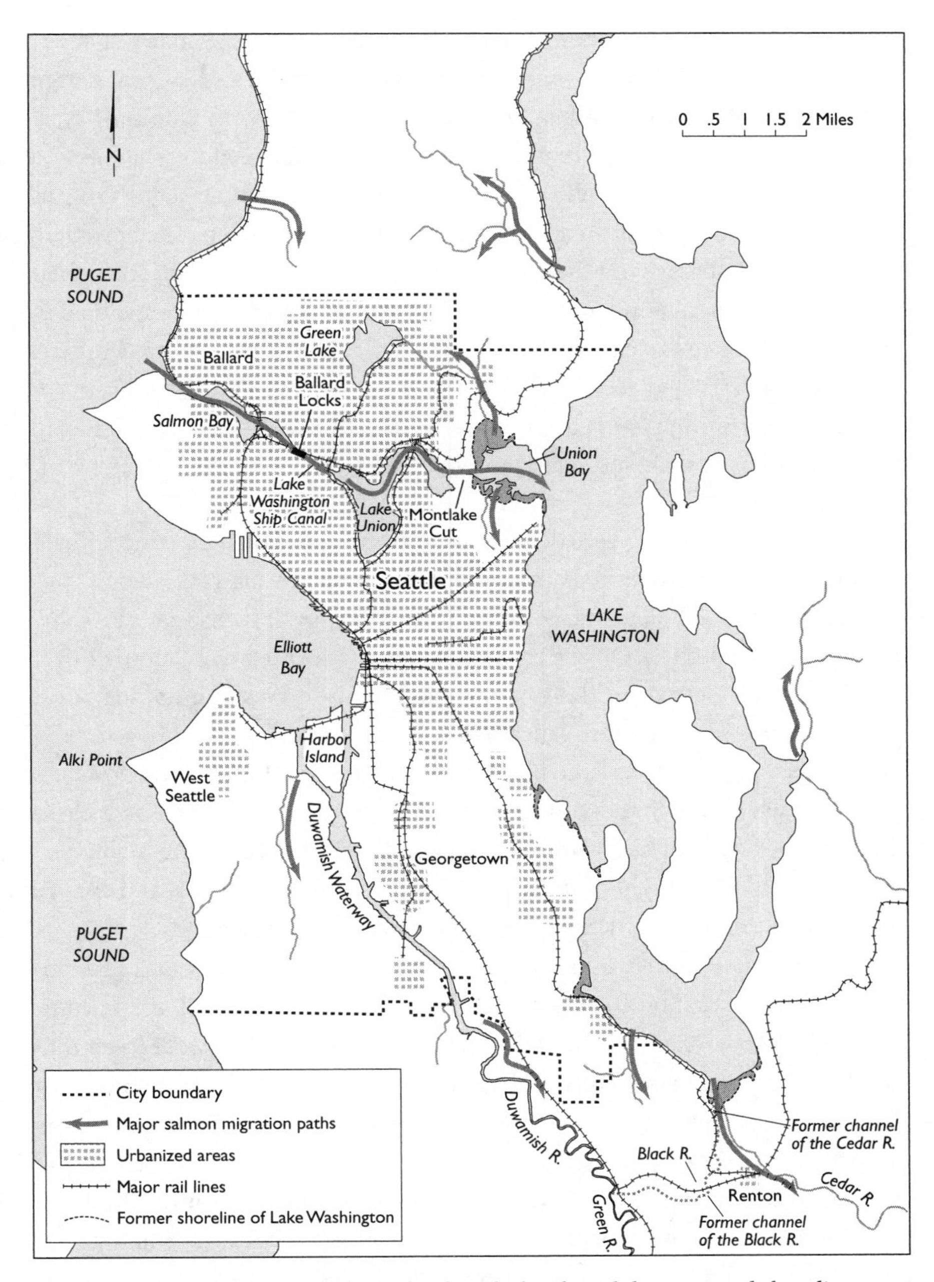

Seattle and its immediate hinterland, with the altered drainage and shoreline of Lake Washington and the Duwamish River, c. 1925.

All of a sudden, the investments that Gilman believed would spawn a boom were endangered. He had reason to worry. The problem was the canal design itself. The Seattle Cedar Lumber Manufacturing Company in Salmon Bay, at the western end, had argued that Chittenden's canal damaged its wharves and millponds when the water level was raised there to accommodate ship traffic. Lumber companies at the canal's eastern end faced the opposite problem when the canal opened and the water dropped, cutting the mills off from their wharves and ships. As one mill operator, William L. Bilger, had complained in an affidavit prior to the canal's completion, lowering the lake would eliminate "free and unobstructed access" to his docks and depress the value of his property. His predictions proved correct, yet the state Supreme Court, which had to decide who was at fault and who deserved damage awards, did not offer consistent solutions.[52]

On one hand, in two cases decided in 1916 and 1920, the high court upheld the doctrine of private property and ruled that the state and county must pay compensation based on damages linked to specific physical changes caused by the canal's construction. On the other hand, in a 1916 case involving the Puget Mill Company, located on the eastern shore of Lake Washington, the court upheld the doctrine of shore lands as public property. "Our holding here is, and we need go no farther," the justices ruled, "that the state has the power to determine where the line dividing these two classes of lands shall be located upon the ground, and has lawfully done so." In six other cases, the high court further defended the right of the state to levy canal-building taxes and change preexisting waterways in the name of improved public navigability.[53]

Legal wrangling over private property rights and the public good had also delayed construction of the planned Duwamish Waterway as well as the completion of Commercial Waterway District No. 2 to reroute the Cedar River into Lake Washington permanently and put a stop to flooding along the now dry channel of the Black River. In one 1913 case, a shadowy group called the Seattle Factory Sites claimed potential damages to their unimproved shore lands at the lake's southern end if the dredging of the Cedar River continued. The high court ruled against the consortium using the same logic it later employed in the 1916 Puget Mill Company decision. Yet, in a case from 1915, the justices ruled that rapid changes to the physical environment compelled the court to draw clear property lines where none had existed before. The plaintiff, William Wardell, had asked the court to uphold his title to lands along a bend of the Duwamish River, which, he claimed, included the riverbank, so that he would receive damages after the waterway split his property in two. The problem was that original title, including rights to the riverbed itself, was no longer valid in light of the new laws giving the state jurisdiction over navigable waters below the low-water mark. Compounding Wardell's problems was the Duwamish's shifting course after the original survey, in the aftermath of tides and floods. Citing earlier principles of sovereign state rights over navigable waters and shore lands, the justices ruled against Wardell.[54]

Ultimately, the courts proved once again to be unreliable arbitrators of the many meanings of place, bound as they were to legal precedent and the prevailing secular faith in market forces to apportion costs and benefits equitably. In some cases, if defendants could prove that fluctuating water levels yielded specific injuries to property or business enterprises, they were entitled to compensation. The problem lay in setting the dividing line between what was public and what was private. As the courts tried to establish clear boundaries between the two types of property, the line kept shifting in response to the increasingly complex alterations of the region's rivers, lakes, and shorelines. What was land one day could be water the next before switching back again depending on human actions and nature's unpredictability. The effects were costly for property owners, but they would prove devastating for creatures in the waters, marshes, and estuaries that had once surrounded Seattle.

Once construction of the Ship Canal was under way, the falling water table around the Lake Washington basin reduced the volume of water flowing seaward in the Duwamish River and smoothed the rapid completion of the industrial waterway. Dredges operated by the Puget Sound Bridge and Dredging Company completed Semple's work and replaced the estuaries at the Duwamish River's mouth with Harbor Island. The world's largest artificial island at the time, spanning almost 350 acres, it was separated from the mainland by the east and west branches of the new Duwamish Waterway. Filling the tidelands was made easier by additional changes farther upstream. Diversions by the Tacoma Water Department from the Green River in 1912 had already subtracted water from the Duwamish. Turning the Cedar River permanently into Lake Washington, and pushing the White River into the Puyallup River, took even more water away. Without the full volume of its former headwaters, the diminished river, carrying less silt and debris, made the job of dredging and filling far easier.[55]

On a clear July day in 1917, before the waterway's official opening, the editor of the *Duwamish Valley News*, gazing from the balcony of the Smith Tower in downtown Seattle, then the tallest building west of Chicago, remarked how "the winding Duwamish River has given place to the straight Duwamish Waterway." The transformation of the Duwamish was as dramatic to human observers as it was destructive to fish and wildlife. The United States had entered the war in Europe that April, and wartime exigencies had pressed the remade river into military and industrial service. Two decades earlier, the Duwamish River had been a very different place. An 1894 article, written in anticipation of Semple's South Canal project, described the lower river valley as a bucolic paradise. The slow-moving river, flanked by earthen dikes, meandered past lush farms growing produce for Seattle markets and acres of hop vines. Indian and immigrant day laborers picked hops and vegetables in season, or harvested berries and caught salmon to store for the winter or sell to whites in the city and surrounding countryside. This idyllic portrait was more elegiac reminiscence than optimistic advertising; the repeated pairing of the rural with the "hurly burly push of Seattle" pointed to the river's expected and ultimate fate. Now, as the newspaper editor observed, gone were the hop farms and produce gardens from the river's lower reaches, replaced by creosote plants, shipyards, flour mills, meat and fish packers, auto and aircraft manufacturers, lumber mills, and boiler works. The natural riches of Seattle's hinterland—tall spruce trees from the Cascade Mountains, fresh salmon from northern Puget Sound, wheat from eastern Washington—all flowed into the Duwamish before their transformation into aircraft, canned fish, or flour to reinforce the Allied arsenal in Europe.[56]

From his perch atop the ultimate symbol of urban power, the skyscraper, the newspaper editor confidently predicted that he could see Seattle's future. His was the dreamscape imagined by Burke, Semple, and Chittenden, if for very different reasons: the enchanted pairing of technology to environment and of market to location. But this was a phantasmal landscape because, despite the semblance of prosperity, the completion of the waterways and the filling of the tidelands had unmasked the environmental and social complexities of conjuring up property. Like the rivers and estuaries that needed taming, the rising currents of capital coming to Seattle had also tended to flow in unpredictable directions. Boosters for the Ship Canal had hoped that completing the water link would bring industry to Lake Washington. Investors put their money elsewhere. Few factories and shipyards gravitated to Lake Washington, and the steep hills around the lake made extensive construction difficult except at the southern end. Industries concentrated at the head of Elliott Bay and

Native workers in front of Yesler's mill and cookhouse, Seattle, 1866.
(Courtesy Special Collections, University of Washington Libraries, UW 5695)

Central Seattle waterfront on Elliott Bay looking north toward Denny Hill, 1878.
(Courtesy Special Collections, University of Washington Libraries, UW 2296)

Filling Elliott Bay tidelands with sluiced material from Beacon Hill on the South Canal project, 1901. (Courtesy Special Collections, University of Washington Libraries, A. Curtis 1769)

Lake Washington Ship Canal locks under construction at Ballard, 1915. (Courtesy Museum of History and Industry, PEMCO Webster & Stevens Collection 44,652)

Opening of the Montlake Cut between Lake Union and Lake Washington, August 26, 1916. The completion of the Lake Washington Ship Canal lowered the water level of the lake by as much as ten feet. (Courtesy Museum of History and Industry, PEMCO Webster & Stevens Collection 51,503)

Indian migrant laborer encampment on filled tidelands along the Seattle waterfront, just south of the downtown business center, c. 1900. (Courtesy Museum of History and Industry, MOHAI 90.45.14)

R. H. Thomson with Seattle map during his second and final term as the city's chief engineer, c. 1931. (Courtesy Museum of History and Industry, *Seattle Post-Intelligencer* Collection, P-I 1986.5.43455)

Workers building the Cedar River pipeline adjacent to the flood-prone Black River channel, 1899. (Courtesy Seattle Municipal Archives, Item 7258)

Photograph by Asahel Curtis of the first Denny Hill regrade, 1914.
(Courtesy Special Collections, University of Washington Libraries, UW 4812)

Houses demolished during regrading operations on Beacon Hill, 1910. (Courtesy Museum of History and Industry, PEMCO Webster & Stevens Collection 19,023)

Replanting trees on cutover lands above Rattlesnake Lake in the Cedar River watershed, April 11, 1935. (Courtesy Seattle Municipal Archives, Item 50648)

Ruins of the North Bend Lumber Company Mill in Edgewick, looking toward the Snoqualmie River, April 18, 1924. The "Boxley Burst" flood of December 23, 1918, destroyed the mill and workers' housing when water pierced the flanks of the Cedar Lake reservoir. (Courtesy Seattle Municipal Archives, Item 6683)

Lake Washington Boulevard looking south toward Seward Park and Mount Rainier, 1914. (Courtesy Special Collections, University of Washington Libraries, A. Curtis 31258)

Traffic and visitors' sign on Lake Washington Boulevard and Madrona Drive, 1910. (Courtesy Special Collections, University of Washington Libraries, Lee 534)

Family on a park bench in Schmitz Park, West Seattle, one of several city parks closed to automobiles, c. 1910. (Courtesy Special Collections, University of Washington Libraries, W&S 17682)

Washington Department of Game officials releasing hatchery trout reared at ponds in Seattle's Seward Park into the lower Cedar River, c. May 1948. (Courtesy Museum of History and Industry, *Seattle Post-Intelligencer* Collection, P-I 23045)

the Duwamish Waterway, prompting city engineer R. H. Thomson to call it the canal "of the greatest benefit."[57]

Engineers and entrepreneurs, convinced that they had divided water from land and put each in its proper place, now discovered that maintaining mimetic rivers and lakes required constant vigilance. Once the Lake Washington Ship Canal locks were opened in 1917, heavier salt water from Puget Sound pushed beneath the locks into Lake Union, and only the building of siphons and flushing mechanisms by the corps prevented further intrusion. With the Cedar River tapped as Seattle's municipal water supply and the Duwamish River prevented from flowing backward into Lake Washington, freshwater was not always available to resist the ocean's advances and to operate the locks. The problems with flooding throughout the entire remade watershed were even more severe. A U.S. Geological Survey report in 1913 found that rivers draining into southern Puget Sound had become more unpredictable and violent than before. Logging and railroad development in the Cascades stripped the river headwaters bare of the forests that held precipitation in check, worsening seasonal runoff and sending volleys of mud and debris to clog channels and compound flooding downstream. Timber cutting in the city's Cedar River watershed, for example, had intensified overflow into Lake Washington now that the river emptied directly into the lake. The corps had to open the canal locks during heavy storms to prevent water from pouring over the top. Along the Duwamish, floods in the winter of 1917–18 reprised the destruction of 1906 even as the *Duwamish Valley News* argued, halfheartedly, that damages could have been far worse.[58]

The reclaimed tidelands turned out to be just as unstable as the remodeled and artificial rivers. Building terminals, laying railway tracks, and erecting bridges required standard grades and solid earth, but filled tidelands were often more liquid than solid. As the fill subsided, sewer lines cracked, water pipes bent, and tracks and trestles warped. In digging a new slip at the base of the old Yesler mill in 1901, Northern Pacific engineers dislodged chunks of "slab wood and large rocks" from beneath the mud. This inferior fill was, the engineers claimed, responsible for the collapse of the White Star Line dock into Elliott Bay that same year. Worse still, cheap tideland properties had inundated the real estate market. When Semple's dredges had moved across the Duwamish delta, speculators flocked behind like seagulls trailing fishing boats, fighting to grab the lands left in the wake. After Semple's machinations failed, the momentum that impelled tideland development continued unabated. Realtors

Cover of C. B. Bussell, *Tide Lands, Their Story* (Seattle, 1903). (Courtesy Special Collections, University of Washington Libraries, UW 23445)

bought pages of advertisements in local papers, feigning an impending tideland shortage as a way to move property sales. One real estate agent, V. Hugo Smith, warned that in hilly Seattle smart buyers would "occupy the level land contiguous to the business center, the tide lands." "Why pay big rentals?" asked H. H. Dearborn, another real estate mogul, when buying tidelands made businessmen "their own landlords." Others played upon buyers' fears of missing out. R. Cooper Willis admonished readers to act promptly lest "the bogy man of despair and the green-eyed man of envy" afflict them. Another seller, C. B. Bussell, produced a pamphlet, *Tide Lands, Their Story,* to help readers "acquire the TIDE LAND HABIT." "Values do not come at once, but gradually, as these lands are improved and used," Bussell wrote, and "those who buy now will reap the profits." He was willing to sell lots "of any size to suit purchasers" and "bargain hunters."[59]

By the eve of the First World War, the choicest properties were already in the hands of the railroads and their land companies, and the railroads held onto their parcels and waited for the market to reduce the wild prices demanded by smaller speculators. In 1907, as the Great Northern Railroad was building its switching yard in Salmon Bay, speculators paid daily visits to agent L. C. Gilman with offers of still more land. "There are a few cases in which we have been able to get options without the payment of any money," Gilman told Louis W. Hill, son of the Empire Builder and the Great Northern's president. And in those cases, the most worthwhile plots "have been taken." All Hill needed to do was wire the order and agents would close on those options while "picking up any straggling pieces that may be had cheaply."[60]

Seattle business and political leaders, who had staked so much on attracting the transcontinental railroads to their city, were now living on railroad time and with a corporation's sense of place. The magic the railroad had bestowed upon Seattle's waterfront had made many wealthy, men like Thomas Burke and Daniel Gilman, but it had robbed others, like Eugene Semple, of their own fortunes. The railroads and the boom they spawned had closed off the city from its most valuable resource, its spacious deep-water harbor, even as the public-minded engineer, Hiram Chittenden, pushed to complete the Lake Washington Ship Canal in the name of municipal interest. All the canals and waterways built to enhance that harbor turned out to be liquid chains confining Seattle's shoreline commons and even the powerful railroads, which, in their quest to shove out their rivals, had lost money and were now burdened with acres upon acres of mud to their name. The business expenses were only part of the accounting facing Seattle's residents. Over the coming years, as Seattle's less fortunate were excluded from the shoreline they had once looked to for sustenance, or had to eat shellfish and fish contaminated by industrial pollution, another line on the ledger would call for attention.

And No Trespassers Are Allowed

The railroads that so many had yearned for now held Seattle's waterfront in their grasp. Their hold proved to be weaker than expected, however. In the first decade of the twentieth century, the struggle for municipal port ownership consumed Seattle as local businesses and civic leaders discovered that the railroads had them at their mercy, not only controlling shipping rates but also restricting physical access to the waterfront. It was a phenomenon not unique to Seattle and, by 1910, citizens in San Francisco, New York City, Philadelphia, New Orleans, and Los Angeles were fighting to regain partial or complete public control of their port facilities. The same individuals and organizations that had lured the railroads to Seattle now pushed to create a municipal port, one of the nation's first. In 1906, George F. Cotterill, a former assistant city engineer and state senator from Seattle, introduced a bill in Olympia based on Virgil G. Bogue's recommendations for tideland development, to create publicly owned port authorities. Five years later, in 1911, after a protracted battle against the transcontinental railroads, port advocates prevailed. Seattle voters approved in a landslide the first slate of port commissioners, including former corps engineer Hiram Chittenden, but victory at the polls did not translate into victory along the waterfront. Until the 1930s, the new port authority would be in a near-constant struggle against the railroads, private wharf and dock operators, and the city's two major daily newspapers, to whom municipal ownership was akin to socialism.[61]

The railroads' hold on the waterfront loosened not only because they faced new political opposition. The pell-mell construction unleashed by the earlier tideland boom and explosive real estate market had hemmed in the railroads, and, because of the previous actions of rivals, no one line could act as it pleased. The Northern Pacific had kept the Great Northern from building a second set of tracks along the waterfront on Railroad Avenue, and the Great Northern had to accept the suggestion of city engineer Thomson, offered back in 1893, to dig beneath the city, a plan James J. Hill had termed "utter madness." A crew of 350 men finished the mile-long tunnel at a cost of more than $1.5 million in 1904. Once Hill reached the tidelands south of downtown, he found maintaining and improving the filled tidelands to be costly. Unsupervised construction had left behind dangerous and expensive grade crossings, and building spurs to warehouses and docks would require negotiating with rivals for right-of-way. He would have to accede to city engineers' demands for proper sewerage, drainage, and street access, and deal with state and federal officials' complaints that dock repairs interfered with navigation. Local land agents for the Great Northern as well as the Northern Pacific called tidelands terrible investments and encouraged their ultimate sale at an undetermined date. As the railroads sat on their sour investments, investors drifted away, leaving behind a mosaic of open spaces and abandoned properties, interspersed among busy wharves and thriving industries, on the former mudflats and estuaries of Elliott Bay.[62]

As the new Port of Seattle was fighting to reclaim some of this space in the name of public utilities, others continued to view these landscapes as part of the former waterway commons. The oldest users of the former commons, the Natives of southern Puget Sound, persisted in visiting former hunting and fishing sites on their seasonal subsistence rounds. Indians pitched tents and beached canoes in accustomed locations, like the remnants of Ballast Island where Henry Yesler's former wharves once jutted into Elliott Bay. The whites who owned the waterfront lands were no longer as accommodating as the old lumberman had once been. Bitter conflicts were already commonplace because whites saw Indians as trespassers. In early March 1893, an angry landowner in West Seattle burned an Indian longhouse that had been standing on his waterfront lot, sending the inhabitants fleeing in their canoes. The lesson was clear: Indians could use places that whites did not yet want—like unoccupied tidelands—to provide things whites wanted, including shellfish, fish, and labor, so long as they did not stay.[63]

With many acres of tidelands unused or underdeveloped, squatting became endemic by the turn of the century. The itinerant laborers who had built early Seattle were now a persistent presence: Greek and Italian fishermen, African American stevedores, Swedish and Japanese loggers, Russian Jewish merchants, farmworkers from the rural Northwest, and Indians, all clustered along Elliott Bay and the Duwamish Waterway. Unlike the first inhabitants of Puget Sound, none of the new residents had prior claims rooted in deep tradition, yet like Indians, they, too, asserted their right to the tidelands and waterfront through subsistence use. "From the street car windows and railway coaches," the *Seattle Post-Intelligencer* reported in 1899, "glimpses are caught of Shacktown, but the daily lives of the cosmopolitan inhabitants, their tales of adventure, of grinding sorrow, of domestic unhappiness are not revealed." Like the Indians on the cover of C. B. Bussell's pamphlet, *Tide Lands, Their Story*, standing on the water's edge watching as factories belching smoke took over their ancestral home, the denizens of Shacktown were tragic figures who seemed destined for history's ash heap. An article in 1904 underscored Shacktown's impermanence: "Every inch of its gray sand is valuable, and will sooner or later be covered with business blocks."[64]

Immigrant squatters and Indians alike faced a new world where the political economy of private property circumscribed their lives and in which state and federal authorities exerted pressures to assimilate or disappear. For the Indians, the Dawes Severalty Act of 1887 codified what had been a tenet of well-intentioned white reformers for years and was reflected in the treaties of 1854–55: civilizing Indians was impossible until they became property owners, preferably ones who were self-sufficient farmers. Splitting communal holdings into individual parcels had been part of an evolving federal Indian policy. The Dawes Act made it standard practice until the act's repeal, in 1934, almost fifty years later. The legislation empowered federal Indian agents to shatter community property and provide individual land allotments, but it did not provide sufficient materials or training to help individual Natives make the transition to farming. In many cases, whites used the act to file spurious land claims, stripping still more acreage from already inadequate reservations. The results were often dire: whole bands reduced to begging and starvation. Puget Sound Indians, however, unlike many other bands elsewhere, were often spared the most corrosive effects of the Dawes Act because they could augment federal assistance with wages, foraging, and fishing. As Indian agent D. C. Govan noted in 1896, with "their beautifully fashioned canoes and necessary nets for fishing" the Indians of Puget Sound were "armed and equipped for the battle of life."[65]

Govan's observation drew attention to what had made reservation policy, until then, fairly easy to implement in Puget Sound: the region's ecological fecundity had enabled Indians to travel beyond their reservations to feed and support themselves when necessary. The Indians knew this because the treaties they had negotiated with territorial governor Isaac Stevens promised signatories the unrestricted right to fish, hunt, pick berries, and collect shellfish on unoccupied and unclaimed lands. By the beginning of the new century, this arrangement was under attack. As some white observers noticed, Indians who received no allotments of land—or opted not to live on reservations—faced a far more uncertain struggle than those dependent on federal agents and government charity. By the turn of the century, the growing urban economy had further constricted the remaining locations beyond the reservations still open to Native subsistence. Charles M. Buchanan, superintendent of the Tulalip Agency, the Bureau of Indian Affairs branch responsible for those Puget Sound Indians living north of Tacoma, warned in 1901 that white fishermen trespassing on reservation tidelands jeopardized the ability of Indians to be "self-supporting." The following year, he implored state and federal officials to help him protect his charges from the "aggressive competition of the white man, with his combinations of capital and craft."[66]

Alarmed by such reports, the commissioner of Indian affairs dispatched Samuel Atkins Eliot, a member of the Board of Indian Commissioners and president of the American Unitarian Association, from Boston in 1915 to survey the state of Northwestern Indians. Although Eliot reported that Stevens had "acted with absolute fairness and candor" in his original treaty negotiations, the state had since imposed fishing restrictions, limiting Indians to fishing only within reservation boundaries. Other tribal groups complained that they had never received compensation from the federal government for land cessions. Eliot was most disturbed by "homeless vagrants," those Indians who had not received allotments because reservations were too small to apportion lands to all enrolled tribal members, or because they were not members of federally recognized tribes. One to three thousand of these "landless and homeless" Indians were wandering "up and down the Sound, living on the beaches and constantly evicted or ordered to move on by their white neighbors." They had nothing but "squatter's rights." Eliot was concerned that "industrial and moral improvement" was impossible under such conditions. What he observed did not shake his benevolent and paternalistic faith in Indian reform through landownership: "They must be given some sense of security and permitted to cherish the ambitions that come only with the possession of property."[67]

What Eliot and other white Americans failed to admit was that, for the Indians, all of what was now Seattle had once been theirs. Now, they were forced to accept, without compromise, a new property regime from which they were effectively excluded. As Charles Buchanan explained, in a 1915 address sent to the state legislature in Olympia, this exclusion was achieved "under the cover of law and the appearance of legal right," or indirectly through "the exploitation of the great natural resources" by whites. As a result, the options facing Natives were limited. They could continue to migrate from encampment to encampment, which as Eliot pointed out was a problem, and one that was becoming less of an option. They could take an allotment, if one was available, and relocate to nearby reservations. Or they could fade into the city in ways that were invisible to outsiders, working as seasonal laborers or finding other employment as best they could. All three options often spelled poverty, especially the decision to keep on moving, and well-meaning white reformers often confused Indian destitution with backwardness and an inability to modernize.[68]

In 1910, after the late Seeathl's nephew, Billy Phillips, and his wife, Ellen, were driven from their shack on the shores of Elliott Bay by a fierce gale, whites started a relief fund. Charity only added pathos to the tragedy. The *Seattle Post-Intelligencer* reduced Billy's plight to strictly an environmental problem. "In the old days Billy was a good provider," the paper reported, but "the camping places along the shores of the Sound are now privately owned and no trespassers are allowed; the game has been killed; even the fish are hard to get." The landscape had changed and Billy had not. What the newspaper's reporters had done was to downplay, even erase, the social context behind Billy's plight. Construction along the waterfront had made modern Seattle possible even as it confined or destroyed the lives of the original inhabitants. Seen from another angle, however, Billy's very presence emphasized the tenuous endurance of common property along the enclosed waterfront.[69]

The presence of immigrant squatters and Indians challenged the faith that Seattle's builders could completely develop property they considered their own, let alone master the environment that made it possible. Tideland and waterway backers were convinced that a remade environment would deliver valuable and durable property, but their own power to make real estate was incomplete. Tidelands and waterways were subject to the same physical forces that engineers and speculators had hoped to master and did not behave with the predictability of a machine. All of the principal actors, from the transcontinental railroads to the local investors, had found it almost impossible to impose order, whether physical or social, on their new properties. The same physical forces undermined any hard and fast division between public and private, and the complexities of the marketplace amplified this unpredictability.[70]

Semple did not build his great canal; his tideland filling business barely kept him ahead of creditors even as he invested in ever more dubious enterprises. He sank his remaining energy and savings into a failed plan to dig a canal near Astoria, Oregon, to bypass the treacherous mouth of the Columbia River, and explored sending his company's dredges north to the Yukon River to prospect for gold in the busted claims of the Klondike district. He paid for both by liquidating the Lake Washington Canal Company, piece by piece, and died in a San Diego rest home in 1908, leaving his family no estate. Gilman and Burke did much better, selling off or leasing properties along the Lake Washington Ship Canal and the right-of-way of the former Seattle, Lake Shore, and Eastern Railroad line. Both became prominent investors in the region's construction trades, mining, and electrical and gas utilities. Burke further transformed himself into a leading cultural figure, championing educational reform and serving on the board of the Carnegie Endowment for International Peace. Chittenden, the engineer-scholar, settled in Seattle, served as a member of the new Port of Seattle Commission, three times as its president, and continued to work as a consulting engineer.

The successes of Burke and Chittenden, repeated like parables in so many histories of Seattle, obscured how unstable a foundation they had left their adoptive city. Seattle's waterfront and waterways had been utterly transformed, but no matter how attenuated the older idea of the commons had become, it persisted. The presence of individual Indians like Billy and Ellen, or the shanties of Shacktown, built in the shadow of Seattle's growing commercial sector, belied the well-established stories of progress forged through partnerships between private enterprise and civic government. The idea of the commons had incredible staying power because it was so flexible. The unstable physical properties of land and water around Seattle only added to its strength. Nor was it accidental that conflict over what constituted property and public interest climaxed during the Progressive Era, because similar tensions existed within Progressivism itself. The enduring appeal of common spaces—tidelands and waterways—was that they were not amenable to rigorous supervision from above. Some on the margins of Seattle society regulated commons in their own fashion, marking spaces through use, occupancy, and access. But so did the rich and powerful; railroads and entrepreneurs eluded the scrutiny of civic authority when it suited their interests.[71]

Emerging from these conflicts were three propositions that had future implications for an evolving ethic of place in Seattle. The first was that making property in a western city like Seattle relied on the intercession of two powerful forces—the transcontinental railroads and the federal government—and neither was as powerful nor as accommodating as city boosters had hoped. Therein lay the second proposition, that the non-human world in Seattle could and did exercise its own contingent power. Indians believed that the waters and lands upon which the city emerged had power over humans. The whites who settled among them turned this belief around, convinced that humans, through labor and technology, gave power to the environment. In Seattle, enlightened investors and skilled engineers could, and sometimes did, unleash the potential of the natural world to their benefit. The Indians were right as well; the waters and the lands could be altered and redirected but never fully controlled. Thus, the third proposition, one that Seattleites had yet to learn: manipulating nature to create landscapes of property and profit often amplified human social divisions by yielding an unstable and hazardous new environment. Improvement for some too often spelled despair, even ruin, for others. This would be the legacy of creating a political economy of private property.

CHAPTER 3

The Imagination and Creative Energy of the Engineer

Harnessing Nature's Forces to Urban Progress

Engineers, along with city planners and landscape architects, emerged as the nation's new builders and system makers at the end of the nineteenth century. Engineering was as much about improving upon human nature as it was about improving upon physical nature. Reforming one was tied to reforming the other. Just as engineers ordered the natural world, they sought to reorder society and put people and their activities in their proper place as well. Their philosophy of place, then, did not consider particular locales as anything more than parts in larger abstract systems. In Seattle, this concept of engineering took two forms—reshaping the city's topography and claiming an abundant, free source of clean water—and its leading practitioner was Reginald Heber Thomson, Seattle's most famous municipal engineer.

A newly powerful class of white-collar professionals, engineers remade the city in North America and Europe at the turn of the twentieth century into a metropolis. They threw steel across open chasms or straight up into the sky; they held back rivers and made them do their bidding; they vanquished disease and made the seasons irrelevant to commerce and construction. Municipal leaders, they built not only new things like bridges but new societies as well. Engineers came from an American romantic tradition that venerated nature as the font of national identity but added to it a more austere tradition of efficiency and mechanization.

Historians have paid considerable attention to how landscape architects and city planners reshaped urban form in turn-of-the-century America, but comparatively little to municipal engineers beyond their building of infrastructure and utilities. Architects and planners are portrayed as the designers of the American metropolis, engineers as the mechanics tinkering with a city's machinery. Yet these roles were often reversed, especially in the fast growing cities of the West, with engineers addressing the most basic of needs, from sanitation to utilities but also employing the latent power of the physical environment to drive civic advancement. Historians who have taken municipal engineers seriously have nonetheless tended to identify too strongly with them, lauding them for solving environmental problems but failing to account for the ways in which their uses of power and politics were sometimes counterproductive.[1]

During the Progressive Era, city builders and city leaders wrestled with the moral engagement of their work, especially the municipal engineer, who, in thought and practice, fused manipulation of the physical environment with socially responsible action. Engineering was always about turning the city into an engine for progress by harnessing the physical energy of nature to urban design. Their vision was one of efficiency and precision. But benefits of the engineers' ethic, in practice, were not evenly distributed across the cities they transformed or among their diverse, and often antagonistic, inhabitants.

For the Relief of the Common People

Seattle came of age in the era of big plans and big engineering, and Thomson became its master builder. He embodied the stereotypes of American engineering in appearance and in behavior. From the neatly trimmed beard that framed his gaunt face to the small wire-rimmed glasses suspended on a sharp nose, he was all right angles and straight lines. Even his name conveyed economy; he used only his two first initials, R. H. An article in the *Seattle Post-Intelligencer* of January 16, 1916, recalling his years of achievements, said that Thomson was more than "a skilled technician" because engineering suffused his "whole personality." He could "win equally well in a tussle with earth or with man." He usually fought both because, as the newspaper reporter intimated, he never separated the two.[2]

Like many white Seattleites, R. H. Thomson was from someplace else. An emigrant from an earlier frontier, southeastern Indiana in the Ohio River valley, he was born to a Scots Presbyterian colony there in 1856, then moved to California to work as a surveyor after training at Hanover College. He found a place in ruin. Decades of hydraulic mining had choked rivers with debris and destroyed farmland, throwing thousands out of work and into the arms of the demagogues who blamed the railroads, mining conglomerates, and alien Chinese laborers for their troubles. Unable to find work as a surveyor, he took a job for three years teaching science at Healdsburg Academy, north of San Francisco, and then moved to Puget Sound, lured by reports of its rich coal mines. In 1882, at twenty-six, Thomson was hired as Seattle's city surveyor, the equivalent of city engineer, for the next decade. After ingratiating himself with the local mining and railroad magnates, Thomson worked throughout the Pacific Northwest, learning the craft of engineering just as it was about to become an established profession. He built Seattle's first sewers and plotted a canal between Lake Washington and Lake Union. He moonlighted as a mining assayer, designed railway bridges, and surveyed a railway route around Lake Washington and over Snoqualmie Pass so the Seattle, Lake Shore, and Eastern Railroad could haul coal and timber from the Cascade Mountains. He learned on the job to plan, to build in future tense, and repeatedly asked employers and associates what would make Seattle prosper. "I generally received this answer, 'Don't worry your head about that. Some railroad will come here sometime and that will make it grow.'" Thomson knew from his early years in Ohio and his several years in California that railroads were no guarantee of progress and that markets were fickle. "Looking at local surroundings," he said later, "I felt that Seattle was in a pit, that to get anywhere we would be compelled to climb out if we could."[3]

In the 1880s, Seattle was a wide-open town hacked out of dense forests and steep brooding hills ruled by speculators and lumbermen who wanted a city that catered to the quick dollar or a good time. For a new arrival, coming by sea, the smoking hills of cutover forest and muddy tide flats must have been a depressing sight, but the built landscape was even more miserable. More than Seattle's topographical and economic shortcomings repelled Thomson. The town's open licentiousness offended his midwestern moral rectitude, and the rapid growth of Skid Row during the 1880s troubled him all the more. The fire fed by the spilled pot of glue that consumed Skid Row and leveled the downtown business district on June 6, 1889, must have seemed like a providential cleansing. Thomson, who was working as a private engineer, had lost nearly all of his maps to the fire, but he soon found work surveying for new buildings and new streets, a financial windfall for the young and ambitious engineer.

Like his fellow citizens, Thomson was eager to rebuild and paid scant attention at first to the consequences of unplanned growth. As in Boston and Chicago, the environmental problems that had sparked devastating fires thwarted attempts to reorganize and rearrange city services, and as the destructive 1889 fire cooled, typhoid struck, killing 166 in that one year alone. The origins of the disease were in the changed landscape of the city. Roads carved out of the rocky, clayey soil, strewn with stumps and exposed rock, with grades on many streets exceeding 10 percent, turned into muddy streams that flowed through outhouses and garbage sumps on their way down the barren hills into Lake Union and Lake Washington. Health officials pumped more water from Lake Washington into water mains, hoping to expel the fever, but this strategy only worsened the epidemic. Fatalities from typhoid declined from 90 in 1890 to 66 in 1891, but even as new water sources and better sewers abated disease, vice never disappeared. Brothels and bars, box-houses (vaudeville theaters attached to the back of saloons), faro joints, and piano halls that were burned out by the fire returned before the rebuilding was complete. Soon the wooden-planked streets of Skid Road were busy once again with the sounds of player pianos, clanking glasses, and an occasional pistol shot.[4]

Seattle remained the destination of choice for the single working men endemic to the industrial Northwest. Rose Simmons, writing for the *Overland Monthly* in 1892, lamented a scene of horrific proportions, the "scores of shanties, lean-tos, and sheds, holding a heterogeneous mass of humanity" huddled amid "sewers pouring down contagion and filth, moral and physical ill-being." Native-born white workers, Indian hops and fruit pickers, Japanese and Filipino loggers, Chinese railroad laborers, and southern European and Scandinavian fishermen caroused in Skid Road's saloons and slept off their revels in residential hotels or lodging houses. City health inspectors railed against these poorly ventilated "death traps," built without plumbing or sewerage, and called upon city officials to offer "the poor and ignorant" the same protections enjoyed by "the more intelligent and wealthy class of our people."[5]

Seattle leaders had relied on a succession of short-term solutions to solve the city's environmental ills, though some had tried to get the city to plan. In 1889, Mayor Robert Moran hired Colonel George E. Waring, a nationally famous sanitary engineer, who later developed modern sewer techniques for New York and other cities, to devise a plan for Seattle. Waring's comprehensive report advocated a separate system of underground tunnels emptying into Elliott Bay to keep sewage and storm or surface water apart. Moran and the city council had balked at the cost, but after the 1889 fire, they turned to Benezette Williams, a sanitary engineer from Chicago, for a less costly solution. Williams's 1891 report rejected Waring's separated system for a combined system that carried both sewage and storm runoff. The city's surrounding waters, he concluded, provided ample room to dilute waste. Although Thomson, who had resigned as city surveyor five years earlier to work in private practice, predicted the Williams plan would turn Elliott Bay and Lake Washington into cesspools, the city council accepted Williams's findings and commissioned the first trunk sewer, the North Tunnel, to run from Lake Union to Elliott Bay.[6]

Williams had no understanding of the region's geological properties, and as a result, the North Tunnel workers hit a small underground lake, flooding the tunnel and driving the project over budget. This unseen subterranean landscape, the legacy of the massive glaciers that had scoured the region repeatedly during the Ice Ages, would frustrate city utilities workers in the decades to come. Thomson, however, was not intimidated. After Albro Gardner, Seattle's city engineer at the time, resigned in frustration, Mayor James T. Ronald, elected as Robert Moran's successor in 1892, remembered Thomson's warnings about the Williams plan and summoned Thomson from his mining work in the Cascades to give evidence at public hearings. Thomson testified that the tunnel had pierced a lens, a glacial deposit of clay and gravel muddied by groundwater. Solutions, he said, were clear—use pumps and breast boards to hold back the waters while lining the bore with vitrified brick—but, he said, only an expert in mining could handle the work. An engineer recommended by a patron on the Board of Public Works turned down an offer of the job, whereupon Ronald appointed Thomson city engineer. He would serve in this position for almost twenty years.[7]

R. H. Thomson was a proselytizer as much as a public servant; he believed in the sanctifying power of reason and technology to redeem society. A devout Presbyterian, he confided to a friend that engineering resembled missionary work: "Having put my hand to the plow, I can't look back. There is so much crying to be done for the relief of the common people, and they are so blind themselves, and there are so few who care, that it seems impossible for me to quit." Thomson's almost Puritanical zeal was consistent with the profession's definition of itself. Engineers, especially those who worked for government, were cultivating their new power as elites who wielded technology and science. The engineers, who declaimed their good works on behalf of society, turned out often to be profoundly anti-democratic moralists, distrustful of an uneducated public.[8]

The engineer was a symbol of America's quest for reform in the Progressive Era, between the Gilded Age and the First World War, when new professions and new bureaucracies promised to tame the excesses of political sleaze and the consequences of unbridled capitalism. Americans, earlier suspicious of their cities, by century's end in increasing numbers regarded cities as unavoidable, even necessary. In this new climate, engineers held themselves above sordid politics and graft. Their office became as indispensable to the workings of the city as that of the police commissioner or the tax assessor. They were the vanguard of an urban political revolution that saw power devolve from the political boss who relied on face-to-face contacts with kith and kin to the dispassionate expert who could tap into a transatlantic, even international, community of professionals all working in the universal language of engineering.[9]

Thomson was a parent and an offspring of this new order. A cosmopolitan, he had traveled to Glasgow, Antwerp, and cities around North America to learn the latest planning ideas and techniques. He read voraciously, keeping abreast of trends in engineering and urban design. He became an able political strategist as well, while cultivating a reputation for being above political schemes. A reporter for the *Seattle Times* wrote on May 12, 1907, in a review of Thomson's career, that the engineer's power was "well nigh unlimited." Thomson could "make or unmake councilmen to influence public committees, and even bring the mayor of the city on his knees, begging favors."[10]

Thomson had acquired this power slowly, learning as he went, but, even so, Seattle had grown so quickly and so heedlessly that he was unprepared at first to address the overwhelming challenges. Comprehensive city planning had only come to maturity at the end of the century. The nation's largest cities—New York, Chicago, and Boston—served as the models for cities like Seattle that followed them. Commercial and retail activities were concentrated downtown. Residences rimmed the margins, with industry interlaced through the urban fabric, near the rail lines and waterways that held the metropolis together.

Municipal engineers, like Seattle's Thomson, as well as architects and landscape architects, like Chicago's Daniel Burnham and Boston's Frederick Law Olmsted, Jr., learned to take in the whole urban landscape, imagining, then building, an entirely new city from the ground up while the city grew around them. They were like composers and the cities they built were to be their symphonic scores. Order was the rule of the day. But beyond order, they had an artistic and moral purpose, a so-called City Beautiful movement, which sought to emulate and outdo the neoclassical splendor of Europe's great cities, an effort equivalent to the organic struggle for the City Efficient. Still, Thomson, as an engineer, was concerned about utility first and beauty second. If he was a composer, his scores were lean and muscular.

In Seattle, engineers, who staffed every major city agency, began to dominate the city's government and the bureaucracies that made it run, from the Department of Water, led by L. B. Youngs in its various incarnations over the next two decades, to municipally owned City Light, which broke away from the Water Department in 1910 to be run by a former Thomson assistant, James Delmage (J. D.) Ross. Thomson stood at the center of this new political class as head of the Department of Engineering. His first big plans were two daring projects to meet Seattle's most vexing needs as it grew from small town to large metropolis: reliably clean water and level, solid land. To achieve these goals, Thomson set out three immediate priorities: secure a clean and reliable source of water in perpetuity, complete the sewer system, and decrease the grades of the city's steep streets. He saw these all as interconnected and essential to the physical and social well-being of the city. Steep grades had meant higher costs for street building and hauling goods and had divided the city's neighborhoods by class. Continuing to pump water uphill from nearby lakes was expensive and risky. Eliminating the hills would reduce expenses, boost real estate values, and promote social harmony. Level sewers and water lines built on flattened streets would save money and provide better drainage. Thomson would embark on lifting Seattle out of its "pit."

Because territorial laws limiting municipal indebtedness (still in effect after Washington became a state in 1889) prevented the city from assuming bond liability, Thomson lobbied the state legislature in Olympia to raise the debt ceiling. A home rule bill in 1893 gave Seattle the ability to exercise eminent domain and issue bonds—provided it would pay just compensation for property taken and put strict limitations on indebtedness. The home rule bill applied to projects within the city limits, a frustration for Thomson's first priority of bringing a clean, endless supply of water (plus electricity) from the Cedar River, one of the region's largest, some thirty miles distant in the Cascade Mountains.[11]

Unlike most western cities, damp Seattle abounded with springs and seeps. Freshets gushed from rocks and the sides of hills after heavy rains, flowing even in the driest of months, so there had been few incentives to build an expensive municipal water system. Water belonged to whoever could grab it first, pay to pump it up the city's steep hills, and sell it at the best price. The largest and most aggressive of these suppliers, the Spring Hill Water Corporation, formed in 1881, controlled the entire supply for the southern end of the city by 1886. Mayor Moran, at the urging of Thomson and others, in 1888 proposed building a gravity-driven pipeline from Rock Creek, a tributary of the Cedar River, south around Lake Washington and into the city, but a clerical error delayed the special election that September. Eleven months later, in July 1889, an outraged populace, blaming the city's poor water system for the devastating fire the month before, voted for the pipeline. The city council, which had no authority to raise the requisite funds, instead had to buy out private competitors, an expensive process. The aqueduct promised by Moran remained just that, a promise.[12]

While Thomson waited for the legislators in Olympia to define the powers major cities of the state could exercise, he turned to his next task—the city's sewers. He started construction on the two main sewer trunks—the North Tunnel and the South Tunnel, running from Beacon Hill to Elliott Bay. To the trunks he attached the branches—storm drains, houses, businesses. A necessary bond measure had passed but the Board of Public Works pushed aside Thomson's preferred contractor, opting to use the project as patronage for constituents who had lost their jobs in the Panic of 1893. Disgusted, Thomson resigned in early 1894, only to be reappointed a month later by a chastened Mayor Ronald, who fired the offending board members and gave Thomson the expanded powers he demanded in a revised city charter.

Armed with his new authority, Thomson now sought to redo Seattle completely, beginning with its sewers. Before he could rebuild the sewers, he needed to start with a new plan. Thomson rejected Benezette Williams's earlier proposal for a combined sewage and storm runoff system emptying into Lake Washington and Elliott Bay; dumping effluents so close to the city, he believed, would create an unsafe nuisance. He looked instead to the deep mixing waters of Puget Sound and decided to follow George E. Waring's previous recommendation instead: to built sewer outfalls on the sound and on the Duwamish River that emptied into it. Thomson was no purist in the end and combined elements from both plans because he agreed with Williams on one key point—that a combined system would better accommodate Seattle's seasonally heavy rainfall and save a great deal of money over Waring's complicated but more hygienic separation of sewage and storm overflow. Thomson's decision, at the time, was an expedient remedy to the city's pressing problems, but it had far-reaching consequences for Seattle over the next century.[13]

Thomson's earlier lobbying efforts to acquire the Cedar River for Seattle's water supply now paid off, thanks largely to legal providence. In 1895, the Washington Supreme Court affirmed the right of Spokane to build a water system beyond city limits and issue bonds for its construction as long as the city redeemed the bonds through utilities' revenues to avoid the ceiling on municipal indebtedness. The decision gave teeth to the 1893 state law, handing Thomson the two instruments he needed to acquire the Cedar River: money and power. His proponents quickly grasped the significance, and a city council report in October 1895 stated that "the time and opportunity has come when the City of Seattle is in a position to say 'Hands off—the inhabitants of Seattle, now and to come, have a paramount claim upon the Cedar River water supply.'"[14]

Winning the Cedar River for Seattle was about more than water alone. Water catalyzed social reform and moral uplift, just as it had for earlier reformers in Boston, who employed the campaign for municipally owned water as a crusade to transform city politics. Clean, abundant public water was, according to Thomson, Seattle's lifeblood, but the Cedar River was also a device to achieve his third goal—to lay low Seattle's hills. Thomson planned to use the kinetic energy of the river, channeled through pipelines and nozzles, to reengineer Seattle's topography and address two of its principal evils—its infamous reputation as a den of iniquity and its persistent unmanaged development. As he explained to a colleague, with a dose of hyperbole, there were "few cities in the world more broken or irregular in their topography" than Seattle. Erasing those hills and gullies would provide a "fair foundation for municipal solidarity" by uniting the city's divided neighborhoods. It was a bold proposal, but it ran against the freewheeling spirit of Seattle's realtors and speculators, who, in Thomson's opinion, had built with little regard for the "broken and inaccessible position in which nature had left" their property. Without regrading, he said, most of Seattle's real estate would soon have "no commercial value whatsoever." In a medical analogy, Thomson compared regrading to an operation enlarging "insufficient arteries" (the city's streets) and removing the blockage (its hills and ravines) that stopped traffic flow and endangered the city. The engineers' near absolute power over city planning enabled them to make connections between more than environmental conditions and urban development; they proposed to save the patient, body and soul. Nowhere was this more apparent than in the justifications offered for Seattle's famed regrades.[15]

As a relative newcomer to America's late-nineteenth-century urban growth spurt, Seattle wanted to catch up with the rest of the nation, and big engineering provided the boost. Beginning in 1898, and for almost two decades, regrading consumed the city. In these regrades, totaling almost sixty separate projects, Thomson's engineers changed the elevation of more than twenty streets, removing several high hills entirely and using the spoils to fill in the remaining tidelands along the waterfront. Since changing Seattle's topography required immense quantities of power, Thomson planned to use the rerouted Cedar River's kinetic energy to drive hydraulic cannons to rub out the city's steep hills. Erasing hills would make new property, remove unwanted residents, and cleanse neighborhoods of filth and blight, while delivering endless water that would also eradicate illness and pollution. To Thomson, clean water and level land were as much ethical as economic or political goals.

The target of Thomson's regrade surgery was, not surprisingly, the infamous Skid Row and the polyglot slums of the city's dank ravines and waterfront tidelands that fronted Elliott Bay. Although the deep depression that followed the Panic of 1893 had temporarily cleaned out the box-houses and saloons when their customers left for better-off locales, the lure of lucre farther up north brought the brothels and card shops back to the waterfront. The 1897–98 Klondike Gold Rush turned Seattle's social landscape upside-down even as it generated huge profits for legitimate city businesses. As Seattle merchants, from the clothier Nordstrom to local hardware moguls, the Schwabacher Brothers, plowed the profits they made from provisioning miners streaming north to Alaska and the Yukon back into new facilities, Skid Row proprietors did the same. By 1898, a major turning point in Seattle's history, downtown saloons outnumbered dry goods stores, and a market in prostitution and liquor connected Seattle to Alaska, parallel to the legitimate trade in groceries, shipping, and passenger service.[16]

Thomson wrote to a friend in Alaska that "the streets are so full that you can hardly walk through them, but the countenances you gaze upon are not those which are going to impress a solid householder with the belief that these are the people [whom] he wishes to make his neighbors." All of Seattle was now "wide open," stricken with "Klondicitus," a malady that attracted vagrants and pimps, and corrupted local politics. The proof was in the streets: busted miners who stayed in Seattle when their dreams up north had evaporated and erected shacks along the city's waterfront wherever wharves and jetties were not; waves of new immigrants from Japan, the Philippines, southern Europe, and Scandinavia who had come to work in the region's extractive industries; and Indians who persisted in making Seattle part of their seasonal subsistence rounds. The influx swelled the city's population from 3,553 in 1880 to 237,194 in 1910, one of the fastest rates of growth in the Far West at the time.[17]

Thomson believed that decay and depravity emanated from damaged and overtaxed landscapes and that changing those landscapes would change the people afflicted by them. He adhered to late-Victorian-era ideals that linked bodily, mental, and moral health to community well-being, and all of these to where one lived. Despite revolutions in late-nineteenth-century science and the rise of germ theory to explain disease, many experts continued to associate poor health with poor landscapes and poor people. The shifting demography of Seattle's downtown and waterfront districts only confirmed their suppositions. As more and more immigrants poured into the residential hotels and flophouses along Jackson Street, Skid Row brothels and watering holes, seeking to capture more of the post-Klondike trade, began to shift their operations north and west toward the central business district to better serve Seattle's growing working-class population. As Skid Row's original denizens left, Japanese and Chinese immigrants and African Americans began to move in and take their place. The new arrivals found an instant city with an inadequate infrastructure. Nagai Kafū, a visitor from Japan in 1903, found Jackson Street "like a Japanese city street, with signs advertising tofu, Japanese soup, raw fish, and noodles" built amid "mounds of horse manure, and an acrid smell of smoke." To affluent white Seattleites, having a street reminiscent of Tokyo in their city was an abomination. To them, Jackson Street was now another "dirty, filthy region that had naturally become the immoral center of Seattle."[18]

Many middle-class residents, reacting with their feet, moved to the new streetcar suburbs of Fremont, Wallingford, and Green Lake, lured there by real estate agents playing on racial and class anxieties. Reverend Mark A. Matthews exacerbated such tensions. The minister of Seattle's First Presbyterian Church, Matthews was an imposing six-foot-five, a giant in the city's reform circles and a confidant of Thomson. Born in northern Georgia's Hill Country, Matthews had come to Seattle in 1902, fired by the Social Gospel movement led by Walter Rauschenbusch and Protestant clergymen of the Progressive Era, intent on employing Christian principles to temper the ill effects of industrialization, unrestricted immigration, and rapid urbanization. Although Matthews was not a strict adherent of the Social Gospel movement, since he believed firmly in the importance of admitting sinfulness to prompt personal salvation, he soon turned First Presbyterian into one of the nation's leading Social Gospel churches. From the pulpit he called for changes necessary to the very fabric of

Seattle, arguing in one sermon from 1906 that "a favorable environment renders indispensable service in the development of character."[19]

Thomson shared Matthews's embarrassment over Seattle's decrepit physical and moral state, and both blamed a corrupt political system that had prevented any sweeping reforms, particularly the municipal ownership of important utilities, such as light and water. But, now, in the wake of the Klondike boom, Seattle's political bodies, as in other cities during the period, were coming under the control of the emergent middle and upper-middle classes. In Seattle, reformers pushed municipal ownership of utilities and ports, and promoted referenda and initiatives for prohibition and against prostitution. Most reformers, largely recent newcomers from the East and Midwest, were white-collar professionals (a few were moderate labor leaders) who advocated efficiency and economy in city government and expulsion of all vice. Their enemies were manifold and powerful, ranging from Alden Blethen, publisher of the *Seattle Times,* who opposed municipal ownership of utilities, to Hiram Gill, elected mayor in 1910, who favored open saloons and an open town as means of promoting social peace.[20]

In 1911, in response to Gill's election, reformers founded the Municipal League of Seattle and King County, an organization of professionals modeled after similar societies around the nation that lobbied for greater efficiency and morality in local government. Prominent members in Seattle included, besides Thomson and Matthews, A. H. Dimock (an assistant city engineer who succeeded Thomson in 1911); George F. Cotterill (another former Thomson assistant and future mayor); Robert Bridges (the first Seattle port commissioner); and J. D. Blackwell (yet another assistant and future chief engineer, and a former superintendent of the Seattle Electric Company, operator of the city's streetcar system). In its campaigns against political corruption and for municipal ownership, the league forged a tenuous alliance with moderate labor leaders against those forces that were "bad for smaller business, bad for morality, and bad for Seattle's reputation."[21]

Until the formation of the Municipal League, however, the most powerful reformer in Seattle city government was Thomson, and he exercised his authority forcefully. Reengineering Seattle's landscapes was inseparable, he maintained, from cleansing its social and political structure, because clean water, proper sewerage, electrical power, and a renovated topography could yield more than good hygiene and efficient planning. He would instruct the people of Seattle how to build and live in a great city. The Cedar River was the key. "I have to have a specific end in view," he wrote to a friend, "and what I am teaching the people now is, that with the cleanliness of water, through God-given Cedar River, and with cleanliness in light, also through God-given Cedar River, that we have just begun the question of municipal cleanliness."[22]

We Order the Parts of Our Modern City

Thomson was an unabashed reformer but he was also a political pragmatist. To move mountains and rivers, he would need more than faith; he would need the power of the purse and the law. And he would need backing from the businessmen and from the public he distrusted. His political strength came not from his quasi-religious zeal but from his effectiveness as a charismatic bureaucrat. He was the quintessential Progressive, never so fanatical as to lose sight of his goals.

Thomson faced his first challenge as the city moved to acquire the Cedar River, thanks to the 1893 decision of the Washington Supreme Court. The Cedar's watershed lands were a crazy quilt of parcels, big and small, owned by the federal government, the Northern Pacific Railroad, homesteaders, small mill operators, and the Weyerhaeuser Timber Company, which had quickly become one of the nation's largest woodland owners and lumber producers after its founding in 1900. The court had required Seattle to pay owners just compensation, provided they wanted to sell or be reimbursed through the legal condemnation process, but the acquisition of land to build the pipeline aroused cries from opponents of municipal ownership. They backed an 1894 plan by a shadowy investor from New York City, Edward H. Ammidown, to buy the lands for the city in exchange for retaining title. In response, municipal-ownership advocates argued that the doctrine of prior appropriation (the principle in western American law that allocated water rights on a first-come, first-served basis), as a result of the original surveys by Williams and Thomson, was the source of the city's claim. Voters turned down the so-called Ammidown plan at the end of 1895, approving instead, by nearly two to one, the $1.25 million bond for Thomson's Cedar River project.[23]

Thomson, chastened by the potential expense of purchasing all of the lands in the upper Cedar River basin at once, now struck a deal with the federal government, the largest landowner in the watershed, to cooperate with the city, and in 1899 he persuaded the General Land Office to withdraw all federal lands in the drainage from sale, settlement, or other disposal in the expectation that

Seattle would build a pipeline to the Cedar River. Workmen fanned out across the countryside to lay wooden and cast-iron conduits from the intake at the town of Landsburg, on the lower Cedar River, across the Renton Valley and the Black River at the south end of Lake Washington, and into the city.[24]

Building the system was a technical and environmental challenge. Workers assembled custom-designed wooden stave and iron pipes, built in Seattle to Thomson's specifications, along a precisely measured grade in order to use gravity to draw the water to reservoirs on several ridges throughout the city. But geology conspired against the workers. Unstable glacial deposits turned to mud, or settled suddenly and snapped pipes in two. After three years of back-breaking work, and after fending off lawsuits charging backroom deals and fraud, Thomson's careful calculations paid off, and, in January 1901, the pipeline opened.[25]

Seattle had acquired its pure water at last, and not a moment too soon, because as Thomson put the finishing touches on the Cedar River pipeline he had already begun his regrades to level Seattle's center city hills. Here, too, he needed to cobble together support from real estate speculators and downtown businesses. Local improvement districts, an innovation borrowed from eastern cities adopted in Seattle in 1893, put the onus of paying for developments like streets or regrades on individual property owners, thereby linking individual benefits to costs incurred by the entire city. Thomson considered the local improvement concept a godsend, telling an editor in Portland, Oregon, that without it "Seattle would have continued until this day to have been designated 'the stump town'" that he wanted to leave behind. Thomson and his subordinates, meeting with neighborhood improvement clubs and businesses, took them through flipcharts and blueprints, pointing out how regrading would deliver significant "financial, hygienic [and] aesthetic benefits" by reducing costs for shipping and time to travel, increasing property values, and containing or eradicating unhealthy slums. The technological management of place was his view of the world. It was a very persuasive view, to a point, after which the engineer relied on political shrewdness. Thanks to Thomson's campaigning, although the city used eminent domain to bully recalcitrant residents and businesses, most supported regrading. Thomson had appealed to particular groups of Seattle residents through their pocketbooks because he soon realized what other reformers had discovered decades before in cities like Chicago: the public interest often ended at the neighborhood's edge or the business doorstep.[26]

The first regrade, completed in 1898 along First Avenue at the foot of Denny

Hill, set the agenda for subsequent regrades: remove the earth and wait for the benefits to follow. Steam shovels and horse-drawn carts hauled off nearly 120,000 cubic yards of earth, some of it used to fill tidelands along the city's waterfront. Property values along First Avenue, now flattened to a reasonable grade, soared in an already hyperinflated market. By 1902, businesses and residences along Second Avenue clamored for their own regrades and property owners elsewhere in the city quickly followed suit. Given the volatility of the real estate market before regrading, determining its effects on property values is difficult, but evidence suggests that some projects yielded healthy returns, at least in the short term. V. V. Tarbill later wrote in the *Harvard Business Review* that, after the Second Avenue regrade in 1906, street-front lots sold for $2,000 per front foot at the corner of Pike Street. One block to the north, where Second Avenue still ran into Denny Hill, similar frontages were worth less than $300 per front foot. Two years later, Thomson estimated that regrading had boosted business real estate values 400 percent and residential real estate 1,000 percent in the affected neighborhoods.[27]

Even Thomson, who distrusted realtors, made them allies in the "struggles to up build this city." The businessmen and real estate agents who opposed Thomson on the municipal ownership of utilities supported the regrades in return. George Emerson, a local lumber magnate, told the stockholders of the Metropolitan Building Company in 1907 that regrading was one step "along the only path leading to Seattle's commercial heights." Emerson extolled Seattle, which had devoured its hinterlands, for now cannibalizing its hills for a higher cause. Thanks to capital and labor, with government and business working hand in hand, all under the guidance of the municipal engineer, Emerson concluded, "we order the parts of our Modern City."[28]

As Emerson noted in his speech, technology would channel human labor, but technology led back to the Cedar River. Demand for regrading began to mount, as Thomson expected that it would, and horses and steam shovels became inefficient. Contractors hired by the city engineer turned to mining technologies instead, brandishing hydraulic cannons, called "Giants," against the hills while burrowing tunnels beneath city streets or running sluices above them to carry the tailings or "slickens" to the waterfront for disposal. At first, workers used water pumped from nearby lakes or Puget Sound, but engines failed and the salt water corroded pipes. The arrival of the Cedar River water gave Thomson a more efficient scouring agent. Using gravity to build the nec-

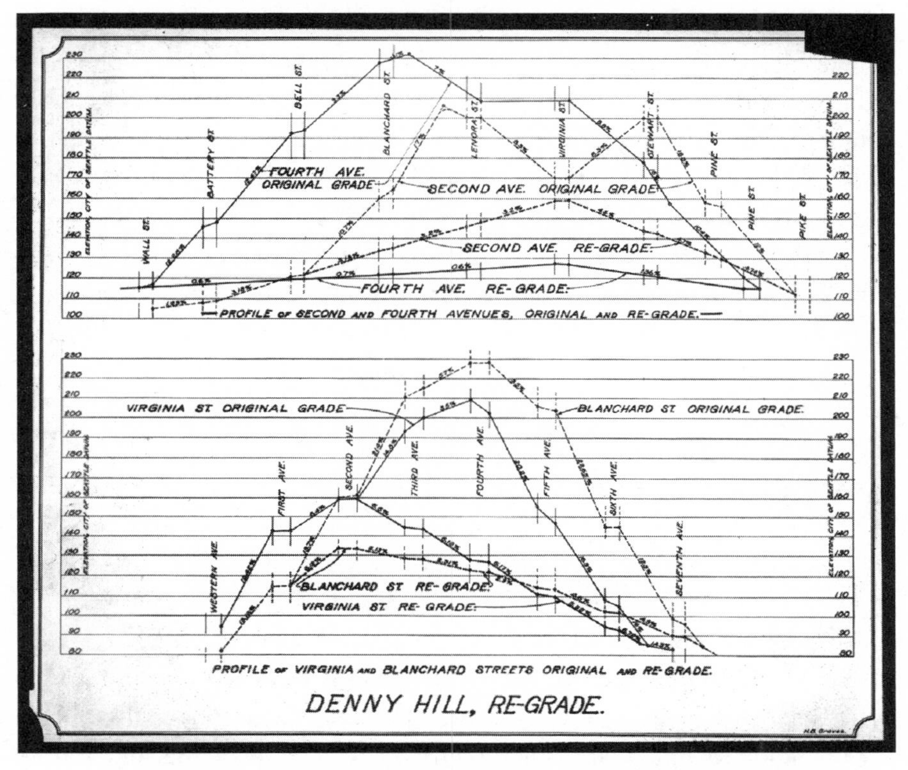

The first Denny Hill regrade, side profile showing the original and regraded slopes, c. 1910. (Courtesy of the Museum of History and Industry, SHS 16,998)

essary pressure, engineers could accomplish the work in the most economical fashion. Nearly all of the city's large-scale regrades were undertaken only after the pipeline was finished.[29]

The sheer power marshaled to reverse eons of geology captured the attention of residents and visitors at first, even as regrading, as Thomson had envisioned it, began to erase the city's most blighted neighborhoods. On the first Denny Hill regrade, engineers and contractors shoveled or sluiced nearly 5.5 million cubic yards of dirt and rock into Elliott Bay, removing over half of the 250-foot mount, at the same time displacing squatters' shacks along the shoreline. On the Jackson Street regrade, workers carted off or washed away approximately 3.4 million cubic yards of soil as city inspectors condemned whole blocks of homes and apartment buildings in the largely immigrant and working-class neighborhood. And these were only the largest. Statistics citing the total amount of earth moved in all of the regrades varied, but estimates went as high as 50 million cubic yards. Others calculated that more than one-eighth as much earth had been moved as had been excavated in construction of the Panama Canal, the yardstick for big engineering at the time. In the words of one engineering report, hydraulic sluicing was the "application of a natural world-making force applied to . . . municipal improvement."[30]

It was the Pleistocene played all over again. Workers used explosives to dislodge large boulders from the glacial hardpan, and set abandoned structures ablaze before undermining them with steady streams of water, washing the remains out to sea. Along the tide flats fronting Jackson Street, settling basins made with earthen levees and wooden walls captured the dissolved hillsides to create twenty-seven new city blocks where slums, "awash with the usual debris and garbage," once stood, adding to the man-made land already created by private developers at the southern end of Elliott Bay since the 1890s. On the Denny Hill project, the process was far less complex—contractors merely dumped debris along the shoreline, a move welcomed by the Northern Pacific, which was expanding its piers and improving its track beds.[31]

Since many regrades proceeded simultaneously, city engineers also had to think systematically in order to coordinate the energy and labor vital for effective operations. Regrading was not a novel idea in city planning, and Thomson may have had Boston's experience in mind, but unlike Boston's assault on its hills in the antebellum period, where workers used horse-drawn carts and shovels, Seattle's attack was thoroughly industrial. Not surprisingly, the hydraulic cannons consumed tremendous volumes of water. On the Jackson Street regrade alone, the giants spewed between 9 and 12 million gallons each day, nearly one-third of the city's daily water capacity. Contractors worked with the Water Department to ensure sufficient supplies without disrupting residential or industrial use. Often operations ran at night, lit by floodlights, with electricity supplied, in part, from a second hydroelectric dam on the Cedar River, approved by voters in 1902 and completed in 1904, which gave Seattle one of the nation's first municipally owned electrical systems and endless lighting for workers. Giant operators were as specialized as the machinery they used, usually "Alaskan veterans . . . experienced and expert in hydraulic mining operations" who had the masculine "skill to 'squirt' water" at high pressure in order to "'herd' earth and mix water" without losing time to accidents.[32]

Thomson's contemporaries saw technology as an instrument of beauty, be-

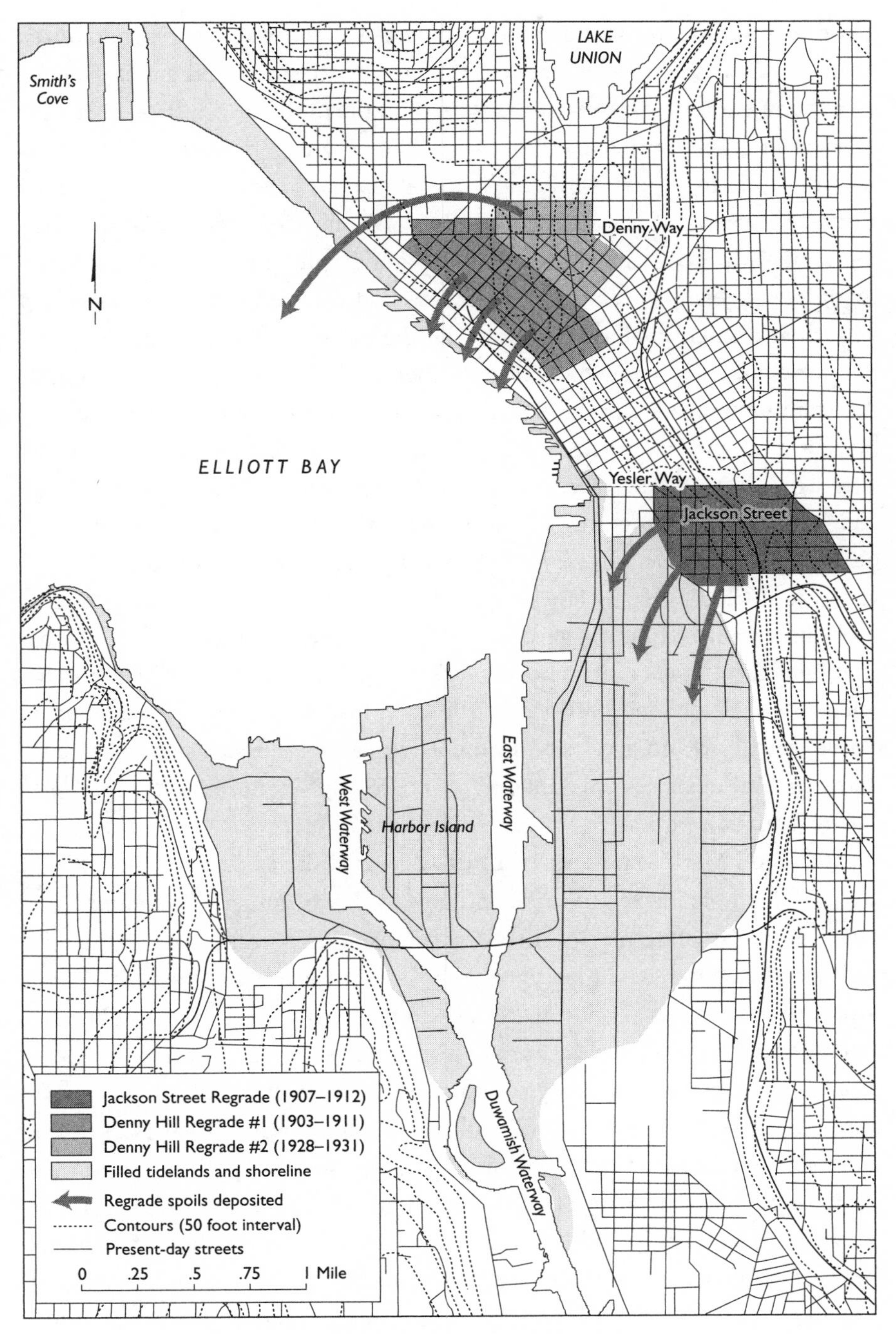

Major regrades in Seattle, c. 1900–1930.

cause beauty arose from pure function and from a careful assessment of nature's needs. It was an aesthetic of place congruent with Progressive Era ideals of efficiency and equity. Many were impressed with the modern industrial creativity of regrading. Phillip Kellar, of *The World Today,* wrote, for example, how "this regrade work, in addition to being stupendous, has been picturesque." Asahel Curtis, a member of the Seattle Chamber of Commerce, brother of the photographer Edward S. Curtis and a skilled photographer himself, chronicled the regrades, capturing spectacular landscapes that dwarfed individual observers with their immensity. His elimination of the individual human presence was not accidental. Engineers had created, albeit temporarily, an urban version of the sublime with modern industrial technology. Often such representations played upon regional pride. A. H. Dimock, a Thomson assistant at the time, later wrote in 1928 that the regrades were "the product of Western conditions" and "pioneering engineering methods." Editorials in the leading engineering periodicals, from *Engineering News* to *Scientific American,* extolled Thomson's vision, and Charles Evans Fowler, a noted New York engineer visiting Seattle, pronounced that regrading gave "full play . . . to the imagination and creative energy of the engineer." If the engineer's imagination made Seattle, then that imagination demanded hard work. The local historian Clarence Bagley later commented how "no great city on the American continent has overcome so many natural obstacles encountered in its growth." Seattle was "virtually one vast reclamation project."[33]

Reverend Mark Matthews, in a 1911 address at the exclusive Rainier Club, the watering hole for Seattle's rich and powerful, proclaimed, "The building of a city cannot be left to men of mediocre brains or determination . . . it must be built by men who have broad views, whose perspectives are those of giants." Invoking the famous sermon by John Winthrop, who, in turn, had taken the metaphor from the biblical account of Jesus's Sermon on the Mount, Matthews concluded: "We are truly a city set upon a hill; we cannot be hid." Matthews was likely talking about his friend and parishioner. It was not hard to see that Thomson, in the minds of some white Seattleites, was perhaps their version of the Changer or Transformer that Puget Sound Natives believed had prepared the world that whites had stolen from them. But at his zenith, three months before he left office in December 1911, Thomson the transformer would discover that there were limits to what brains and machinery could do. Power had limits he could not transcend.[34]

We People Who Live in Your Regrade Districts . . . Are Martyrs

By the end of the twentieth century's first decade, Seattle had become a major city, outstripping its rival to the south, Portland, and surpassed in size only by San Francisco and Los Angeles in the West. In 1900, Seattle was the nation's forty-eighth largest city; two decades later, it was the twentieth largest. Annexations of outlying towns between 1907 and 1910—tree-lined Ravenna, industrial Georgetown and Ballard, bucolic West Seattle and South Seattle—further boosted the city's population and added thirty-seven new square miles of territory. Meanwhile, the reformers, consolidating their power, succeeded in taking City Hall, thanks to the efforts of the Municipal League. George F. Cotterill, who served as an assistant under Thomson from 1892 to 1900, ousted the corrupt Hiram Gill as mayor in the 1912 election. The next year, he reported to the city council about the tremendous work undertaken by his former boss: in a twelve-year period, beginning in 1900, nearly 200 miles of streets had been paved, 400 miles of water pipe laid, 525 miles of roads built, and 800 miles of sidewalk poured.[35]

Cotterill had little time to celebrate this achievement, however, because he had inherited a city where organized labor, energized by Seattle's position as a leading shipping center on the Pacific Coast, was now challenging the industrialists who ran Seattle's shipbuilding companies, steamship lines, railroads, mills, and factories, insisting on better wages and working conditions. Some joined the Industrial Workers of the World (IWW), a militant union, founded in 1905, that espoused socialism and advocated the abolition of the wage system. IWW organizers and members, known as "Wobblies," were busy in the extractive industries of the Far West at the time, and Seattle soon became a Wobbly stronghold. Others, fearing the red tide rising around them, struck back against the labor radicals. A mob attacked the suffragist speaker Annie Miller outside the IWW office during the Potlatch Days parade, a new citywide boosters' celebration, in 1913. The center of reform politics in Seattle could not hold. Voters removed Cotterill from office in 1914, electing none other than the tainted Gill, who claimed the mantle of Progressivism even as the language of progress was beginning to lose its luster.[36]

Many historians have chronicled these political disputes as evidence of Seattle's devolution into social turmoil, but none considered how the equally massive changes to the physical environment reflected and amplified urban conflict. Regrading and municipal waterworks were more than mere infrastructure; they were the apotheosis of Progressivism manifested in concrete and steel, water, and earth. Yet Thomson and his associates, on the hillsides of Seattle and on the mountainsides of the Cedar River, instead of offering stability, had crafted hybrid landscapes, neither fully natural nor fully under human control. Engineers had produced, very often, landscapes more dangerous and less reliable than those they had altered.

The first troubles began along the Cedar River, which the city had tapped for its water but did not control completely. Most of the watershed, totaling nearly 142 square miles, remained largely out of the city's hands. The mill towns of Barneston, Cedar Falls, and Taylor, all above the city's intake at Landsburg, were home to hundreds of loggers and homesteaders who could contaminate the city's water supply. Thomson faced a challenge—how to acquire the entire watershed without plunging the city further into debt. The first pipeline soon proved inadequate as the city, from 1900 to 1910, nearly trebled in size from 80,600 to 237,194, prompting construction of a second pipeline in 1908. Thomson was convinced the city needed jurisdiction over the entire watershed even as citizens had become suspicious of spending still more money to feed Seattle's public works binge. Three bond measures, beginning in 1895, enabled the city to control, on paper, nearly 38,000 acres by 1914—only a fraction of the area. Still, landowners refused to sell unless the city agreed to reimburse them for the market price of harvestable timber in addition to the value of the land. With timber the unchallenged mainstay of the regional economy, city attorneys could not easily persuade other watershed occupants to put up their land and trees for sale. Ironically, the key to the watershed was to control the trees, not the river, forcing Seattle to pursue a pragmatic strategy by exchanging timber rights for land titles while endorsing logging on its own properties to pay for further land purchases or system construction. Seattle's dilemma was unique among cities at the time. In the Northeast, cities turned already logged-off lands into municipal watersheds, or bought out small logging concerns and individual landowners using eminent domain to acquire protected water supplies. In California and Utah, forested lands in the Sierra Nevada and Wasatch Mountains did not produce timber as commercially valuable as that in the coastal Northwest. Thomson had to navigate a more complex social landscape. Major companies, like the Northern Pacific Railroad, the largest in the watershed, or Weyerhaeuser signed contracts with the city that included stipulations on logging or transportation to protect water purity. Local logging companies removed the trees for the timber sellers, and afterward Seattle would purchase the logged-off lands at a fraction of their timbered price. With smaller owners, city attorneys combined intimidation and persuasion to compel them to sell their lands with the timber rights, or else face the legal condemnation process.[37]

Although separating the land from the trees could give the city what it wanted at a price it could afford, Thomson and other impatient officials wanted complete control sooner, fearing disease or unanticipated water shortages. Events soon sharpened their concerns. In 1906 the Milwaukee, Chicago, and St. Paul Railroad applied for a right-of-way across city-owned lands in the lower watershed. The resulting controversy—some residents worried that trains would dump raw sewage as they sped along the tracks—led to a special commission, chaired by a national expert on sanitation, William T. Sedgwick of the Massachusetts Institute of Technology. Sedgwick's report simply emboldened city officials to confiscate even more of the watershed because, as the sanitary engineer warned, if "even the spit of one come down with typhoid" entered the water supply, an epidemic was likely.[38]

Acting on these recommendations, the Department of Water tightened regulations over its watershed lands. The Milwaukee Road was required to follow strict sanitary guidelines; watershed guards locked out hunters and unapproved loggers; and the city began a three-decade fight to condemn and eradicate the towns of Barneston and Taylor. In the interim it needed more money to fund watershed expansion, especially after voters in 1910 approved the construction of a permanent masonry dam to replace the timber crib at the mouth of Cedar Lake and create a larger reservoir for water and power. Ignoring grave concerns about the stability of the dam's location—the northern wing abutted a large glacial moraine, an unstable ridge of gravel and mud—the city served notice that year to those landowners whose trees would be flooded: the U.S. Forest Service, Northern Pacific, and Weyerhaeuser. The Forest Service filed a counter-lawsuit, but city health commissioner J. E. Crichton argued in an affidavit that, for safety and sanitary reasons, the city required "absolute power and dominion of all lands within Cedar Lake Basin."[39]

Seattle officials had hoped that the city might ultimately get all of the federal watershed lands through condemnation as well. However, the U.S. Forest Service, which now administered the Snoqualmie National Forest, set up in 1905, stiff-armed the city and refused to surrender its lands. In frustration, the city appealed to Washington's Senate delegation, and Congress passed legislation, in 1911, requiring the Forest Service to sell to Seattle nearly 20,000 forest acres at a discounted price, the appraised value of the timber plus $1.25 an acre. Forest Service officials were dismayed. An assistant supervisor complained that the bill was equivalent to a "gift of $600,000 worth of timber" to the city. The Forest Service's concern was understandable: a report in 1910 estimated that Forest Service lands contained as much as 270 million board feet of mature Douglas fir and western red cedar, ready for the axe and mill. If Seattle prevailed, the Forest Service concluded, it would set a dangerous precedent by subjecting an important forest reserve "to the varying vicissitudes of municipal politics." The Forest Service started a campaign to block the city's plans.[40]

Seattle tried to play tough, but the city's engineers and sanitarians were at a loss when it came to managing timber. Forestry, like engineering a new profession born in the Progressive Era, was to the forests what engineering was to the city, a rational and scientific enterprise designed to improve old practices. Realizing its ignorance, the Board of Public Works, which was responsible for managing the watershed lands, hired Hamilton C. Johnson, a graduate of the Biltmore Forest School, as the first city watershed ranger in 1912. The next year, Johnson called for a far-reaching solution: cutting trees on city watershed lands and then replanting them to generate revenue without endangering water security. Although sawmills were eager to buy the timber, the Board of Public Works, in an effort to streamline city management, put all watershed management under the command of the Department of Water. L. B. Youngs, the department's superintendent since 1895, another civil engineer and an ally of Thomson, deemed reforestation too expensive and shelved Johnson's report. Youngs appealed to the Forest Service to compromise, as it had with numerous other cities throughout the western United States. In 1915, the city abandoned its fight to seize national forestlands. Although Youngs believed that the entire watershed should eventually be put under city jurisdiction, that goal could wait, and he focused on acquiring more private lands. In the name of saving money, he contracted with one firm, the Pacific States Lumber Company, in 1917, to clear timber from city and Northern Pacific Railroad properties. After loggers cut the trees, Seattle would get the land. It was a shortsighted policy to ignore Johnson's warnings and not put watershed management under professional foresters. Soon inspectors were complaining that reckless operations were endangering both public health and timber quality.[41]

As the Water Department was logging and gaining still more land, city engineers marched into the watershed to domesticate the Cedar into a reliable source for water and power. Engineering would marshal the river to turn turbines and fill pipelines, and to remake the Cedar into a reliable hydraulic machine. The engineers' ideal river, however, had already disappeared because logging had changed the hydrology of the basin even as the Cedar remained an undisciplined watercourse, ebbing and cresting with precipitation and snowmelt. Engineers discovered that freshets from the logged-off slopes above Cedar Lake could raise the level of the now-impounded lake as much as forty feet in one week. Downstream, the problems were equally dire. Dredging and construction by the Milwaukee, Chicago, and St. Paul Railroad along its new right-of-way had shifted the Cedar River dangerously close to the city's Number One pipeline. The railroad assured a worried Thomson that no harm would come to the city's water lifeline. Storm surges in November 1911 proved them wrong when floods exploded the original timber crib dam at Cedar Lake, washing away the pipeline, drenching the town of Renton, and depriving Seattle of potable water for almost a week. Afterward, public pressure mounted to build the new masonry dam with all due haste.[42]

Meanwhile, another calamity was unfolding upstream. Logging was altering the local environment to the detriment of the forest and the city dwellers downstream. To improve taste and clarity, in 1914 Water Department crews removed alder trees and surface soil lining Cedar Lake and secondary reservoirs in the watershed to reduce turbidity. They also planted exotic deciduous species—black cottonwood, hickory, and eastern white ash, among others—to shade the new reservoirs. Although reforestation efforts were piecemeal until the mid-1920s, such plantings transformed the forest and its waters. Removing mature trees exposed streams and lakes to direct sunlight, elevating temperatures and encouraging algae growth that imparted unwanted tastes to the water. Migrating salmon could not reach their historic spawning grounds in the upper Cedar River, above the city's intake, because health officers feared that decaying fish carcasses would contaminate water supplies. The salmon got no break below the intake, however. Logging along the lower reaches of the river also harmed salmon by exposing the gravel nests, called redds, built to protect eggs and juvenile fish, to fatally warm waters and predators. Moreover, in response to citizen complaints, Water Department crews began to apply copper sulfate to lakes and rivers in the early 1920s, and chlorinated water entering city pipelines to eliminate noxious flavors.[43]

Aggressive logging also made fires more common in the Cedar River watershed. Embers and hot exhaust from the donkey engines used for winching logs, combined with the abundance of slash left from logging, repeatedly led to burning during the dry summer and autumn months. Huge blazes had swept across the watershed before. One consumed most of the drainage about three hundred years before Seattle's founding, but industrial logging increased the danger. Before 1920, only two fires grew large enough to threaten marketable timber; after 1920, big fires occurred with frightening regularity, sometimes as many as nine in one season. The Department of Water, now dependent on timber revenue for its operations, vainly tried to prevent fires, but like a toy top losing its momentum, the Cedar River forest was spinning out of control environmentally.[44]

A change in administration brought a bleak new perspective that logging was now harming, not helping, water quality. George F. Russell, who succeeded Youngs in 1923, did not share his predecessor's monomaniacal focus, and in a letter to the city engineer of Vancouver, British Columbia, in 1924 he outlined the double bind facing Seattle. "At the time," Russell admitted, harvesting timber to purchase land "seemed a wise thing" to pursue. The timber was ready to harvest and the city could not afford to buy land outright, but "time has demonstrated this was a serious error." "In our Pacific Northwest we are particularly fortunate in having supplies of water of great purity, which gives us a fine standing in the world for healthfulness," he concluded, but "the purity of the water is in great measure preserved by standing trees." Russell's confession had shown that for all of its energy and optimism, the Progressive ethic of place, extolled by Thomson and its colleagues, could and did go wrong.[45]

As the regrades had dramatically shown, no utility was as important for progress as securing reliable water supplies, but unexpectedly, along the Cedar River, controlling the terrain and its vital watercourse began with controlling its trees. In trying to control one resource, water, by transforming it into a common form of property, the Water Department ran headlong into another form of property, timber, which proved harder to control. The city valued water; logging corporations and the Forest Service valued the trees. Reconciling the two had proved impossible. Downstream, city engineers discovered that some of the problems that confronted them along the Cedar River—an unpredictable and disturbed environment and an intractable array of political opponents bent on protecting their real estate—now confronted them in Seattle.

The *Seattle Mail and Herald* in 1903 compared regrading to "grasshoppers in a Kansas cornfield." "Nature has been worked over from a thing of beauty to an angular checkered surface," the writer bemoaned, "with scarce one feature left to know it by." Others attacked regrading as profane and destructive to property. A 1910 editorial in *Engineering News,* which had supported the regrades at first, now called them a mistake: "It is a pity that a city, which has such a magnificent natural site, with such commanding views as Seattle, would have lost such a great opportunity to lay out its streets to conform with the natural features." Thomson brushed off such criticisms. "Some people seemed to think that because there were hills in Seattle originally, some of them ought to be left there, no difference how injurious a heavy grade over a hill may be to the property beyond that hill."[46]

What Thomson forgot was that avarice, anger, and natural forces could thwart his splendid plans. He and his engineering associates first had to confront the unpredictable properties of water and earth. The same glaciers that had created Seattle's hills had also armored them against removal. Glacial erratics—large boulders, some the size of small houses, carried in the ancient ice streams—could be dislodged only with explosives. Hydraulic sluicing mimicked the power of rivers or glaciers in the city's center, unleashing landslides that swept away houses and turned unfinished streets into muddy, impassible quagmires. Streams of slickens scoured holes in iron and steel flumes. Sluice intakes, called "grizzlies," clogged with debris, and men had to dislodge the chunks with crowbars and gaffing hooks against water running at thousands of pounds of pressure per square inch. When slickens reached Elliott Bay, the city had to spend time and money to obtain separate permits for each project from the U.S. Army Corps of Engineers to prevent interference with maritime navigation. One beleaguered engineer on the Pine Street regrade of 1904 listed every possible delay in his weekly reports: clogged water lines and sluices, tired workers, landslides, rain, more rain, and mechanical malfunctions.[47]

Although Seattle residents were used to living in a city under constant construction around them, the regrades taxed their tolerance. The same new laws that gave engineers and their contractors the authority to regrade also gave individual citizens the ability to protest. Local improvement district files at city hall soon bulged with scores of letters complaining of blocked streets, broken water and sewage mains, shattered windows, collapsed foundations, mudslides, debris, noise, dust, and snarled traffic. James Love, pastor at the Reformed Presbyterian Church, located along the Olive Avenue regrade, remarked that with the autumn rains "patience, even the Presbyterian kind, ceases to be a virtue" because "nearly impassable" streets and sidewalks prevented parishioners from worshipping. When contractors on the Jackson Street regrade filled the tide flats before raising buildings to the new grade, they worsened sanitary conditions by blocking sewer outfalls, which in turn spilled noxious waste onto city streets. Neighborhoods that supported regrading at first now implored contractors to finish early, and in turn the Engineering Department often blamed contractors for the subsequent mess. The fair foundation for municipal solidarity envisioned by Thomson turned out to be both broken and unfair.[48]

In the escalating cycle of recrimination and reaction that consumed Seattle during the regrades, those who seemed powerless, those who lived in the tideland waterfront shacks or in the apartments along Skid Row, fought back. An embittered resident told the *Seattle Post-Intelligencer:* "We people who live in your regrade districts . . . are martyrs. Sure, it's a good cause, and all that, but we're martyrs anyway." But they, as martyrs, did not go resignedly to their fate; they often went to court to claim that the regrading had diminished the value of their property. Thomson and his engineers believed regrading would ultimately benefit everyone, and it was no accident that several of the largest regrades, especially the Jackson Street and Dearborn Cut projects, targeted the city's poorest and most dilapidated neighborhoods. Local property owners often felt differently, however, and found the courts sympathetic to their plight. American courts during this period were struggling to define the proper relationship between property as protecting personal wealth and property as advancing social propriety. The Board of Eminent Domain Commissioners reviewed every regrade and then held condemnation trials in King County Superior Court to determine how much property owners would receive in compensation.[49]

Complaints over taxes and condemnation awards generated the most protest because city officials relied so heavily on the advice of real estate agents. Scott Calhoun, the city corporation counsel, in a page-long editorial in the *Seattle Post-Intelligencer* in May 1907, defended the process for the Denny Hill regrade, claiming that in the 350 verdicts entered, which affected nearly 800 property owners, the city had been "more equitable and just than in any other large condemnation proceeding tried during the last two years." What Calhoun did not apologize for was the inherent inequity in asking those who stood to benefit most from regrading to evaluate the worth of other citizens' property. Nor did he mention what regrading had really cost the city as complaints clogged city and superior court dockets, with more than forty separate cases reaching the state Supreme Court on appeal.[50]

The high court cases that involved personal injury and property damage were the most harrowing, especially for residents of modest means living in the poorest neighborhoods targeted by Thomson's regrades. One homeowner, H. F. Povine, charged the city and its contractors, Hawley and Lane, with negligence after regrade fill settled beneath his simple house and wrecked it. J. W. Coffer, a timber cruiser living in an apartment on Fourth Avenue, sought compensation when cinders from elevated trains blinded him in one eye, costing him his job. Peter Casassa and his neighbors sued twice; blasting and sluicing had undercut the slopes beneath their homes, sending the houses tumbling into a ravine. Julius Johanson complained that regrades burst his water pipes and loosened the earth beneath his house, pulling the foundation out beneath him. In all of these cases, the high court found that assessments were below fair market value and ruled in favor of the afflicted residents. In other cases, the justices defended the city against regrading claims that would "render that means of improving streets impracticable." Even though the high court also ruled against unfair assessments, it did not overturn the city's ability to create local assessment districts, pursue municipal improvements, or levy taxes.[51]

Thomson was especially frustrated when lawsuits by the city's more powerful real estate developers stood in the way of what he considered the common good. In one famous case, James A. Moore, owner of the new Washington Hotel, the former Denny Hotel, atop the south summit of Denny Hill, fought hard against repeated attempts to dislodge the imposing Victorian edifice that towered over the downtown district. Moore, a land speculator from Nova Scotia who had arrived in 1886 and quickly bought up property in downtown Seattle, had acquired the hotel from a band of squabbling developers, who, after the Panic of 1893, had left the hotel unopened and unfurnished. Moore had planned to make it the premier watering hole in Seattle. At first, he refused to budge. Finally, in 1906 the city council and the Engineering Department persuaded him to sign a special agreement, named for him, whereby he and other private owners in the neighborhood would pay to remove their own earth in exchange for no special assessment on their property. According to Thomson, the resulting boost in real estate values on nearby lots persuaded Moore to join the city. Smaller property owners, left out of the "Moore Agreement," were not impressed and filed suit. The state Supreme Court eventually sided against the city and Moore, calling for a reassessment of the condemnation awards. In addition to sacrificing his grand hotel, Moore eventually paid over a hundred thousand dollars out of pocket. It was only a temporary setback. After the regrade, he opened a long-planned theater, one of the finest in the Northwest, in the shadow of the former hill, and soon made back his fortune.[52]

Moore had the resources and connections to turn Thomson to his will. Most other citizens acquiesced to the engineers. Those who resisted faced expensive, sometimes embarrassing, consequences. Since regrading operations had often preceded judicial decisions, contractors removed earth and destroyed buildings before court proceedings had concluded. Under local improvement district guidelines, individual owners arranged for the regrading of their own property upon approval of a given project. Without an owner's consent, contractors could still remove earth to the edge of the lot but no farther, leaving lots and homes atop pinnacles of earth, popularly known as "spite mounds," towering above the new streets. In the resulting mess that the regrades caused, the courts tried to validate the sanctity of private property while upholding the common good. Former owners who sold their property during regrading escaped assessment taxes, leaving the new owners responsible for unanticipated debts. Renters ejected from their homes still had to pay their landlords who were, in turn, responsible for assessment taxes.[53]

One story among hundreds captured the human cost of Thomson's war against Seattle's hills. Sandy Moss, who migrated with his family from Topeka, Kansas, to Seattle in the early 1900s, was one of the hundreds of African Americans and Asian immigrants who had settled on Beacon Hill, taking advantage of the then inexpensive prices at the south end of Seattle. The Moss home was along the Dearborn Street regrade, which would cut through Beacon Hill to connect Rainier Valley with downtown, and his father refused to vacate. The hydraulic cannons tearing into the hillside became unbearable, and the family took rooms across the street where they could keep an eye on their home. One morning their house had vanished. "So we walked over to the brink of the hill and looked," Moss recalled, "and the house was three blocks down below, and it was about two hundred feet lower and it was just a bunch of matchwood down there. It was all broke up . . . like you would step on a apple box and crush it." The Moss family had to move, like so many others, because his family received some compensation from the city "but not enough to pay for what the homes and the property was worth."[54]

Local resident J. W. Charlton had warned Thomson in 1910 that regrading projects had "piled up so fast" that many residents, overtaxed and overwhelmed, would "lose their homes." Many citizens did. A spiraling cycle of municipal debt and unfinished projects, lawsuits against recalcitrant contractors for malfeasance, grievances from more and more citizens, and uncollected taxes from property owners unable or unwilling to pay had trapped the city. Thomson tried to speed up the regrades in order to finish them as quickly and cheaply as possible. On the Denny Hill project, the city's largest, Thomson's strategy, according to his assistant O. A. Piper, had led to "cooperative disregard for street lines, the paramount idea being the rapid prosecution of the regrade as a whole." Decaying public support and shaky financing had ended Seattle's regrading mania by the beginning of the First World War. By 1911, Thomson had left the post he had occupied for nineteen years to help found the municipally owned Port of Seattle. Even as Thomson moved on to his next project, the battles raged over how he and his engineers had left behind them in the heart of the city acres of newly hazardous spaces.[55]

Conditions were at their worst in the former Jackson and Dearborn regrade districts, where the hydraulic cannons had punched the hole through Beacon Hill (and Sandy Moss's home) and engineers filled the gap with bridges linking the southeastern districts of the city to downtown. In their speed to finish, contractors had left huge portions of the now-exposed hillsides barren. Rainstorms pounding on the exposed clay soil triggered damaging mudslides that swept away homes and knocked out the Twelfth Avenue Bridge—one of the city's main north–south thoroughfares—repeatedly for the next decade. Eight lawsuits made their way to the state Supreme Court between 1914 and 1927; in all but one, the justices ruled against the city. As Justice Tolman wrote for the court in 1927 in *Wong Kee Jun v. Seattle,* the last of the so-called Jackson Street regrade cases, city engineers were to blame for the catastrophe they had created by "causing slides" through their "permanent invasion of private property."[56]

Empty lots now dotted the space where half of Denny Hill once stood beneath the broken summit of its jumbled remains; to the south, denuded slopes of raw earth loomed above Jackson and Dearborn Streets. Real estate agents and civic boosters tried to put the best face on a bad situation, churning out flyers and promotional materials, touting the affordable real estate north of downtown in the newly minted Denny "Regrade District." A birds'-eye map drawn by architect Dudley Stuart in 1917 proclaimed the area as "Seattle's Coming Retail and Apartment-House District," detailing imaginary businesses and brownstones against the scenic backdrop of the Olympic Mountains and Elliott Bay, but his map was a fantasy. Fearful of larger assessments, residents had reached a compromise with city engineers in 1906 to regrade only the portion of the hill nearest Elliott Bay. The remainder loomed over empty lots waiting for a boom that disappeared when Seattle slipped into a depression after the First World War.[57]

Only a year after Dudley Stuart's boastful map, city engineers faced a real crisis. After the ruinous 1911 flood, the Water Department had thrown itself into completing the proposed concrete dam at the mouth of Cedar Lake, ignoring warnings that the basin's porous glacial soils were unstable. From the time that workers poured the first column of concrete in 1914, the dam leaked, sometimes profusely, at its edges. Engineers proposed sealing the lake bottom with asphalt, but that proved too costly and risked endangering water quality. At the recommendation of consulting engineers, Thomson, now in private practice, and William Mulholland, architect of the famed Los Angeles Aqueduct and the head of that city's Department of Water and Power, workers ran a kind of reverse regrade, sluicing mud and gravel onto the submerged lakebed to stop the hemorrhaging. Their efforts staunched the leaks until the dam was finished in autumn 1918. It was a temporary fix.[58]

In the evening of December 23, 1918, Charles Moore, the night watchman for the North Bend Lumber Company in Edgewick, a small logging town in the Cascade Mountains, was making his rounds along the company's millpond on Boxley Creek. His job was important—to keep an eye on the creek's waters, whose tempo rose with heavy winter rains. Almost two miles to the south, on the other side of a high ridge, Cedar Lake lay behind a new $2 million masonry dam, now leaking on its northern wing and pushing its way through the glacial moraine that separated the reservoir from Boxley Creek below. Only days before, W. C. Weeks, president of the North Bend Lumber Company, had complained that the banks above the creek were seeping. Seattle's city engineers had assured him that the fissures would disappear with time. Moore thrust his lantern through the rain and observed that the creek was rising, a foot every two minutes. The engineers were wrong.

Moore ran to the mill, pulled hard on the steam whistle, tied it down and then sped from door to door, shouting, "Out of your beds, the dam is going to go!" Residents scrambled to put on their clothes and dashed out into the night and onto high ground, with the muddy waters reaching their ankles, then to their knees, then up to their shoulders. Those who hesitated almost lost their lives. A wall of water and mud a story high had pushed through the moraine, swallowing over a half million cubic yards of earth as it rushed into Boxley

Creek. The mill and bunkhouses held back the torrent briefly, long enough for the dawdlers to climb to safety. They huddled around a bonfire until daybreak. Amazingly, no one died.

The next morning, Christmas Eve, the residents of Edgewick picked through the remains. Automobiles were perched atop broken homes or jammed into ruined kitchens and living rooms. The surge carried most of the town downstream, clogging the Snoqualmie River with debris and cutting the main line of the Milwaukee, Chicago, and St. Paul Railroad to Seattle in two. Initial damage estimates were more than a quarter million dollars for the North Bend Lumber Company alone. The company and the town were totaled.[59]

Who was to blame for the "Boxley Burst"? The city's two principal newspapers called the project a "white elephant" and said there was "no practical way to make a water container out of a sieve." Both lambasted the city council for ignoring the advice of its paid consulting engineers, especially Thomson, but the papers also noted that the same experts had promised that leaks could be fixed. "The present predicament of the city," the *Seattle Post-Intelligencer* concluded, "would seem to be a refutation of those optimists." Engineering confidence was part of the problem. It was a lesson that many citizens of Seattle already knew all too well.[60]

The challenges of building a workable, efficient city had tantalized men like R. H. Thomson, driven to surmount the physical environment's shortcomings by improving upon nature's flawed design. Men like Thomson tried to write new social rules and physical processes onto what they saw as a clean slate, and in many ways they were successful. Engineers, business elites, entrepreneurs, and planners—the paragons of Progressive virtue—tried to rationalize the social and environmental workings of the city all at once. Their idea of the public was a people policed, regulated, marked out, and defined. Drawing social boundaries on the city served the same purpose as drawing physical boundaries: it provided for the efficient delivery of goods, government, and services. After all, the goal of the engineer, as Timothy Mitchell notes, is "to simplify the world," to rise above the exigencies of place, and thereby to gain "the powers of expertise by resolving it into simple forces and oppositions": nature and technology, public good and private interest, efficiency and waste, order and chaos. This is not necessarily a mission devoid of ethical purpose. Without Thomson's efforts, Seattle would not have had pure water, cheap electricity, or clean surroundings. For Thomson the engineer, the physical environment was a source of energy that, properly exploited, could vanquish disease, repel vice, and fortify the common good.[61]

It was a principle that Thomson would cling to after he left his office to pursue other public-minded endeavors. He remained consumed by his work, a human perpetual motion machine, serving on the Seattle city council from 1916 to 1922 while continuing to design the first floating bridge across Lake Washington and facilities for the Port of Seattle. He even returned in 1930, for one year, to oversee the completion of the Diablo Dam on the Skagit River for Seattle City Light and his former protégé, J. D. Ross. In his autobiography, Thomson wrote that while he had accomplished much, there was a great deal left to do "because a city is a growing thing, and what might satisfy today will be insufficient tomorrow."[62]

Thomson had failed to anticipate that physical processes, such as water pushing against a dam, or frustrated citizens, such as angry homeowners, could undermine the boldest plans. The regrades and the waterworks sprang from a deep desire to reconcile cities with their environment in search of social order. That was the promise Thomson had in mind when he imagined lifting Seattle from its "pit," the promise he made, along with other reformers, to remake Seattle's environment, hoping to achieve efficiency, equity, and sometimes beauty. Thomson's ethic was an engineers' ethic, one fixed to a place through his mastery of technology and knowledge deployed in the pursuit of usefulness for the greatest number. In the minds of many citizens at the time, Thomson's philosophy saved Seattle and its residents from disease, fire, and pestilence plus the fraudulent machinations of corporate monopolies. Through bending rivers and moving hills, the city engineer fortified Seattle against its human and non-human enemies while turning it into a fit place to live. Yet by linking social progress to the ability to improve nature, Thomson had, unintentionally and ironically, further inscribed inequality and instability into the landscapes of Seattle and its hinterlands. When his aspirations fell short, when some people and some places suffered more than others did, many wondered if the idea of improvement had failed. The battles that followed over civic justice and environmental protection had their beginnings in Thomson's anodyne philosophy, and in the coming years they would grow ever more hostile.

CHAPTER 4

Out of Harmony with the Wild Beauty of the Natural Woods

Artistry Versus Utility in Seattle's Olmsted Parks

Seattle's reformers wanted a city that worked but a city that was also beautiful. For that beauty they sought what every great metropolis at the time aspired to build: splendid, verdant parks. And for that desire they turned to another kind of expert, the landscape architect, and to the acknowledged leaders in the field, the Olmsted Brothers. By bringing the rural into the city, landscape architects like the Olmsteds imagined that parks would not only provide beauty and pleasure but also promote the health and the morals of Seattle while simultaneously creating a more democratic society. The question was whether beauty would come at the cost of efficiency, or if the impulse of a growing and striving city could coexist with the yearning to tame the excesses of that impulse with artistry.

The new American park, as imagined by the Olmsteds and their disciples in Central Park in New York City, Prospect Park in Brooklyn, and the "Emerald Necklace" of Boston, was an amalgam of city and countryside. It was the culmination of a revolution in urban design and planning that started in the early nineteenth century with the work of the architect and horticulturist Alexander Jackson Downing. Historians have praised as well as criticized the Olmsteds for offering an alternative democratic vision of American life to preserve open space in an urban setting without dissolving into agrarian nostalgia. By the turn of the century, the Olmsted solution was part of a larger contest between conflicting ideas of what American urban life should be. This competition was both a clash of thought and a collision of practice. At its core, it was a debate over the meanings of place and society. On one side was what Thomas Bender has called a "lost tradition" of urban life that "emphasized the ideas of organic social relations, community, and natural beauty." This was the original Olmstedian legacy, the viewpoint of the landscape architect. On the other was the Progressive philosophy of efficiency and systems building; this was the perspective of the municipal engineer. Beyond the realm of ideas, however, urbanites fought over what functions parks should perform. Some agreed with the Olmsteds and wanted parks as landscapes designed for the contemplation of nature's beauty. Others resisted the notion of parks as scenic time capsules and imagined them instead as public spaces to be used as one wished.[1]

Both were false dilemmas yet ones that have become enshrined in how Americans think of urban parks. Parks are always both natural landscapes and social spaces, though a park with a padlocked gate and reserved for the privileged few is not a democratic space. The Olmsteds themselves realized this in designing parks that were scenic and naturalistic as well as socially useful and often functional. The Olmsteds knew that park design and politics were inseparable from dynamic physical landscapes in an ever-changing urban environment. In blueprints or maps or park landscapes, the muddy, leafy, noisy, disputatious world those plans embodied was often absorbed or lost. So, too, were the disputes that parks engendered.

The same tensions between public and private power that had shaped the quarrels over Seattle's waterways and tidelands, real estate and water supply, now plagued the design and use of its parks. Intended as refuges from the perils and clamor of urban life, parks, ironically, became ecological and social extensions of those perils. They magnified tensions, too, between landscape architects, who portrayed themselves as the arbiters of good taste and design, and city engineers, who depicted themselves as the champions of cost-effective management for the public. The irony was that the landscape architect of the early twentieth century employed the methods and philosophy of the engineer—careful planning, a reliance on expertise and modern technology, and a desire to mold society along particular, often rigid lines—to achieve their organic effects. Cultivating Seattle's urban pastoral would prove to be neither peaceful nor bucolic, in part, because the contest between John C. Olmsted, the landscape architect, and R. H. Thomson, the city engineer, would be just as messy and just as fraught as other urban issues. The legacies of their dispute would drive subsequent conflicts between park users and park managers over how best to balance the needs of the many with the wants of the few in an increasingly unstable urban environment.

Man and Nature in Happy Harmony of Effort

Walking along a high bluff above Elliott Bay in May 1903, near the U.S. Army post at Fort Lawton, the famed landscape architect John Charles Olmsted rejoiced in the panorama before him. He had arrived only a few days before with his assistant Percy R. Jones to begin work on Seattle's new park system, and the visit had been a whirlwind of meetings and walks and drives. After touring "the highest point in the city," Queen Anne Hill, one of Seattle's finest neighborhoods, and a "beautiful park with plenty of natural trees and shrubs growing on it," Olmsted's hosts, the Seattle Park Board, took him through "a very crude unfinished part of the town" at the foot of Queen Anne, the Interbay district. Here, railroad "switching, freight, and shop yards," operated by the Great Northern, were "outlined by some cheap houses, a store or two, and a cheap hotel" built upon "filled tidelands now unoccupied," the residue of "the land boom in that district having come to grief some years ago." After lunch, Olmsted and the Park Board members strolled along a bicycle path to the high point at Fort Lawton, passing through "woods that had only small trees" remaining because "all the best firs and cedars [had been] cut and taken away for logs."

Once Olmsted reached the summit, his skepticism melted after he saw the "great tall trees and lots of great alder trees and maples and a dense undergrowth and ground covering of ferns, brake, salal, and roses," with "occasional outlooks over Puget Sound and the hills beyond and the snow clad Olympic Mountains." "The sound is not so very unlike Penobscot Bay," he wrote to "Fidie," his wife, Sophia White Olmsted, "except in having very few houses in sight outside of Seattle itself and West Seattle." "The opposite shores of the sound," he noticed, "are almost continuously wooded on low hills, the fir of course giving the character." Perhaps Olmsted was reminiscing about the summertime vacations at his family compound on Deer Isle, off the coast of Maine. The sun was warm, the air cool. "I wish you were along," he told Fidie.[2]

The day after Olmsted's arrival, he told the *Seattle Post-Intelligencer* he was yet "unable to tell you what plans would be best to beautify the parks," but "I do not know of any place where the natural advantages for parks are better than here." "You have taken time by the forelock and purchased parks when the city was young," he remarked, and if the city followed his suggestions, he predicted the parks "will be in time one of the things that will make Seattle known all over the world." Olmsted had girded himself for the challenge, coming to

Seattle with experience no one in his field could match and carrying the promise and burden of his family's pedigree as a nephew of Frederick Law Olmsted, the acknowledged founder of American landscape architecture. He had also become a stepson when, in 1859, his uncle married his mother, Mary Cleveland Perkins Olmsted, after the death of his father, John Hull Olmsted. He had followed his stepfather, in 1867, to Mariposa Estate, a gold-mining operation in California's Sierra Nevada, where he learned first-hand geology, botany, and natural history. In 1869 and 1871, as a member of Clarence King's surveys of the 40th Parallel in Nevada and Utah, he received further lessons in reading landscapes, a skill he had honed at Yale's Sheffield Scientific School and, after graduation from Yale, as an apprentice to his stepfather. Together with senior partners Henry S. Codman and Charles Eliot, the young John C. Olmsted designed, among other projects, Boston's "Emerald Necklace" of linked metropolitan parks. In 1898, after his stepfather's retirement and the deaths of Codman and Eliot, John Olmsted co-founded Olmsted Brothers with his younger half-brother, Frederick Law Olmsted, Jr., and the following year became the inaugural president of the American Society of Landscape Architects, the first professional organization of its kind in the United States.[3]

The process of building urban parks had changed significantly since 1858, when Frederick Law Olmsted and Calvert Vaux submitted their "Greensward" plan for New York's renowned Central Park. The elder Olmsted had worked within an older tradition of great artists competing for prized commissions, but by the time his stepson arrived in Seattle in 1903, landscape architects and the Olmsteds had become modern business executives. No longer seeking commissions or entering competitions, the Olmsted Brothers instead fielded requests from hundreds of clients each year. While his brother, graced with the famous name, became the public face of the firm, John Olmsted worked mostly in the background, building a company that would make the family name last. He was the hidden intelligence behind the firm's business practices. He crisscrossed the nation by train and automobile, planning parks and residences, sometimes juggling as many as 250 jobs at one time. The drawers and closets in the Olmsteds' office in Brookline, Massachusetts, bulged with maps, blueprints, plans, and correspondence from cities throughout North America. John Olmsted drew or commissioned the bulk of them. By the beginning of the new century, the Olmsted Brothers' park plans were an industrial product for sale across the continent. Just as prospective homeowners could, by the 1920s, order kits from companies like Sears, Roebuck to build beautiful yet affordable houses, municipal officials could now turn to the Olmsteds or other professional landscape architects to buy the talent to make their cities lovely and green.[4]

John C. Olmsted represented a new kind of professional, one whose primary allegiance was to a skilled class like that of R. H. Thomson, Seattle's municipal engineer. His training shaped his critique of Seattle's needs and limitations. Unlike Thomson, Olmsted did not identify with a locality; he was a booster for his profession, not his hometown. As president of the American Society of Landscape Architects, he helped to develop, with his half-brother, a code of professional ethics that advised landscape architects on how to behave with public officials and their clients. As did city engineers, landscape architects saw themselves as a class apart, as specialists endowed with great abilities. But unlike engineers, whose business was efficiency and economy, landscape architects purported to labor in the name of beauty—and on behalf of something more. They wanted to redraw and rebuild the entire city as an organic entity, sometimes literally from the ground up. In this sense, landscape architects saw themselves as aspiring to a higher calling than workaday city engineers. Only they could see and understand the city in full detail. As Olmsted explained in an 1894 address to the Boston Society of Civil Engineers, the municipal engineer was to the city what the family physician was to the family. "Just as the family physician calls a specialist into consultation in cases requiring expert advice which he may not feel competent to give, so the city engineer should, in all cases of sufficient importance, secure the services of the specialist in sewerage, in water supply or in park designing."[5]

Over the second half of the nineteenth century, parks had served not only as genteel pleasure grounds but also as open spaces for democracy and public protest, as agents for reform and improved public health, and as arenas for vigorous outdoor recreation. By the dawn of the twentieth century, parks had also become symbols of urban sophistication. A modern park system was a civic adornment for a city that had come of age, and Seattle business and political leaders, eager to project cosmopolitanism, were willing to pay for beauty. Years of park planning in Seattle preceded Olmsted's visit. Eugene O. Schwagerl, Seattle's second Superintendent of Parks, had proposed an ambitious system, inspired by Frederick Law Olmsted's Boston Emerald Necklace, to unite Seattle's private and public greenswards of land with a latticework of parkways and boulevards. Schwagerl quoted almost verbatim from the senior Frederick Law Olmsted in his 1892 report: "Parks are the breathing lungs and beating hearts of great cities." And, like Olmsted, he saw parks as great public spaces where "rich and poor mingle to inhale the unalloyed, God-given perfumes to body, mind and soul." Parks were to the body politic what hearts and lungs were to the human body—vital organs pulsating with the life force of nature. A Seattle with magnificent parks would have vigorous citizens. Without parks, Seattleites would languish. When Schwagerl left office in 1895, his plan never implemented, an inspired city council, by 1902, nonetheless acquired almost 500 additional acres for parks.[6]

In 1900, the editors of the *Seattle Argus* had objected to the plan: "Seattle does not need the kind of parks that are so necessary in the East" because "she has so many natural beauties that artificial ones are superfluous." The city council, the year after Schwagerl's departure, had a similar perspective and amended the city charter to abolish the post of parks superintendent and create in its place a Park Committee reporting directly to the council. The hope was to begin saving Seattle's natural beauty from the onslaught of urbanization. By 1902, the *Seattle Post-Intelligencer* was complaining that the perpetual construction of Seattle had put "insurmountable difficulties in the way of enjoyment of these natural beauties." Dense undergrowth and fallen trees, the legacy of logging, were choking Seattle's open spaces, and "the hand of man must . . . aid nature to fit these beauties for daily life." The *Seattle Mail and Herald* declared that same year that "parks in and around Seattle are the wonderful evidences of what man and nature in happy harmony of effort can do." Earlier Seattle boosters had advocated beauty as a means of enhancing the city, just as efficiency had been invoked to justify regrading, the mechanical removal of steep hills in the middle of the city to spur business growth. The Park Commission's first annual report, in 1891, advised that "park cities" attracted investment as effectively as any factory or railroad. It took a decade of inaction, but in 1902, the *Argus* editors changed their minds and, in an editorial, advised that parks could induce easterners of "wealth and culture to move out West" and "enjoy all the luxuries which their money bought for them in New York and Philadelphia, or Boston."[7]

The *Argus* had felt the political wind at its back. Portland and Tacoma had already approached the Olmsteds. Frightened now at the prospect of losing potential migrants to rival cities, Seattle's neighborhood improvement clubs, the same clubs exploited by R. H. Thomson in support of regrading, were now calling for more and better parks as well. Local business leaders, too, were pushing for parks. J. D. Blackwell, superintendent of the Seattle Electric Company, owner of several private city parks, appealed to the Olmsteds in a way that mattered—the "natural park features" of Seattle, he warned, were in peril "of being butchered by persons unskilled in park work." Charles W. Saunders, secretary of the Park Committee that had replaced the Park Commission, concurred and explained to Olmsted that Seattle's magnificent scenery provided "opportunities . . . for a Park System . . . beyond the average of other cities." Seattle's "soil, atmosphere and growth," he explained in a 1903 letter, "may be found to be different from what you have heretofore studied in connection with park work." John Olmsted, who handled most of the firm's correspondence and negotiations, intrigued by the challenge, agreed to begin work later that spring.[8]

When Olmsted and Jones arrived in Seattle in late April 1903, they found obstacles as well as the "opportunities" Saunders had promised. From Captain John F. Pratt of the U.S. Coast and Geodetic Survey, stationed in Seattle, Olmsted learned that it was a small city. It had limited finances and a boom-and-bust realtor mentality that would make land acquisition difficult. Some of what he discovered in person depressed him. Seattle's existing parks, a hodgepodge of unkempt parcels of woods and fields scattered throughout the city, offended his standards of refinement. In his tour of Denny Park, the oldest and most developed, Olmsted sniffed at the work of Eugene Schwagerl, whom he called "a local landscape gardener, Swagel, or Swaugel," whose "walks are very crooked often and his banks steep and high and his plantings very mixed but pretty much the same selection for every place." Olmsted's design aesthetic was more naturalistic. As his stepfather, Frederick Law Olmsted, had written in his 1866 report for Brooklyn's Prospect Park, "a mere imitation of nature, however successful, is not art." The purpose was not "to imitate nature, or to produce an effect which shall seem to be natural and interesting," but to assess "the capabilities and limitations" of a given site, then choose the appropriate plantings, layout, and design to mesh with the local environment and not against it. Schwagerl's style was cruder, and in a tour of several private residences that the former parks chief had designed, John Olmsted criticized the "large boulder banks for picturesque effect no matter whether the house was Old Colonial or American or Italian." "So he seems to be no considerable artist in his line," Olmsted concluded.[9]

Between meetings, Olmsted walked everywhere, learning Seattle by foot, and what he saw in places was outright desolation. Homes and streets clung to impossibly steep slopes, leading to erosion and landslides whenever the rain fell. "The utter disregard of land grades shown by almost all land speculators," he explained to his wife, "is very noticeable here." Loggers had cut down some of the finest trees, leaving behind the stumps and charred slash. The Seattle Electric Company's noisy streetcars ran through Volunteer Park, rundown shanties ringed Green Lake, a popular bathing spot at the north end of the city, and warehouses and tenement houses on the downtown waterfront to the south left no room for public beaches. Everywhere Olmsted looked, he saw a mess: acres of burnt stumps, muddy roads blocked by landslides, thickets of fast-growing alder trees and of Himalayan blackberry, a noxious alien species from Eurasia. Yet he also saw great potential, especially in Seattle's natural scenery, and he peppered his correspondence with uncharacteristic superlatives recounting Seattle's climate as a horticultural blessing, like England or Ireland in its ample moisture and soft sunlight. Its subtlest features animated his imagination: moss clinging to boulders or downed trees, and gigantic ferns flourishing in the city's nooks and ravines. When he closed his eyes, he saw dogwoods replacing aging firs, and rhododendrons mixed among indigenous greenery. Like his stepfather, the junior Olmsted used both native and non-native species in his parks and was not above altering the course of streams and reshaping the contours of hillsides. Olmsted sculpted parks in his mind before turning over the first shovelful of soil.[10]

After nearly two months of surveying and walking, observing and talking, Olmsted wrote a report outlining a grand vision; and the city council accepted it in October 1903. Despite rugged and damaged terrain, which posed significant design difficulties and financial costs, unparalleled opportunities existed, he said, in Seattle's abundant views of the distant mountains, and the "valuable remains of the original evergreen forests," some within the city itself, sheltering beneath them "dense and beautiful undergrowth." Olmsted's plans for Seattle were big. Beginning in the southeast corner of the city, still covered with forests, he proposed as a keystone park the entire Bailey Peninsula. He recommended connecting that new park, via a spacious shoreline parkway, to Ravenna Park, a private reserve north of the new University of Washington campus, which he planned to have Seattle purchase. Two other parkways would link Green Lake, at the north end of the city, to parks along the Puget Sound shoreline in the former town of Ballard. On top of Queen Anne Hill, a boulevard would encircle the ridgeline and attached parcels along the Magnolia Bluffs, adjacent to Fort Lawton, towering above Elliott Bay. In West Seattle, yet another shoreline drive, along Alki Beach and through the still-undeveloped Duwamish River valley, would complete the circuit. As his family had done in Boston and New York City in the 1860s and 1890s, Olmsted hoped to ring Seattle, too, in garlands of green, a truly emerald city.[11]

Olmsted realized that private landowners along Lake Washington's shores, who anticipated a huge windfall once the long-delayed Ship Canal was completed, could demand high prices for land left uncovered by the receding waters. He told property owners and city officials alike to remember the symbiotic effects of "long, wide and handsome parkways on the value of adjoining land." Public parks and private property were not incompatible but complementary. Parks could benefit landholders on the city's hills by turning their slide-prone slopes into valuable and scenic real estate. Olmsted urged the city council to buy all of the ridgeline above Lake Washington, which he called the "Rainier Heights Landslide Section," because leaving unclaimed land would encourage "cheap houses" next door to "one of the best residential districts of the city."[12]

Olmsted needed the support of the city's moneyed elites who backed Seattle's park movement to make his plan work, and he had learned from his stepfather that currying favor with realtors and landowners, who had their hands on the levers of government, was the most effective way to turn blueprints into realty. Elbert F. Blaine, a powerful developer and scion of one of the city's founding families, J. D. Lowman, a prominent printer and businessman, and J. W. Clise, another leading landowner—all sat on the Park Committee. But Olmsted's wooing of Seattle's landed elite was not a resounding success. Many considered his initial estimate, nearly $1.2 million, too expensive for Seattle's limited tax base. After extensive negotiations, Olmsted agreed to trim those portions outside of the existing city limits—the Bailey Peninsula and the southern portion of Lake Washington Boulevard—as well as those, like Green Lake Boulevard, that interfered with development plans in newer residential districts. To reduce local property taxes, he agreed to modify those parcels that abutted established neighborhoods.[13]

Those same landed elites, developers like Clise and businessmen like Lowman, now worried that a weak Park Committee could not enact Olmsted's sweeping plan, set about changing the city charter yet again to create a more powerful Board of Park Commissioners that would not be beholden to the political whims of the city council. The gambit was more than a simple grab for power by Seattle's real estate and business interests: it was a question of civic modernization. Olmsted, who supported the move, reminded his clients that

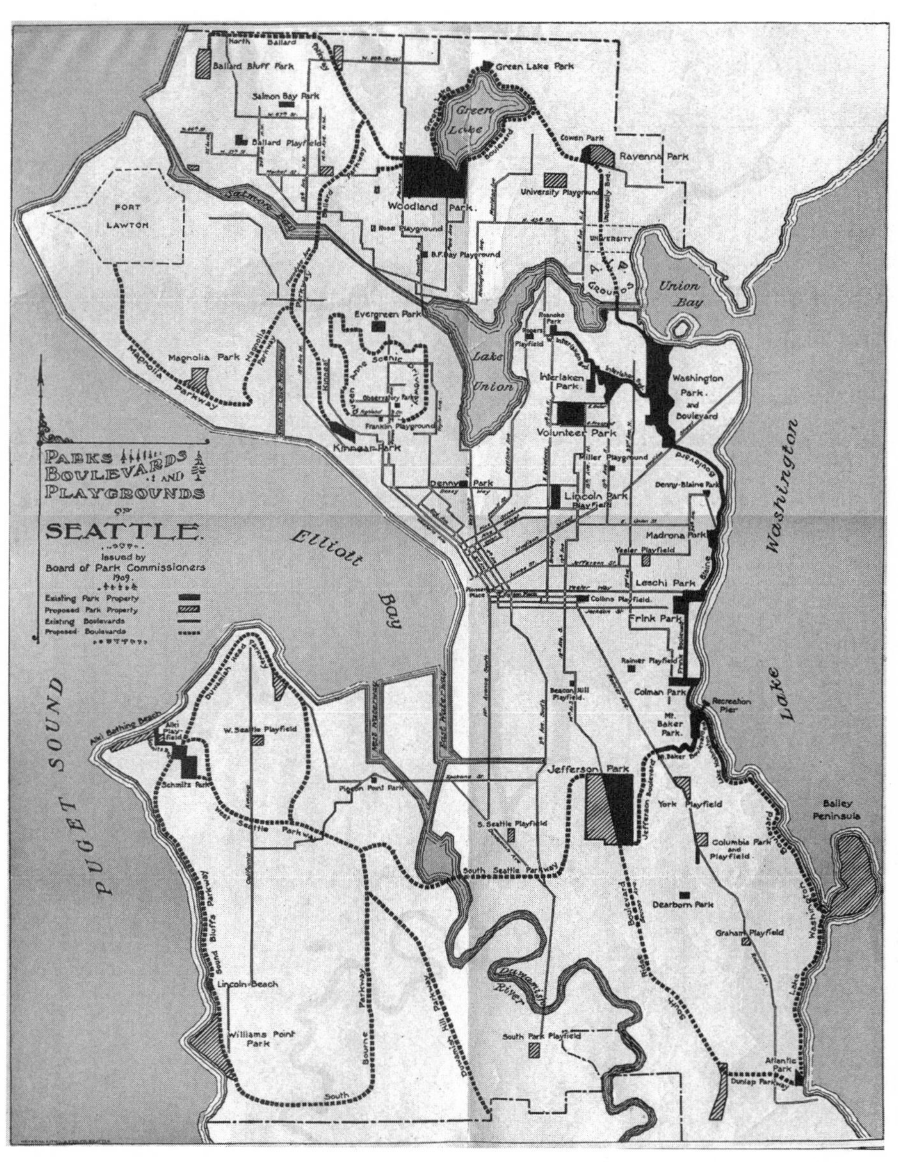

Map of the original Olmsted Brothers park system, 1909. Several of the proposed parks and boulevards, like the Duwamish Hill Parkway, were never built. (Courtesy Seattle Municipal Archives, Item 607)

park commissions "working independently of the Common Council" were "customary in most of the larger cities of the country." He further suggested that the new board include "a competent real estate man in connection with the purchase of land; a competent lawyer as to legal questions; a competent designer as to plans; and a competent superintendent for the execution of the designs and for maintenance." The existing Park Committee, serving at the pleasure of the city council, had no authority over employees or property. Olmsted feared that without a strong agency devoted to parks, elected officials, as they had in other locales, would botch his designs and mismanage his creations. So, with Olmsted's imprimatur, the city council proposed a charter amendment in 1904 to create an autonomous park board, but it met stiff resistance from other reform-minded citizens and political officials who wanted more parks but not the concentration of power in the hands of those entrusted to build them. R. H. Thomson was one of the most vocal critics, and, in an editorial just before the March election on the charter amendment, he pronounced the amendment "repugnant to good morals . . . to a democratic form of government." His reaction was curious given the more common tendency of Progressive Era reformers to empower independent boards and commissions in order to remove technical decisions from an uninformed and easily influenced electorate. But Thomson had a personal stake in this campaign: a sovereign park board could operate without the consent of the Board of Public Works, which he controlled as city engineer.[14]

This time, proponents for a strong park board outflanked Thomson by using the engineer's own tactics against him in a direct appeal to neighborhood improvement clubs. By early February 1904, amendment supporters had collected more than three thousand signatures. Even that was barely enough. The city's principal newspapers endorsed it, as did numerous neighborhood improvement clubs, but the measure passed by the slimmest of margins in the March election. The tallies broke down along class lines. Voters in the blue-collar Second, Sixth, and Ninth Wards, which were home predominantly to semiskilled and skilled workers, trounced the amendment; in the Ninth Ward, the margin of defeat was almost two to one. Electors in more economically mixed or affluent districts, which often included more parks, supported what the *Seattle Times* called "the gardens of the people." Advocates for a strong park board had achieved victory but no ringing mandate.[15]

Olmsted had pushed hard during the election for the reforms he considered necessary, but now he was concerned that politics had corrupted his plan.

He aired his misgivings to Charles Saunders, the outgoing committee secretary, unconvinced that the average citizen understood "the vital necessity of having parks controlled by a board of citizens selected for their taste and knowledge of successful park work." Olmsted's parks, despite their commitment to beauty and the purported well-being of all citizens, had become another function of urban politics.[16]

Exceedingly Ugly, Awkward, and Undesirable Roads and Playgrounds

Olmsted had earned himself the enmity of the city's most powerful public servant, R. H. Thomson, the voluble and headstrong engineer, the polar opposite of the reserved and patrician landscape architect, who, dressed in fine suits, with piercing eyes set in a high furrowed forehead sloping above a natty beard, seemed out of place in rough-and-tumble Seattle. Thomson relished a good public fight; Olmsted, a sensible businessman in his own right, preferred the quiet of his drafting table. Thomson's aesthetic was industrial and modern, a stark contrast to Olmsted's naturalistic artistry. Both were skilled politicians, but the edge in this fight went to Thomson, though in uncharacteristic self-deprecation, the engineer downplayed his political acumen to a friend "because of all things, I [am] that the least." Olmsted was a private person who, when he was on the road, longed to be at home with his wife. He could be aloof, even rude with clients, forgetting names and speaking tersely, a shortcoming he ascribed to his natural shyness. "I tend to always keep to myself and make no friends," he wrote after his first visit to Seattle, a place that made him feel especially uncomfortable. He complained about the rich food and its effects on his digestion, and his shoddy accommodations at local hotels. The ruggedness of the Northwest chafed his body; confined to his hotel room by his labors, he complained to Fidie that he was "so out of practice in outdoor work that I have got somewhat 'soft' as athletes say." The bad weather and rugged topography made what exercise he did get difficult as he stumbled through underbrush, got winded climbing hills, and shivered in the incessant drizzle.[17]

Seattle's social milieu, particularly along the waterfront and on Skid Row, repulsed him, too. "Cheap amusements are abundant," Olmsted grumbled in 1907, describing Seattle's dance halls and saloons, filled with transient immigrant laborers and Indians, the denizens of Skid Row and Chinatown. "I fear they don't save much money in Seattle—the laboring people, I mean," he observed, and "they spend as fast as they get money, sometimes faster." The year before, he noticed how "a large number of toughs" packed the streets and speculated whether there were "enough police to keep them in order, especially in the winter when more come in from Alaska and the Yukon." "I've often wondered how many hotels there are in this city," Olmsted wrote later, in 1910, commenting on the temporary homes, perhaps, for the toughs he feared. "I should say over a thousand!" Residential hotels, which housed largely single men and the working poor, were the bane of Progressive Era reformers, portents of vice and social disorder. The number and concentration of Seattle's residential hotels shocked Olmsted, who said they "seem more numerous than in any other city I know. It spells race suicide!"[18]

A popular belief among many turn-of-the-century elites, race suicide was the fear that the birthrate among non-white or Catholic immigrants would dilute the racial stock of native-born white Protestant Americans. Olmsted's allusion to unwanted race mixing in his letter home could not have been clearer. The day before, Olmsted "saw a lodging house with a big transparent sign over the front door: THE RED LIGHT. It may have been a sort of joke, as it was in the 'red light' district so-called. It doesn't look very tough by day but I'd rather not go there at night." The tight connection between race, place, and decadence in Olmsted's mind was typical of many landscape architects as well as other urban experts of the era. He had learned the lesson from his stepfather: poor, debased places tended to produce poor, debased people, and vice versa. Changing the land would thus change the people, elevating their condition and sorting them by their natural social status. Indeed, if there was ever a city in need of the Olmsted touch, it was uncouth, mongrel Seattle.[19]

If Olmsted found Seattle provincial and its residents uncultured, many Seattleites considered Olmsted an eastern snob, and none more than Thomson. The city engineer had little patience with the effete landscape architect and his approach toward the complexities of city engineering. Olmsted may have designed the parks, but Thomson's engineers, working in concert with park board employees, had to build them. He butted heads with Olmsted soon after the March 1904 election when construction commenced on the new greenswards, beginning with the major anchor parks in Seattle's four corners. One disagreement erupted when Olmsted's landscape architects began mapping and planning Washington Park, set in a narrow valley along the western shore of Lake Washington, just below Interlaken Boulevard. Thomson's engineers, who were busy finishing the expansion of the city's new sewer system, extended an outlet that drained into the park's small creek. The engineers' attention to sanitation left little room for scenery, and James F. Dawson, Olmsted's chief assistant in Seattle, wrote that August in disgust, complaining that Thomson's crew had destroyed "the brook for all purposes and beauties we anticipated." When Dawson confronted Thomson, the engineer offered to build a septic tank to impound the sewage, prompting the landscape architect to dismiss the idea: "the brook would be of no value to us for the park" and instead would present "a severe detriment" to visitors. Dawson was unsatisfied with leaving the creek in its natural state because Olmsted's plan called for removing the swampy thickets of ferns and downed trees that surrounded the stream to create spacious lawns and manicured woods. A septic tank, he feared, would ruin the natural effect Olmsted wanted to create, and, eventually, the park commissioners agreed to encase part of the creek in a storm sewer to drain the valley. Such a solution, Olmsted claimed, would make the grounds "useful to visitors as well as beautiful to their eyes," while allowing a segment of the creek, running through the newly minted meadows, to fill the park with sound and light.[20] Engineering had its charms after all.

Olmsted and Thomson were, at first glance, polar opposites, but in practice and in philosophy they had much in common. When they collaborated successfully, the two built some of Seattle's most scenic and popular parks. One was Green Lake Park at the north end of the city, a large glacial tarn ringed by second-growth forest and rolling hills, a summertime destination for city residents. Olmsted saw the potential for the small lake to be one of Seattle's most beautiful bathing spots and, in his 1903 plan, embraced the lake in beaches and bicycle paths. He relied on Thomson's engineering crews to lower the lake by ten feet, and to fill in the muddy shoreline to create sandy beaches and parkways. At Olmsted's urging, city workers turned Ravenna Creek into the lake's sole outlet, and built bucolic artificial islands to attract waterfowl. The engineer did not assent, of course, to the landscape architect's plans for aesthetic reasons. Lowering Green Lake furthered Thomson's own objective to complete an extension of the North Trunk sewer, which ran beneath the new shoreline. The combined efforts of engineer and landscape architect, no matter how at odds they had been before, helped make Green Lake one of the city's park gems, much to the delight of nearby homeowners who anticipated a spike in their property values.[21]

The parkways and boulevards that connected the Olmsted parks to one another were the primary battlefields between the landscape architect and the city engineer, and the object of their struggle was the automobile. Cars opened parks to the masses and won approval from the health commissioner, J. E. Crichton, who considered automobiles more healthful and less dangerous than horses and their tracks of disease-riddled manure. Olmsted was less enthusiastic. He wanted visitors to stroll through his parks but understood that they were just as likely to drive through them, and he emphasized building pleasure drives to choreograph the motorist's driving experience. Olmsted designed roads, like trails, to let the driver ramble and follow the terrain like footpaths because, otherwise, traveling by car would be "exceedingly ugly, awkward and undesirable."[22]

For Thomson, a road was all about quick and economical transportation at the lowest grade possible, and Olmsted's work often interfered with that goal. When Thomson proposed to extend Stone Avenue, a major north–south arterial, through Woodland Park and around Green Lake in 1907, the North End Improvement Club and newspaper editorials joined Olmsted in attacking the plan. Olmsted feared it would kill the remaining stands of hemlocks and cedars in that district of Seattle. In frustration, he complained to park board commissioner A. B. Ernst. While Olmsted had "the greatest admiration for Mr. R. H. Thomson in his capacity of City Engineer," he wondered if Thomson "might welcome the suggestions of a competent landscape architect when it comes to a plan affecting the development of the city's parks, without implying any doubt as to his ability as a civil engineer." If Olmsted meant no offense, it was hard for anyone other than himself to see the compliment. Thomson refused to relent, preferring the ease of grade and the linearity of design to the defense of scenery. Olmsted's arguments prevailed, and the city council's street committee, in February 1909, recommended building the road along the eastern edge of Woodland Park and northward along the lake's shoreline.[23]

For all of their mutual disdain, Olmsted and Thomson, at first, respected each other professionally. In an 1894 address on the construction of public parks, Olmsted implied that city engineers served "as the intelligence and brains of the municipal government in all physical matters." Thirteen years later, in a 1907 address to the Congress of Horticulture, the landscape architect began to encroach upon engineers' territory. "The ability required to successfully design important municipal railroad, river, canal, and harbor works," Olmsted argued, demanded "a logical solution of each problem." Landscape architects, he implied, were just as able as engineers were to tackle the "complex problems of fitting land for human use." While he did admire Thomson's work in Seattle on the hydraulic regrading used to flatten the city's hills, calling the technique "so easy and sensible" in contrast to a "big excavation done by men with pick and shovel and horses and carts," he did not consider the engineer his equal when it came to thinking of the city as a whole.[24]

Olmsted's chauvinism was more than professional conceit. He and his step-brother, Frederick Law Olmsted, Jr., had helped to devise the McMillan Plan for Washington, in 1902, the first redesign of the nation's capital since Pierre L'Enfant's original plan in 1791. Named for Senator James McMillan of Michigan, its principal backer, the McMillan Plan took L'Enfant's Baroque-style design, with its wide boulevards flanked by ceremonial structures, and made it workable for a modern city, balancing the needs of streetcars and railways with a system of scenic parks and recreational spaces. Afterward, Frederick Law Olmsted, Jr., wrote that, as understood and practiced by landscape architects, city planning acknowledged "a city's organic unity, of the interdependence of its diverse elements, and of the profound and inexorable manner in which the future of this great organic unit is controlled by the actions and omissions of today." Municipal engineers and their champions expressed doubts. In an address to the Engineers' Club of Philadelphia, in 1911, Thomas W. Sears suggested, "landscape architecture as applied to cities is first and foremost utilitarian; after this, if one can introduce beauty, so much the better." Perhaps, he argued, "cities strive for 'beautility,' the combination of beauty and utility," but, above all else, "it should be remembered that a city must be designed conveniently and economically."[25]

Beautility, like economy, was in the eye of the beholder, and when homeowners along the proposed Lake Washington Boulevard suggested shrinking the avenue to retain more of their private property for sale, Olmsted lambasted Edward C. Cheasty, the park board president, for bowing to the pressure. "Between economy of land and construction, and trying to suit the ideas of individuals," Olmsted complained, "the boulevards are narrow, crooked, and with unscientific boundaries." The parks builder may have said he believed in beauty, but what he really aspired to achieve was an amalgam of beauty and utility. The Park Board had ignored the advice of its own superintendent, John W. Thompson, an Olmsted pupil, and bowed before demands for "efficiency and economy." Olmsted decried the outcome: a boulevard built "in a stiff and formal manner distressingly out of harmony with the wild beauty of the natural woods." Property owners wanted parks to promote a bubbly real estate market; engineers wanted to build efficient roads; and both wanted to eliminate the graceful curves along Interlaken Boulevard, which ran uphill from Lake Washington through one of Seattle's largest stands of tall conifers. Olmsted's confidence in Thomson and the Park Board fell apart. "It is all very sordid and lacking in any aspect of patriotism on the part of the landowners," Olmsted wrote in a letter home. After the homeowners along Lake Washington further protested the condemnation of nearby hillsides for a scenic overlook, Olmsted objected to the Seattle city council that "in this instance the greatest good of the greatest number . . . ought to prevail over the personal and individual opposition of a few interested landowners." Seattle homeowners and politicians wanted the Olmsted cachet, but they did not always want the doctrine that came with it.[26]

Roads had been Olmsted's path to conflict with Thomson; playgrounds would be his quarrel with Seattle's citizens. The neighborhood associations that had so enthusiastically supported the original 1903 parks plan now reversed themselves, criticizing Olmsted for ignoring playgrounds and outdoor recreation. He bristled at the charges. Seattle did not need playgrounds. "I do not think the city is yet so large and the working population so crowded," he maintained in a 1906 report, "as to make it vitally essential to provide play grounds . . . as it is so vitally essential in cities two or three times the population of Seattle." Like a stern schoolmaster lecturing unruly pupils, Olmsted set about educating Seattleites as to the virtues of landscape contemplation. Parks were meant to be pastoral refuges; the more cluttered they became with unnecessary playfields, concessions, and haphazard plantings—a sandlot here and a bandstand there—the more they would reflect the chaotic state of city life outside. Olmsted urged the Park Board to encourage genteel sports—basketball, tennis, croquet, or other games better suited to the true purpose of a park. He presumed to speak on behalf of women and children who might be offended by the antics of sporting boys and men. "The interests of the public at large," he concluded, "should be regarded as of far more importance."[27]

That public began to pull away from the Olmsted catechism, just as it had also started to question the city's engineer and his mania for all things efficient. As Seattle began to annex more of its outlying neighboring towns in the years after the 1904 election—Ballard, West Seattle, and Georgetown—the need for more parks with playgrounds became politically acute. With three of the new annexed towns predominantly working class and populated by recent immigrants, subsequent elections for park property acquisition bonds in 1905, 1906, 1908, and 1910 demonstrated that a wide spectrum of citizens supported parks but with one condition: creating a multiple-use system designed for both scenery and recreation. Local concerns began to propel park and playground construction, and Seattle's blue-collar voters were partially responsible for this shift. The 1905 and 1906 elections illustrated the growing influence of Seattle's working-class electors. In the first election, a majority of city voters approved a $500,000 bond measure, but it failed to garner the required three-fifths vote. Residents in the Second, Sixth, and Ninth Wards, all predominantly working class, had voted against the bonds. The next year, in 1906, the measure came before voters again with the endorsement of twenty neighborhood improvement clubs, the Central Labor Council, and the Seattle Real Estate Association. This time, the Seattle voters approved the bond—7,501 to 4,188—despite failing to carry the Ninth Ward, which turned down the measure, though by only 69 votes.[28]

Olmsted, horrified at the dismemberment of his original plan, called for an even more aggressive acquisition of scenic properties, choosing to ignore public sentiment or hoping to shift it in his favor. His 1908 report proposed a new forty-five-acre park in Ballard, a working-class neighborhood of Scandinavian immigrants in northwestern Seattle, overlooking Puget Sound, with a parkway connecting the parcel to Woodland Park, and the still-uncompleted parkway running through portions of the Duwamish River valley between West Seattle and Beacon Hill, plus the purchase of the Bailey Peninsula. He also noted the uneven distribution of parks throughout the city, and that "South Seattle appears to have been overlooked in the locating of small parks." Olmsted's proposal quickly became fodder for elected officials in the city's outlying districts, the same blue-collar neighborhoods that had voted against the bond measures in 1905 and 1906, and who claimed that the plan ignored their needs. Eugene Way, a council member from the Second Ward in southeastern Seattle, chair of the city's parks and boulevards committee, threatened to withhold funding for parks on the north end until there were more parks for the south end. "You buy Woodland Park for the north and you give us a new pest house," he complained to the *Seattle Post-Intelligencer* in 1909, "and you build a rat incinerator in the south end." Now the Park Board wanted Ravenna, "a privately owned park" north of the University of Washington campus, famed for its beautiful trees, "and for the south end you promise a new restricted district." Way wanted equity for his constituents and forced the park commissioners to capitulate to a compromise: acquire Ravenna Park, the beaches at Alki Point, and the Bailey Peninsula while cutting the proposed park in Ballard. (The commissioners also axed the proposed parkway through the Duwamish River valley, something that Olmsted had foreseen. He reluctantly noted that the district was probably "more valuable for factory land being flat land near navigable water.") The commissioners, responding to public pressure, also added still more playgrounds to Olmsted's park system.[29]

Olmsted saw playgrounds as a threat because designing parks was not a task "left to public opinion," but a highly "technical one involving aesthetic principles which few even of the most cultivated citizens would be competent to pass upon." He now had to fight Austin E. Griffiths, an English immigrant and local attorney, organizer of the Seattle Playgrounds Association, who had encouraged blue-collar voters and middle-class improvement clubs to lobby for more recreational spaces. Founded in 1908, the association promoted playgrounds as correctives to the ills of modern life, so that children with nowhere to play would not become adults without jobs and without hope. Griffiths was but part of a larger movement. Playground activists in cities across the nation, drawing from prevailing theories in psychology and sociology, championed organized play as the path toward stronger bodies and higher morals. Playgrounds were to some Progressives what the original Olmsted parks had been to reformers a generation earlier: arenas for democratic advancement where children rich and poor, immigrant and native-born, could come together. "Play and play traditions suited to modern conditions in a great city," Griffiths argued in a booklet describing the playground movement, depended on "opening . . . existing public parks, squares, or open spaces" to encourage a vigorous "outdoor life."[30]

Griffiths's labors did not go unrewarded. In the 1908 special election, voters approved a $1 million bond issue for parks and playgrounds by more than a two-to-one margin citywide and in every ward. Two years later, in 1910, the Park Board asked for twice as much, a $2 million bond, and almost 60 percent of voters again agreed to pay the price, regardless of which ward they cast their ballot in. The addition of more playgrounds to both bond measures likely helped to ensure their approval, but not without objection by Olmsted. He acknowledged that any city needed playgrounds, "surely necessary as a matter of public health and good morals," but Olmsted, ever the vigilant aesthete, believed they were the proper responsibility of "the School Board as a matter of right education of the physical, mental, and moral natures of children." If playgrounds had to be included in parks, it was a "question of balance of advantages." "No matter how meritorious in itself," any improvement, like a playground, he maintained, "should be subordinated very completely to the landscape design." Unable to stop the momentum for more playgrounds or quell the city's compulsion for more roads, Olmsted turned to the physical terrain he had come to know so well. Since Seattle's rugged hills and deep ravines severely restricted where safe playgrounds and economical roads were possible, he advised the Park Board to acquire "larger and more park-like" parcels "on the steep hillsides or on the shores of lake, harbor, or river where they will afford the pleasure and refreshment due to their command of views."[31]

In the fast-growing city, Olmsted's larger plan to put scenic parks above all else was impractical, politically and financially. It ran headlong into Seattle's class politics and the need for, in the words of the *Seattle Argus,* writing in 1903 against the original park plan, more "sewers, water mains, sidewalks," and other improvements promised by the city's indefatigable engineer, R. H. Thomson. Yet Seattle remained divided along economic lines drawn by altitude, with more affluent residents enjoying the best views and latest amenities, living atop the hills or next to the most scenic parcels, and the less prosperous clustered in the lowlands with inadequate utilities. Faced with these challenges, Seattle could neither afford all of Olmsted's recommendations nor ignore citizens' demands for play fields in or adjacent to city parks. Landowners along Lake Washington's shores were unwilling to relinquish properties for parks and miss the anticipated land boom following the completion of the Ship Canal and the lowering of the water level. Around Lake Union and Elliott Bay, as well as in the Duwamish River valley, industrial expansion and tideland filling had ruined suitable lands or priced them beyond reach.[32]

On a trip to the Pacific Northwest after the 1910 election, Olmsted lamented that the mayor of Portland, fixated on finances, was squandering his city's future. According to Olmsted, what was at stake was nothing less than the ability of city governments to protect and nurture greenswards for subsequent generations. "Economy is all right in its place but to stop acquiring land for parks," he wrote, "when the city is bound to have parks for sometime" was shortsighted "because the land is rising in value besides being improved." Boston would have "served millions by buying parks when it was the size Portland is now," and the same held true for Seattle.[33]

The Toot of the Automobile and the Odor of Gasoline in Sacred Precincts

John Olmsted never lived to see if his predictions came true, but he lived long enough to face yet more rejection by Seattle and Thomson. As the craze for urban planning spread across the nation following the adoption of the 1902 Mc-Millan Plan for Washington, Seattle's business and political leaders, not wanting to be outclassed, pushed for their own comprehensive plan. Their efforts culminated in the Municipal Plans Commission, approved by voters in March 1910 by almost a two-to-one margin. That summer, the new commission met to find a leader worthy of the task they had set for themselves. Olmsted desperately wanted to be the lead planner and had good reason to assume he was the obvious choice. He had designed the grounds of the Alaska-Yukon-Pacific Exposition, Seattle's first world's fair, which opened in 1909, and the new campus of the University of Washington, site of the exposition. He classed himself among the nation's best city planners, the equal of Daniel H. Burnham, creator of Chicago's Columbian Exposition of 1893 and that city's 1909 plan. Olmsted's fame did not impress Thomson, who wanted someone familiar with Seattle's particular problems, preferably an engineer like himself, and it was Thomson's opinion that mattered most to the new commission. Thomson recommended five finalists, Olmsted among them, but endorsed only one: Virgil G. Bogue, the former railway and mining engineer who had surveyed Seattle's unclaimed tidelands in 1895. The commissioners offered Bogue the position, and he promptly accepted. At a banquet later that year, they feted Bogue as an "engineer of national repute to advise it upon its greatest undertaking—the building of a world metropolis." Thomson had exacted his revenge.[34]

Bogue threw himself into his work and, by the summer of 1911, proposed the biggest city plan, as measured by area, for any western city until Frederick Law Olmsted, Jr., and Harland Bartholomew's doomed 1930 plan for greater metropolitan Los Angeles. "Greater Seattle," Bogue predicted, would one day encompass Lake Washington and sprawl from Everett to Tacoma, including nearly 150 square miles and more than a million people—nearly quadruple Seattle's 1910 population. To meet this challenge, Bogue's plan called for a list of major improvements: a new Civic Center of six neoclassical buildings north of downtown where Thomson had removed Denny Hill by hydraulic cannon and steam shovel, more arterial streets, municipal shipping terminals on Elliott Bay's filled tidelands, and an additional 57,600 acres of parks and boulevards stretching from Lake Washington to the Cascade foothills. "Seattle should have these mountain roads for the education, the joy and recreation of her own people," he wrote, "and for her guests who come from afar—even from beyond the sea—to gain health and strength, both of body and soul, amidst these scenes of wonder and beauty." Lifting rhetoric from the Olmsteds, he crowned his plan in the mantle of nature's beauty. With "encircling hills, the lofty, snow-capped Olympic peaks closing the westward view," Bogue decreed that a greater opportunity "for high and permanent distinction never fell within the privilege of a municipality."[35]

Seattle voters never warmed to Bogue's extensive and far-reaching plan, and, once the cost became known, fiscally conservative members of the Municipal League of Seattle and King County formed a Civic Plans Investigation Committee to oppose it. The league, in 1912, published a scathing critique, "The Bogue Plan Question," speculating that the ornate Civic Center alone would cost nearly $3.5 million, well in excess of the city's bonded debt limit, and might require removing part of Queen Anne Hill. Spirited support from Bogue's allies, including Thomson, who scolded "oratorical engineers" and "advisory geniuses," could not save the plan. The Municipal League's leader, Paul B. Phillips, head of an organization that was no foe of civic improvements, was exasperated. "The people are in the position of a man whose architect has designed a dwelling for him," Phillips wrote, and now that the owner had questioned the work, it was "up to the architect to defend the plans." Of course, the city was already engaged in a costly public works binge, led by none other than Thomson, of regrades and waterworks, sewers and roads, waterways and canals. According to the terms of the 1910 ballot, when the electorate had approved the Municipal Plans Commission, voters also had to accept any subsequent plan by a simple up or down vote, and, in March 1912, Seattleites of all classes turned down Bogue's vision by a decisive majority.[36]

Opposition to the Bogue Plan spanned the city's political and economic divides, with working-class residents and white-collar businesspeople alike concluding that the Municipal Plans Commission would wield too much power or take money away from local needs, like playgrounds. The largely professional and middle-class Municipal League opposed the Bogue Plan, but so, too, did the Central Labor Council. Yet the defeat was not a categorical rejection of city planning. Seattle eventually completed many of the improvements Bogue envisioned, with the exception of the grand Civic Center in the Regrade District, including arterial streets, regional highways, and scenic parkways stretching into rural King County. The waxing power of neighborhood improvement clubs and civic organizations in Seattle politics, coupled with voters' fatigue over spending still more money for more public works projects, were the likely reasons for the plan's failure. The 1911 *Plan of Seattle* lost at the polls, like so many big plans before and since, because it came before voters at the wrong time and without sufficient attention to other immediate needs that commanded political and fiscal attention.[37]

If Seattle voters had no interest in big plans, they continued to support parks and playgrounds, albeit not in the fashion that Olmsted had intended. Even if Olmsted might have disparaged the tastes of the people, the people took pride in what the landscape architect had designed for them. Roland Cotterill, secretary for the Park Board, wrote in a 1909 article that Seattle's Olmsted-designed parks were no "after-thought" but an integral part of the greening metropolis. "There is probably no other city in the United States of the size of Seattle," he boasted, "which is better provided with parks, playgrounds and boulevards." The varied topography, with lakes and hills and ocean, allowed park builders to blend "the natural with the artistic . . . to a degree little short of perfection." That same year, Seattle had almost seven hundred acres of parks and boulevards, a remarkable achievement for a city barely more than fifty years old. A sizable majority wanted still more and, in 1912, voted to add three key parcels to the city's system: privately owned Ravenna Park, Alki Point in West Seattle, and the Bailey Peninsula, which became the site for a new Seward Park. All three had been in the original Olmsted plan of 1903, and, with these additions, Seattle had twenty-eight parks, plus more than twenty-four miles of scenic boulevards, twenty-two playgrounds, four fieldhouses, and a salt-water bathing beach at Alki Point. Over the next eighteen years, the number of parks reached forty-two, playgrounds almost doubled, and nine more bathing beaches and sixty-two small "vest pocket"–style parks dotted the city with patches of green.[38]

Boosters like Cotterill did not disclose the new problems caused by the hard work of implementing park plans, and among the most vexing challenges were automobiles and the roads they traveled. If Olmsted saw cars, begrudgingly perhaps, as moving vistas that framed his landscapes, motorists considered cars foremost for transportation to and from those vistas. As more and more visitors drove to city parks, more and more cars spewed fumes and noise, struck wild animals, threatened pedestrians, and destroyed roadbeds, exacerbating soil erosion. Gus Knudson, head keeper at the new Seattle Zoo at Woodland Park, warned that exhaust was choking the deer and elk in the small zoo corrals. "It will be just a matter of time" until all the animals were dead, since "this is all artificial to them." Parks officials began to ban cars from certain parks, beginning with Schmitz Park in West Seattle. "Suggestions that drives be constructed within the park have been resisted," the Park Board wrote in its 1911 report, "and the toot of the automobile and the odor of gasoline have not permeated the sacred precincts." But keeping cars out of all parks was impossible because Olmsted had designed many with automobiles in mind. L. Glenn Hall, a landscape architect, said, speaking of plans to build still more roads, "I do not believe that the natural character can be preserved very long if automobiles are permitted." One proposal, to encircle Seward Park in asphalt, would destroy "the finest large natural forest of any park in the whole system."[39]

John Olmsted's earlier advice against building too many roads in Seattle's parks was more than an aesthetic objection. "The wild beauty of the natural woods" served a practical purpose: to keep unstable soil out of gravity's clutches. Environmentally sound engineering grounded Olmsted's worries. Seattle's steep hillsides made for breathtaking views so long as the trees and groundcover kept the rains from chewing into the earth. Once city engineers or private developers removed the forests to build roads, they let loose the processes of erosion. Since city parks and boulevards often abutted some of the city's best real estate with the best views—the high ridges above Lake Washington, or the commanding summits of Queen Anne and Capitol Hills—and since many of those sites were adjacent to steep hillsides and ravines, landslides were not happenstance. Along Interlaken Boulevard, a parkway designed by Thomson and criticized by Olmsted for its sharp turns and steep grades located at the northern end of Capitol Hill, mudslides had become routine. Olmsted had predicted this and now, after every heavy rain, the falling water loosened the slopes and streams of earth spilled onto the streets, jumping over retaining walls and closing the road for weeks at a time. In 1912, fifteen local residents petitioned the city to build concrete abutments and bridges in the name of safety and access. City engineers tried to shore up the crumbling roadbed, but they were fighting a constant battle on Interlaken Boulevard as well as along nearby Lake Washington Boulevard.[40]

"A disastrous washout" in the summer of 1931 on Interlaken Boulevard again threatened homes and posted a "great danger" for motorists; three years later, another onslaught of landslides damaged the concrete bridge built to withstand the seasonal inundation. Homeowners asked the Park Board and the city engineer to do something. In the middle of the Great Depression, city officials pled poverty and did nothing. "The dirt is still coming down and piling up on my property," wrote local resident J. W. Wheeler, "and I feel that this kind of improvement would be more protection than a span across the entire ravine." In May 1934, three months after his first letter, the city had still neglected to fix the road, prompting Wheeler to remark how "one of the most beautiful drives of the city is practically closed." He continued to complain about the now-constant landslides for the next two decades.[41]

Yet the avalanches did not scare off the road builders. In 1929, state engineers proposed extending a major north–south thoroughfare, Aurora Avenue, through the middle of Woodland Park, the same park R. H. Thomson wanted to slice apart, in 1907, with another road. As before, the engineers faced a coalition of citizens, organized as the Save the Parks Association, led by a former park board head, A. S. Kerry. "Cutting up Woodland Park is only a stepping-stone to cutting through several other Parks," Kerry warned, and "the question is when will they stop and who will stop them?" Kerry had reason to worry. The previous year, in 1928, the Puget Sound Bridge and Dredging Company applied to build a private toll bridge from Seattle to Mercer Island and the eastern shore of Lake Washington. The proposed bridge, built on an earthen levee, would shatter Seward Park and its tall pines and firs. Homeowner associations in the affluent Mount Baker Park neighborhood, worried at the effects of a busy nearby road on home prices, joined forces with the parks commissioners and community clubs across the city to stop the bridge. Road opponents in Woodland Park were not as successful. The Save the Parks Association asked Seattleites if they would pay $2 million to "build a speedway for strangers," but merchants and small businesses surrounding the park, the so-called strangers who stood to benefit from the new road, persuaded the powerful Seattle Building Trades Council, then the city council, to back the expansion of Aurora Avenue. City voters sided with the Save the Parks Association only to have the state overrule them and begin construction on Washington's first highway, State Route 99, in 1932.[42]

The roads that carried speeding drivers and noxious fumes into city parks also carried effluents into Seattle's lakes and streams, and, as early as 1900, Green Lake, named for its summer algal blooms, was slowly going eutrophic—dying, in ecological terms—thanks to an increase in nutrients, namely phosphorus and nitrogen, from leaky septic tanks and sewer lines. Before Green Lake became a park, logging had already hastened runoff. Now, the new sidewalks, roads, and hardened shores sent sheets of polluted water into the lake basin. As more nitrogen and phosphorus leaked into the lake, the nutrients exacerbated algal blooms during the warm spring and summer months. The stink of rotting algae and dead fish, suffocated by the anaerobic or oxygen-free waters, led the Green Lake Improvement Club to propose changing the lake's name to deflect attention away from its algae-choked waters.[43]

The very engineering ordered by Olmsted and completed by Thomson to beautify the lake only made matters worse. Diverting the lake's major tributary, the stream from Licton Springs, once a popular mineral bath used by Natives and early settlers, to feed the North Trunk sewer addition increased the concentration of pollutants. Waterfowl attracted to the artificial islands, built to lure birds there (Olmsted had pushed for even more to enhance the lake's scenery), encrusted the shorelines with phosphorous and nitrogen-rich excrement. As mats of the green scum grew thicker every summer, city health officials vacillated on a remedy. City health commissioner H. M. Read proposed returning the Licton Springs drainage to the lake, then reversed his decision when he learned that the springs' outlet flowed past several leaky residential septic tanks. Frustrated, local homeowners protested that preventing the springs' water from entering Green Lake would wreck the lake itself and the playgrounds at the headwaters. Health inspectors, unmoved, simply sealed off Licton Springs even as the clarity of the water and its quality continued to degenerate.[44]

Illegal logging became another persistent nuisance in Seattle's parks, dating back to the first parks' report of 1892, which reported that few locations were untouched "by the blasting fires of the logging camps and clearings." Building new parks did not end the cutting, because urbanization had splintered the original city forest, leaving the remaining shards vulnerable. Nearly a decade later, the *Seattle Post-Intelligencer* said that illegal cutting of Christmas trees in Volunteer Park was "mutilating the trees and shrubs." Ravenna Park, named for the famed Italian pine groves, nestled in a small canyon between Green Lake and Lake Washington, suffered the most. Its owners, the Beck family, advertised the giant trees and moss-draped slopes as a sylvan paradise "unshorn of [by] axe and fire." It was an overstatement even then, for the Becks' property, like most of Seattle, had been logged and burned, first by Indians, next by white invaders, until the Beck family stopped the logging at the end of the nineteenth century. Yet enough of the original cover remained, complemented by new growth, to impress Olmsted, who marveled at a few surviving cedars nearly twelve feet in diameter at their base. The Park Board rebuffed his pleas to put the park in his 1903 plan because of the high price demanded by the Beck family until voters approved its purchase six years later.[45]

A few months after Ravenna became city property, sometime in the summer of 1909, several large cedar and fir trees disappeared from the park. Some speculated that corrupt park workers had stolen the trees under cover of night, or that local residents had taken them for lumber or cordwood. Enraged citizens demanded an investigation, but no one was ever charged. Vandalism became even more acute when Ravenna Park, once somewhat isolated, joined the public system of more than 640 acres. An ever-expanding road network linked the parks, and it may have inadvertently provided escape routes for timber thieves. The overextended Park Board could not employ full-time guards and could not remain confident that they might do their job. City officials reluctantly concluded that keeping the park's magnificent trees intact would be impossible given the value of its timber and the shortage of reliable park employees.[46]

Seattle's Park Board was becoming overtaxed because every road built and every tree removed threatened to intensify ecological and social volatility, since parks, havens for urban wildlife fleeing speeding cars and hunters, were havens for less fortunate humans, too. The animals usually arrived before the people. Squirrels, raccoons, quail, waterfowl, and songbirds crowded feeding stations provided by parks staff eager to enhance visitors' experiences. By 1936, wardens were distributing almost one and a half tons of seed a year in Woodland Park alone, much to the delight of the Seattle chapter of the Audubon Society. Neighbors and park visitors did not appreciate all animal visitors equally, and no animal was as reviled as the American crow, a persistent nuisance, blamed for scavenging garbage, attacking songbirds and waterfowl, and ruining the quiet with their loud cawing. Mary I. Compton, president of the Seattle Audubon Society, wrote in a letter to the *Seattle Times,* in 1919, "the crow is one of the greatest menaces against which our game birds have to contend, and the same applies as well to all of our smaller birds." "The sportsmen of Washington, especially those on the west side of the Cascades," she pleaded, "should begin at once on removing the crow from the protected list."[47]

Controlling crows became such a concern that, according to W. M. Elliott, the city had paid "experienced gunners," in 1919, "to kill off this pest in the various parks." Elliott, who lived next door to Volunteer Park and had been part of the first hunting posse, complained in 1929 that "time dulled the enthusiasm of the other shooters." He offered his services once again to "go after these devils in the Park." Seattle's crusade against crows was but an episode in a nationwide campaign to eliminate them. Early Americans had loathed crows for raiding their crops and harassing their livestock; by the twentieth century, with the concomitant interest in songbird protection, modern Americans decided that coexistence was impossible. According to wildlife biologist John M. Marzluff, crow hunting is remarkable because, in contrast to other game, "the quarry is hated, rather than revered or respected by the hunter." With the possible exception of wolves, few animal species have engendered as much personal animus as crows, but, unlike wolves, human beings had made the American landscape more habitable for their nemesis. Highly intelligent and adaptable, corvids, the bird family that also includes jays, magpies, and ravens, are aggressive and opportunistic, feeding on a wide array of foods. Urban corvids, in particular, adapt to their environment and favor trash and roadkill for meals. Crows are also highly social and often hunt in flocks, called a murder, ganging up to attack smaller birds or steal food from larger animals, like bears and eagles. The spread of cities and roads into the countryside in the twentieth century only expanded their range and numbers; crows and humans coevolved through time, creating their shared habitat of thinned forests, fragmented fields, and garbage dumps.[48]

Other residents, like P. F. Appel, complained less about annoying wildlife and more about illegal hunting in the city's parks, which had "resulted in the disappearance of practically all of the songbirds and quail." Appel grieved for the once "very prevalent" birds of Ravenna Park, now gone. As early as the 1910s, those living next door to city parks had grown weary of rifle reports at dawn or dusk as hunters, who considered parks a version of the rural commons, pursued their targets. Another resident from West Seattle, P. S. Rowntree, complained that Schmitz Park endured "the mercies of the hoodlums and tough boys." "Boys and men came with rifles, killing ruffed grouse, quail, pigeons, doves and squirrels," he explained, and he had "personally put out four boys with guns and four dogs" the previous winter when park employees were not on duty. Conservation groups like the Audubon Society were especially incensed at poachers and even legal sportsmen who hunted in and near city parks, and, in 1919, Compton, the chapter president, implored the city's Park Board to take action to stop them. "It is deplorable," she protested, "to drive from Lake Washington the game birds which come here in great numbers to escape the hunter's gun. . . . Are not other people entitled to enjoy the pleasure of these wild birds which come to Lake Washington each winter?" Parks wardens, at the request of local birdwatchers, established a refuge at Seward Park, in 1919, on the shores of Lake Washington.[49]

Five years later, the Audubon Society, in exasperation, petitioned the Game Commission of King County in even stronger language. In a letter from 1924, Audubon leaders called for "active supervision of Alien shooters and trappers," primarily Italian men, using "mufflers or silencers" to mask their crimes as well as "snares and traps," and Japanese boys armed with "air guns [and] sling shots." "Considerable poaching is done in Union Bay," the letter noted, "[along] east shores of Lake Washington, and the lake shores near Renton and swamps contiguous." The society's outrage was understandable: all locations were popular and accessible bird-watching spots, adjacent to the string of Olmsted-designed parks stretching from Washington Park to Seward Park, connected by tree-lined Lake Washington Boulevard. "It is believed that much of this indiscriminate shooting has been done to augment a scanty food supply," the Audubon Society concluded, "as their action is not confined to the killing of game birds, but any bird or animal met." Non-white and immigrant hunters were a menace and "generally ignorant of the game laws and restrictions." Nativism and class bigotry were central tenets of the early conservation movement, and Seattle's bird lovers did not mind those residents "who are not aliens" trapping wild quail in the city for food, which they said was "to be expected." The premeditated killing of songbirds, especially by foreigners and the poor, however, was an abomination.[50]

Olmsted had envisioned Seattle's parks as great arboreal arenas for the mixing of diverse citizens to promote democracy just as his stepfather, Frederick Law Olmsted, had imagined for his creations a century earlier. Yet as with parks in other parts of the country, like New York's Central Park or San Francisco's Golden Gate Park, Seattle's parks were arenas for public prejudice. Squatters and the itinerant poor earned the most scorn. Native-born Seattleites' distaste for immigrants, minorities, and the homeless extended to more than hunting because, as with parks elsewhere in the nation, making public space often entailed removing private citizens living there. When the Parks Department acquired lands along Alki Beach in West Seattle, in 1909, with scenic views of downtown across Elliott Bay, it asked the city attorney to condemn rows of shacks fronting the shoreline. Squatters remained a persistent nuisance along Alki Beach, the longest stretch of city-owned public shoreline in Seattle, for the next two decades. In 1929, the Alki Community Club, a group of West Seattle homeowners, said they were "endeavoring to rid Alki Point of shacks and to prevent any others from being placed upon property to be rented." Upon hearing of plans to expand the public beachfront, the club president said "that the moving of these shacks upon property at Alki, and especially near Seattle Park property, is a great detriment to the community and to the city at large." Squatters did not confine themselves to the city's shorelines, because parks were another commons for the indigent and transient to call home. Gus Knudson, director of the Woodland Park Zoo and the park department's sanitary inspector, had to wage "continuous war on rats and mice." In almost every location, the causes of the rodent infestation, as he found over the course of 244 inspections across the city in 1923, were the piles of "garbage and junk" and "old unused shacks and buildings" scattered in the city's parks.[51]

Expelling squatters and razing shacks was easy compared with the mounting struggles to keep Green Lake blue. The capping of Licton Springs in 1919 to prevent further pollution failed to stop the summertime algae blooms, and as more houses continued to sprout around the lake, most remained unconnected to the city's North Trunk sewer. Leaky septic tanks drained into the lake, and, in 1929, the algae problem had returned, prompting William Graefe, who lived nearby, to complain how "the *SCUM OF ALGAE*" made the lake into a stinking cesspool. Parks managers were stuck with a lake that stubbornly refused to remain clean and had resigned themselves to continual maintenance. Several studies proposed changing the drainage system of the lake yet again to flush out pollutants, applying copper sulfate to kill the remaining algae, and armoring the shoreline to retard future algal blooms. Throughout the 1930s, dredges operated by the Works Progress Administration, a depression-era employment program, harvested noxious aquatic weeds, such as elodea, and removed bottom muck thought to promote algae. By the 1940s, Roland Loeff, the city parks superintendent of maintenance, put an end to the "ill-advised" mud removal, blaming the persistent algal blooms on pollution and warm summer weather at "just the time that swimmers want to use the lake." Chemical treatment to remove the algae would render the water "so alkaline" as to "make it untenable to fish" and "undesirable to swimmers," and flushing the lake regularly with large volumes of freshwater was impractical. "To persist in a vain attempt to upset a natural process," he concluded, "will only bring mediocre results, or more probable, none at all."[52]

Attempts to cultivate or design a particular kind of environment, one pleasing to a diverse citizenry, could boomerang as it did with the bird refuge at

Seward Park, established in 1919 "to keep the wild waterfowl, such as mallards, blue bills, geese, and other varieties, in this locality." The refuge had worked too well. According to the 1938 annual parks report, some birds became "quite tame" and provided "a great attraction for people who prefer looking at the birds to shooting them," a problem for local homeowners. As Donna M. Henderson, the Lakewood Community Club's secretary, explained, tame waterfowl spent "most of their time in the yards of the nearby residents . . . [engaged in] the messing up of their lawns, the digging out of the flowers and shrubs and in general making a nuisance of themselves because they have not been given sufficient room to rest and feed." She proposed moving the sanctuary to the opposite end of Seward Park, so both birds and their admirers "would no longer disturb the residents of our beautiful city and parks." What Henderson did not know was that green lawns, the marker of a well-kept yard, attracted Canada geese and various species of ducks, which grazed on the grass instead of diving for aquatic plants. Protests from the local Audubon Society chapter kept the refuge open until the end of the decade, after which parks officials, citing sanitation concerns, stopped feeding the birds to attract them to the site and rescinded the park's status as a designated wildlife sanctuary.[53]

The problems facing Green Lake and Seward Park, among others, compelled city officials, as well as the Olmsteds, to consider making some parks effectively private, as they did with the University of Washington Arboretum, carved out of Washington Park, just south of the college campus. University of Washington scientists, as early as the 1920s, had badgered the Park Board to build an arboretum, and in 1934 the board relented and signed an agreement with the university to create a botanical garden. Two years later, the new University of Washington Arboretum Foundation, with the help of the Seattle Garden Club, hired the Olmsted Brothers to landscape the new site. Basing their design on Harvard University's Arnold Arboretum, the Olmsteds proposed closing off the garden, as well as sections of the park itself, like Foster Island, to the public in order to protect the collections from vandals and thieves. Condemnation was swift and bitter, forcing parks officials to reject the Olmsteds' suggestion and limit fencing to the arboretum only. "The physical development of our children," the Central Federated Clubs protested, "comes before the study of trees." Grace Cadwallader, who lived beside Washington Park, was even more insistent. "Shouldn't those interested in making the right kind of citizens," she asked, echoing the original Olmstedian spirit, "find ways of getting people into the parks, not in ways of keeping them *out?* . . . Shouldn't these words be broadcast: 'Let all come who will, be they motoring, cycling, or walking.' The important thing is that they *come.*"[54]

John Olmsted, who died in 1920 from cancer at the age of sixty-eight, had perhaps anticipated this and other conflicts since he had little confidence in elected officials or anyone else in managing his parks properly. It was a point that his successor, James F. Dawson, repeated in 1929 to A. S. Kerry, the new parks superintendent. Kerry needed help. Presumably, the Olmsteds had not purposely designed a park system to produce pond scum and landslides, to house squatters or breed vermin. Dawson was familiar with the history of Seattle's parks; he had helped John C. Olmsted draft the 1903 plan. Then, as now, Dawson felt that Seattle's Park Board had inadequate powers to manage its properties, so Seattle, not the Olmsteds, was to blame for its messy, polluted, and conflict-ridden parks. Green Lake in particular attracted his scorn. "There is no reason at all why it shouldn't be just as attractive as some of the finest parks in Boston, Brooklyn, Minneapolis, Oakland," he said, "or other cities where the parks are maintained in a satisfactory manner." It was Dawson, after all, who suggested, in 1936, restricting public access to Washington Park lest it fall into permanent disrepair.[55]

Dawson invoked an ideal Olmstedian landscape that forever remained beautiful and ordered. Yet his high-handed reply missed two important points: Seattle's Park Board, like park boards elsewhere, had to respond to a diverse constituency while caring for a complex mosaic of forests and beaches, gardens and lawns, that constantly eluded control. Dawson's cavalier remarks also ignored another fact. Seattle's parks were no more or less unmanageable than many Olmsted-designed systems in other cities at the time. Squatters and vagrants in New York City's Central Park avoided eviction by park wardens well into the twentieth century, and local horticulturists in San Francisco protested loudly when William Hammond Hall eradicated native plants in Golden Gate Park in favor of exotic ornamentals.[56]

Olmsted had given Seattle greenswards that rivaled those in any American city of the period, despite the compromises that had angered him, but the process brought unanticipated costs. Olmsted relied upon artistry and engineering, trying to create landscapes designed for pastoral contemplation, yet the parks, even as he planned them, were hybrid landscapes, part natural and part artifice, and became more so as a consequence of his compromises. He perhaps understood this better than the city engineer, but Olmsted, too, could not transcend the consequences of human avarice or the unpredictability of a non-human world remade by humans.

Olmsted was a nineteenth-century Romantic who lived and worked as a twentieth-century Progressive. The two sides of his personality were often in conflict, if not in the man then in his creations. Thomson and his engineers took physical nature as a kind of workshop, an assemblage of parts ordered properly by experts to ensure their smooth functioning. In contrast, Olmsted's parks were more beautiful than the engineers' urban workshops but less pristine than the romantics' dream of wilderness; they stood in the middle ground of a feasible beauty with practical purposes. Yet, like Thomson, Olmsted believed that only he and a few other landscape architects had the appropriate talent and experience to create beauty within the metropolis. Olmsted, however, could not stay in Seattle to see his creations built and nurtured to his exacting standards. The demands of his business had forced him, in the case of Seattle and other far-flung cities, to rely on epistolary planning instead, and by the time that John Olmsted died, he was still managing fifty jobs at one time.

Over the course of his career as a senior partner at the firm his stepfather had founded, he directed as many as 3,500 separate projects. This was John C. Olmsted's legacy. From the period between 1884, when he joined his stepfather's firm, to his death in 1920, he had a hand in creating parks from Dayton to Detroit, Baltimore to Boston, and Portland, Maine, to Portland, Oregon. It is a legacy lost to many historians because his younger and better-known stepbrother, who had his father's name and later took over what was then the largest landscape architectural office in the world, overshadowed the adopted sibling. Frederick Law Olmsted, Jr., went on to help frame the congressional act that created the National Park Service in 1916, and implemented the key components of the McMillan Plan for Washington in 1902, including the construction of the city's most prominent landmarks: the White House grounds, the Jefferson Memorial, Rock Creek Parkway, and the Federal Triangle. The other Olmsted, John, tended to blend into the business, which had become a modern enterprise. The Olmsteds' art could, and did, suffer as a result. Letters and periodic visits were a weak substitute for a persistent pressure, applied in person, coupled with a deep understanding of the unique properties of each place.[57]

Building parks was now an industrial process, akin to municipal engineering; it required standardized ideas to fabricate multiple new landscapes for diverse needs suited to particular cities. The Olmsteds did revolutionize urban design and planning by offering a new vision for the American city, one in opposition to the mass-produced world of the engineered metropolis. But their vision was not a romantic fantasy. Their goal was purposeful and rational, though it required faith as well, like growing and tending a garden. The Olmsteds' ethic of place was to cultivate better citizens by cultivating parks. They believed that if enough city dwellers visited well-designed parks, strolling along winding paths and taking in the beauties of the natural world, they would emerge stronger and wiser and better able to connect to one another through spontaneous displays of affection and respect. Seattleites appreciated and understood this vision, but only in part, and with an eye to their own perceptions of their immediate needs, from driving real estate sales to opening up public spaces for their children to play in, to embroidering the urban fabric with greenery. The philosophical underpinnings did not interest them. Nor did the two Olmsted brothers grasp completely the full scope of their famous father's legacy; they had come of age in an America that was already urban, industrial, and atomized. Cities, like Seattle, could not easily sustain the harmonized social relations Frederick Law Olmsted wanted to promote with a place like Manhattan's Central Park, because such harmony had rarely ever existed. There were limits to the ideal of the urban garden.

Diverse groups brought their conflicting desires with them to the park gates because building parks, like regrading hills or remaking watersheds, depended on ecological and social processes. Disentangling the two was impossible, and, as park enthusiasts quickly discovered, non-human nature and human nature behaved unexpectedly and perversely in their response to each other, even in the places where both belonged. When people complained about fouled bathing beaches, or when poachers killed birds under the cover of night, there were social consequences that played out to the benefit of the more fortunate; the powerful tried to impose what non-human nature should be on the less powerful. So, too, unpredictable water, earth, biota, and fauna had an unintended and active role in reshaping parks as public spaces, drawing people into cooperation or conflict with one another. Human beings rarely respected property lines or laws themselves in the city's parks. Nature never did.[58]

This harsh dynamic was distressing. Parks were supposed to be free from chaos, independent landscapes of beauty or restorations of damaged lands. Park builders, even the Olmsteds, failed to understand that the new park environments could promote and foster bad behavior and widen social rifts. When park builders and citizens tried to restore the environments they saw as out of balance, they were also trying to erase certain human presences. Seattle's park boosters considered parks static spaces where the opposing forces of urban external life—city versus countryside, individual desires versus the common good, beauty versus utility—could be brought into balance, into an idyllic harmony. But as with a garden, creating and maintaining a park was a constant struggle to maintain a chimerical balance, one that demanded the artistry of the landscape architect and the ingenuity, if not the arrogance, of the engineer.[59]

CHAPTER 5

Above the Weary Cares of Life

The Benefits and High Social Price of Outdoor Leisure

In the summer of 1909, Seattle held its coming-out party, the Alaska-Yukon-Pacific Exposition. Since the Chicago Columbian Exposition of 1893, or even before with the Philadelphia Centennial Exposition of 1876, world's fairs had become a favored device to announce the advent of urban greatness, and Seattle was no exception. Boosters had intended to celebrate the tenth anniversary of the Klondike Gold Rush, but their intentions were frustrated when rival Portland declared plans to honor the centennial of the Lewis and Clark Expedition in 1905, and when Virginia announced a fair to commemorate the tercentennial of the Jamestown colony two years later. The Arctic Brotherhood, a fraternal organization of Klondike veterans, teamed with *Seattle Times* publisher Alden J. Blethen and city businesses to outdo its competitors. The result was the "Alaska-Yukon-Pacific Exposition," a world's fair that celebrated neither history nor past heroes but present location and aspirations. As a fair booster wrote in *The World's Work,* a monthly magazine, since Alaska and the Yukon were not "*known* places," promoters were "compelled to break trail into the lands" much like the sourdoughs who had climbed the snowy Chilkoot Trail on the way to the gold fields.[1]

The Alaska-Yukon-Pacific Exposition, or AYPE, opened on the new grounds of the University of Washington, designed by the Olmsted Brothers, the noted landscape architects who had given Seattle its renowned park system. One reporter called the exposition "the world's wondrous treasure box." Fairgoers toured exhibits of the region's abundant resources—tall trees felled for lumber, gold and silver ingots smelted from Alaskan mines—as well as scores of amusements from reindeer rides to vaudeville shows. The fair was a giant marketplace, celebrating the Pacific Northwest's environment as an endless supplier for eager consumers. Not everyone was convinced it told the whole story, however. The exposition's leaders, eager to save costs, had circumvented the city's building trades with use of open-shop hiring, prompting complaints from city and state union leaders. "To the class conscious Seattle wage worker," said an editorial in the *Seattle Socialist,* the fair was "a great fantastic monument to the brutal avarice of the capitalist class." Other laborers took issue with the very theme of the fair itself. The story that promoters pushed at the fair—labor and technology besting and improving the physical forces of nature for the benefit of all—was not necessarily true. A group of unionists from Fairbanks protested that the rosy portrait ignored "Nature's irrevocable laws of supply and demand." In the real Alaska, farming was difficult, mines were dangerous, and "heartless capitalists" went off with others' hard-earned fortunes. The landscape itself laid bare the contradictions of industrial capitalism.[2]

The class conflicts exposed at the 1909 fair intensified in Seattle over the next few decades as the city's newly powerful middle class, the same group that had supported city engineer R. H. Thomson in his bold regrading and waterworks crusade and John C. Olmsted in his sweeping parks plan, had begun to assert their own entitlement to nature's treasures. Decades of shaping and reshaping of Seattle's physical environment primarily to benefit railroads, industry, and real estate speculators had separated the city's social spaces into those dedicated to production and those dedicated to consumption. Spaces on the tidelands and waterfront were for assigned, productive enterprises essential to progressive urban living, such as manufacturing, retailing, and shipping. The spaces in the city's greenswards and rural hinterlands were set aside for new consumer activities, such as outdoor leisure and recreation. But Seattle's middle class objected to efforts by others, notably poor and blue-collar residents, to call their privileges into question. Those who had greater political and economic authority saw production and consumption as complementary forces working for their benefit. White-collar professionals in the city's major businesses and in Seattle's cultural and political institutions were often the leaders of local conservation groups and the most avid of outdoor enthusiasts. Catching fish, climbing mountains, or picnicking in the city's parks were more than mere idyllic escapes; they were also matters of politics and power. Affection for outdoor recreation in Seattle was not limited to the middle class, but those with the greatest pull as consumers enjoyed the greatest power to control their playtime and their playgrounds, too.[3]

In the newborn geography of outdoor recreation that spanned the city and its environs, Seattle's outdoor-minded middle class sought to impose standards of decorum, often forcefully, upon those they saw as playing or working improperly. Theirs was an ethic of individual entitlement turned outward to remake society and nature alike. Well-intentioned reformers now reversed the equation earlier espoused by both Thomson the engineer and Olmsted the landscape architect—that improving the natural world would improve human nature. They wanted instead to change their fellow citizens' behavior in order to restore and protect the physical environment, or exclude those who did not fit with prevailing racial or political preferences. As with the city's parks and playgrounds, outdoor recreation became another arena in which combatants, with Seattle's scenic backdrop behind them, fought over who had the right to define labor and leisure.

Nature's Own Great Exposition

The drive to promote Seattle's economic and recreational virtues was interrelated with the full integration of Puget Sound into the national mass market, made possible, in large part, by the arrival of the transcontinental railroads and by the Klondike Gold Rush. Nearly three decades of promotion, beginning in the 1890s and rooted in the "back to nature" movement, had encouraged visitors and residents alike to prize Seattle's blend of urbanity and simplicity. Boosters believed that Seattle must become a city to match its scenery because scenery was a resource, like timber and salmon, open for exploitation. Just as the railroads and business elites had touted Seattle's manufacturing and shipping advantages, so now they publicized its leisure possibilities. The Chamber of Commerce crowed, in an 1898 pamphlet full of photographs, illustrations, and descriptions of the city's assets, "Seattle is becoming a health and pleasure resort of deserved fame." It offered tourists "a country full of charms."[4]

The icon that defined the panoramic Northwest was Mount Rainier. Citizens of both Seattle and Tacoma claimed the glacier-clad peak as their own, and fights over who possessed the mountain (as if it could be owned) prompted an ill-fated crusade, led by Tacoma, to rename it Mount Tacoma or Tahoma in putative honor of its Indian heritage. The Northern Pacific Railroad, whose terminus lay in Tacoma, appropriated the mountain in its own promotional literature and worked, in surreptitious fashion, to turn Mount Rainier into a national park and, thereby, exchange the mountain's poorly timbered lands for more valuable property elsewhere in the Cascades. Seattle boosters, working through their congressional representatives, squelched the promotion to change the mountain's name but did join the national park campaign, which attracted support from the Appalachian Mountain Club, the Sierra Club, and the National Geographic Society. Their combined efforts led to the establishment of Mount Rainier National Park on March 2, 1899. Modeled after other national parks in the West, it testified to the power of the railroads and urban elites to make places of leisure beyond the city limits. And the railroads sought a return on their investment. The Northern Pacific, in its annual promotional circular, *Wonderland,* ballyhooed the prospects awaiting recreational travelers in the Cascade and Olympic Mountains.[5]

In an era when Americans still believed that city environments brought ailments, ranging from neurasthenia to hay fever, such appeals accented the region's healthful climate and stress-free way of life. An article in the magazine *Pacific Coast Sportsman* described an automobile trip to Snoqualmie Falls in the nearby Cascade Mountains as "the greatest health-giving experience," claiming that the fresh air "forced around one" would "rehabilitate worn-out nerves and drive out worry, insomnia, and indigestion." The city's Chamber of Commerce accentuated the point; only the most hardened man could fail to "live an ideal life in a Seattle hillside cottage overlooking sea, lake, or snow-capped mountain, amid a refined modern civilization." Indeed, noted preservationists like John Muir were saying similar things in their efforts to bring urban Americans into the outdoors, often working with local businesses in their promotional efforts. Seattle business leaders, by marketing their hinterland as a recreational paradise, were emulating counterparts in New York City and Boston who had similarly appropriated the Adirondacks, Berkshires, and rural northern New England decades earlier, or contemporaries in San Francisco who laid claim to the Sierra Nevada as their backyard playground. Political leaders, who usually came from the ranks of local business, pitched in, as Thomas J. Humes, Seattle's mayor, did by writing an article, in 1903, extolling the phenomenal fly-fishing so close to the big city.[6]

Alaska, too, was part of Seattle's new leisured hinterland; so were the northern Rockies and British Columbia. Seattle business operators dreamed of provisioning the tourists passing through town on their way to the outdoors, in any direction, as they had with the Klondike miners just a few years before. And as working in the outdoors was seen as a masculine pursuit, outdoor leisure was a manly affair. Mother Nature, said a Chamber of Commerce circular in 1905, had "done her share" to make Seattle and its environs a fantasy land for energetic tourists eager to paddle canoes, fly cast for jumping trout, or camp beneath giant trees. "There is an inspiration in scenes like these," the chamber noted, "which make men stronger, and wiser, and better, which raises them above the weary cares of life and adds largely to the sum total of their happiness."[7]

By the first decade of the twentieth century, Seattle was poised to surpass its closest rival, Portland, as the preeminent city in the region. It would soon have more people, more railroads, more industry, and more capital. The 1909 fair had helped to propel Seattle past Portland, a "regional city" focused on its immediate hinterland of the Columbia River and Willamette Valley, to become a "network city" that could claim the Pacific Rim and Alaska as its satellites. Selling Seattle as a network city had begun with the shape of the Alaska-Yukon-Pacific Exposition fairgrounds, themselves designed by John C. Olmsted as part of the master plan for the University of Washington's new campus on the shores of Lake Washington and Lake Union. The main promenade, punctuated with a cascading fountain at one end, parted a grove of tall pines to reveal Mount Rainier looming in the distance when the skies were free of clouds or rain. A reporter for *The World's Work* wrote that "Nature's own great exposition" held "the mind in thrall."[8]

On the fairgrounds in 1909 were dioramas of hardy pioneers building cabins in the gloomy woods, aquaria of gleaming salmon and trout raised in government fish hatcheries, newfangled motion pictures of Alaskan landscapes, model mines and timber mills with working saws and ore carts. Visitors could touch as well as see. They ran their hands over piles of gold nuggets or caressed lush pelts. The buildings and iconography emphasized a simple lesson: that if left undeveloped, the earth's surfeit was wasted. On the cover of the fair's daily program, a grizzled Alaskan miner standing in a stream shoveled gold nuggets that melted into a cascade of coins falling on Seattle streets like rain. The Forestry Temple was a neoclassical structure of bark-covered columns of timber from rough-hewn gigantic old-growth cedar trees shipped from the Olympic Peninsula. It would later house the University of Washington's School of Forestry.[9]

Many visitors raced through the official exhibits (or skipped them altogether) to head straight to the carnival quarter, the Pay Streak, named for the gold-bearing veins of the Klondike. This street party of hustlers, penny-ante arcades, and burlesques was a tamer version of Skid Row, absent the brothels and saloons. Fair promoters who may have scorned the coarse atmosphere and noisy barkers saw the Pay Streak as both a financial and an educational boon. Fairgoers could pretend to be sourdoughs at work in the Yukon, rocking actual sluice boxes filled with fool's gold, or test their muscles sawing logs and splitting cedar planks alongside real lumbermen. Other amusements put native peoples on display as exotic objects of edification and derision. In their Pay Streak village, Eskimos from Siberia and Alaska staged mock seal hunts and dances for curious spectators while, across the way, past the Gold Camps of Alaska vaudeville revue, indigenous Igorrotes from the Philippines, armed with spears and clubs, waged mock raids against imaginary enemies. Indians held fake battles modeled after Buffalo Bill Cody's popular Wild West shows, drawing huge crowds farther down the midway. Native peoples participated willingly in these reenactments, in exchange for pay and board, even as the Smithsonian Institution scientists and Bureau of Indian Affairs agents who helped to entice them there ensured that sideshow promoters depicted their charges according to prevailing racial notions. The racist sentiments behind such displays were obvious. When an Igorrote named Ka-lang-ad supposedly "escaped" from his Pay Streak compound, frightened residents called the police to capture the errant native. Afterward, the *Seattle Times* editorialized "the time is past when tattooed and painted savages" roamed the woods.[10]

Even as living Indians walked the streets of Seattle, selling handmade baskets and buckets of clams or fish from Puget Sound, the Bureau of Indian Affairs exhibit, on the main fairgrounds, had set up mannequins to represent what assimilated Indians might become in the future. One mannequin was a denim-clad "plowman," educated at a government boarding school; another model was a maid, busy weaving "yards of thread lace" for her white employer. Portraying Indians as pliable citizens, willing to work and live as servants to whites, was part of the fair's allegory of progress. It was a story that, according to John Harkman, president of the University of Washington's Board of Regents, began with explorers and loggers and ended with "the church in the valley and the schoolhouse on the hillside." Harkman, in a welcoming speech at the exposition's opening ceremonies, summoned images that the historian Frederick Jackson Turner had employed in his famous address at the 1893 Chicago exposition on the significance of the frontier in American history. It was a message white Americans understood; as the editors of *The World's Work* wrote, the fair exemplified the "noble battle" of remaking the natural world and its inhabitants through the "dignity and joy of labor."[11]

For those who were not white or those not affluent enough to be honored for their labors at the fair, Harkman's allegory of progress was a discomfiting one in a city that was becoming as segregated by race and class as any other in the United States at the time. During the first and second decades of the twentieth century, native-born whites who had emigrated from the upper Midwest and the Ohio River valley or the rural Northwest composed 75 percent of Seattle's population, while non-whites were about 5 percent of the total, with residents of Japanese descent the most numerous and African Americans the next. (The other approximately 20 percent were immigrants from northern and western Europe.) The city was, essentially, segregated geographically and most of Seattle's immigrants and minorities tended to live in enclaves: Scandinavians, predominantly Norwegians, in Ballard; Japanese, Chinese, and blacks south of the central business district along Yesler Way or Jackson Street; Italians in Rainier Valley, "Garlic Gulch" in the local argot. The demographic changes justified, in part, the regrades that cut off tops of hills and filled the muddy tidelands and they, in turn, continued to reshape the city's social fabric. Seattle residents were now increasingly segregated by class, too, with the wealthiest citizens able to move up to the hilltops and beyond to outlying streetcar suburbs. Middle-class homeowners and prosperous unionists concentrated in the middle belt of neighborhoods on the intermediate heights of the city's hills and ridges or in valleys and flats, such as Montlake and Wallingford. The poorest still lived at the lowest level, along Skid Row or on the Elliott Bay waterfront.[12]

The city's civic leaders, the same group that had spearheaded the 1909 fair, at first saw no social divide. Upward mobility among Seattle native-born workers, as well as immigrants, changed their minds. In a speech to the Seattle Chamber of Commerce in 1903, the state labor commissioner, William Blackman, reported that Seattle had "about seventy-five different labor organizations" all agitating for improved working conditions, higher wages, and "striving in turn to save from their earnings something for a home." By 1908, as the city prepared to host its first world's fair, the Washington State Federation of Labor was calling for a boycott because of the plan for open-shop hiring. The *Seattle Union Record,* newspaper of the city's Central Labor Council, argued that the fair, "for the good of organized labor everywhere and for the good of Seattle, deserves to be a frost, and we ask our friends to bend every energy to that end."

The Central Labor Council staged a protest march outside the fairgrounds on Labor Day. More than a hundred Filipino merchant marines visiting Seattle were also offended by the exposition, calling the Igorrote Village on the fair's Pay Streak, in a letter to the *Seattle Post-Intelligencer,* "no more representative of the Filipino than the American Indian is of the average citizen of the United States." (At the same time, scores of Filipinos in Seattle, perhaps eager to demonstrate their loyalty to America, wrote to the *Seattle Times* and the *Seattle Post-Intelligencer* in defense of the fair and its sideshows.) The fair was a cornucopia of consumer goods and of the raw resources used to make them, and the lessons offered were comforting and familiar: that Seattle was a prosperous city, its prospects bright. Still, the fair's promoters could not hide, as much as they tried, the contradictions behind Seattle's good life.[13]

Those contradictions only grew sharper after the turnstiles closed, and the class conflicts that had long tormented Seattle became ever more explosive. Labor protests, strikes, even armed confrontations were commonplace in the Pacific Northwest during the early twentieth century, just as they were across the North American West, from the infamous 1914 massacre of coal miners and their families in Ludlow, Colorado, to the bloody firefight one year earlier between Wobbly farmworkers and National Guardsmen in Wheatland, California. These and other battles were skirmishes in a larger war between those who owned the extractive industries that now dominated the West and those who worked in the forests, beneath the earth, or in the fields to bring the goods to American consumers. At stake was who should control converting nature's raw riches into everyday wealth, but it was also about something else: who would set the terms for enjoying the fruits of toil. Just as the forces of corporate capital at the time were trying to oust organized labor from the places of production, a similar combination of tourist promoters and outdoor enthusiasts busily made efforts to remove working people from the places of leisure and consumption. The politics of play were often just as fraught with strife as the politics of labor.[14]

Unleashing the Spirit of Play

In 1908, Caroline Sheldon, a reporter for *The Chautauquan,* stopped in Seattle on her way to Alaska to witness preparations for the coming world's fair. She had not expected a city set in the wilderness within an amphitheater of "amethyst peaks." Despite Seattle's "apparently chronic condition of uptorn pavements, . . . buildings making a transit of the city on wheels, and stupendous enterprises in regrading," she could think of "no other city in America which makes such an appeal to the imagination." The people caught her attention, too. "The true Seattleite is singularly free from nervousness and worry," she observed, because "he appreciates the magnificent scenery and manages to find some time to enjoy it." "Everybody goes boating or canoeing, and nearly every young person learns to swim," she continued, because "the forest primeval" was within "easy walking distance" of the city, providing "an opportunity for the poorest child to come in contact with nature."[15]

In his classic study of 1899, *The Theory of the Leisure Class,* the sociologist Thorstein Veblen named the culture that Sheldon witnessed on her visit to Seattle. Veblen saw a new consumer society emerging in which "consumption" had become "a larger element in the standard of living in the city than in the country." By 1894, the United States had become the world's leading manufacturing center, almost matching the combined output of Great Britain, Germany, and France, a material advancement that translated into vast cultural change most profound in America's cities. Awash in a sea of possessions, a host of city dwellers constructed new identities as consumers, defining themselves by their dress, their food, their homes, their clubs, and their neighborhoods. They created new consumer spaces—baseball stadiums, municipal parks, bicycle paths, and arcades. Outdoor pursuits in particular alleviated the tedium of commuting to and from the workplace and slowed the breakneck pace of urban social life. To Veblen, leisure paralleled and merged with the rise of consumerism, a means of escape for Americans of means from the strictures and complexities of a new machine-made world. These urban consumers idealized hypermasculine figures like Theodore Roosevelt with his "Strenuous Life" of cow punching, camping, and big-game hunting. Seattle fit well into the broader trends identified by Veblen, and the editors of *Welfare,* a local monthly, wrote in an editorial entitled "The Spirit of Play" that play was now "as important as work." Without the "play instinct," one of "the forces that have led men into the wilderness and hence to the discovery and conquest of new lands" would disappear. Play could achieve "the blending together of the whole man" and salvage what modern life had eroded. Leisure was now a matter of survival. Historians later labeled some manifestations of this phenomenon "anti-modernism"—the pursuit of authentic experiences.[16]

These urban play seekers in Seattle as elsewhere created a new geography of recreation that connected the countryside to the city as tightly as any trade network. Organizations like the Washington State Good Roads Association, founded in 1899, aggressively lobbied county and state officials for more and enhanced thoroughfares, not as much for conveyance as for better access to the great outdoors. One prolific literary commuter in Seattle, W. S. Phillips, a local sportswriter known by the romantic pseudonym "El Comancho," traveled those roads almost every weekend, hunting and fishing, to help publicize Seattle as the "new Garden of Eden." "By all means," Phillips wrote, "if you live in the old, staid, provincial East, pack your suit bags some day . . . and find out what living really means."[17]

Seattleites who could afford to enjoy Eden's fruits began to build exclusive communities to protect their new garden, and the town of Beaux Arts was the exemplar. On the eastern shore of Lake Washington, this model community of up-to-date homes seemed to emerge from the forest, complete with modern utilities like electricity and telephone service, and within easy reach of the beach and the woods. Advances in homebuilding technology made the "Craftsman style" bungalow available for families in the working-class and lower-middle-class neighborhoods of Wallingford, Ballard, and Fremont as well. The steady expansion of pre-milled components, popularized by real estate developers and architects, reduced home construction costs, putting the bungalow style in the reach of more and more Americans. These precursors to the modern prefabricated or pre-cut kit homes, like those marketed by Sears, Roebuck, and Company, quickly became one of Seattle's most popular house designs. Bungalows soon began popping up all over Seattle. These mass-market bungalows had the same aesthetic touches as the hand-built Craftsman homes in fancy Beaux Arts—exposed wood beams and hardwood floors, detailed stonework, fireplaces—but at a price working families could afford. Plus, the Craftsman homes seemed to echo what made Seattle special: its matchless attachment to the outdoors. These homes "seemed to blur the boundaries with nature," as the historian Janet Ore argues, by offering "a crucial respite for those who benefited most from the modern world." An evening at home, sitting next to a crackling fire, the soft rain beating against the windows—all helped to revitalize these homeowners to face "the public world, not to reconstruct it but to implement and manage its transformation."[18]

To the city's boosters, budding neighborhoods of Craftsman bungalows owned by industrious professionals and laborers seemed to confirm that the campaign to advertise Seattle was a success. Housing construction was an indicator of economic progress, and geography was Seattle's strongest selling point.

Publicists for the Northern Pacific Railroad recast Puget Sound as the "Mediterranean of America," a place that, unlike sunny California, had days that were "sunshiny and cool" and "nights that call for blankets." But such appeals to climatic benefits had a more sinister cast, some playing upon prevailing racial stereotypes, based on concepts of inferior peoples of southern origins and eugenics, the pseudo-science of breeding of superior races, in their ploys to lure visitors north. Erwin L. Weber, an engineer and meteorologist, reiterating the well-worn argument that civilization and phenotype followed lines of latitude, wrote in a 1924 Chamber of Commerce booklet that "intense and prolonged sunshine" was detrimental to "human progress." Weber noted that the Pacific Northwest, situated in "the zone of filtered sunshine," possessed "all the basic requirements necessary and desirable for the development of the most virile types of humanity, and the highest attainments of civilization" thanks to "freedom from climatic extremes." Zoë Kincaid, a columnist for the *Seattle Mail and Herald,* observed that Seattleites had no reason to travel to "other climates for amusement and recreation" because bungalows and canvas tents built on the forested shores afforded "a luxurious and independent life." There, the children of Seattle could "grow straight as pines, learned in water craft and wood lore," because living in nature yielded superior people. "Brought up in the shadows of the mountains," she concluded, "they are taught to be true Westerners, men and women of right-living and thinking."[19]

These men and women of right thinking, as adults, joined clubs like the Mountaineers, formed in 1906 to sponsor social and sporting trips in the outdoors. Inspired by similar organizations in other West Coast cities—the Mazamas of Portland and the Sierra Club of San Francisco—the Mountaineers tapped its first members from Seattle's professional ranks. Its founders included photographer and Chamber of Commerce leader Asahel Curtis, electrical magnate W. Montelius Price, University of Washington geologist Henry Landes (husband of the future Seattle mayor Bertha Knight Landes and the club's first president), and University of Washington history professor Edmond S. Meany. By 1907, the club had more than two hundred members, men and women, who were committed "to explore the mountains, forests and water courses of the Pacific Northwest." In keeping with its times and its mission, the Mountaineers was an exclusive association. Although almost half of its charter members were women, later memberships required the endorsement of two standing members before publishing the names of potential inductees in the monthly bulletin. If no one raised objections, the board of trustees admitted the applicants to the club. Such barriers kept the Mountaineers predominantly middle and upper class, overwhelmingly white, and largely Protestant. The same barriers also provided the Mountaineers with political influence, which the club used to its advantage. Meany, who served as president for twenty-seven years, often spoke for the Mountaineers when promoting Seattle's wilderness beauties as a private citizen.[20]

Edmond Meany, the public face of the Mountaineers, suffused the club with his energetic personality. He had come to Seattle in 1877 with his family from Michigan when he was fifteen years old. He enrolled in the Territorial University of Washington, where he was the secretary of the Young Naturalists' Society, an early research organization committed to nature study and conservation and, after graduation, worked as a newspaper reporter and state legislator while pursuing graduate studies in history at the University of Wisconsin. In 1897, the University of Washington invited him to join its history faculty, a position he would hold until 1932, and he became a beloved professor and prolific writer. Meany amassed a huge collection of materials on Washington's past and collaborated with Edward S. Curtis on the research and production of *The North American Indian,* and he founded the state's first scholarly history journal, the *Washington Historical Quarterly.* A relentless booster of Seattle, he raised funds for the Alaska-Yukon-Pacific Exposition, which he later said was "pointing its finger to the place where the finest of the Aryan stock may find its rejuvenation only to evolve a still more robust, vigorous, and brainy type." He also supported conservation and political reforms to protect the Pacific Northwest, the seedbed for future robust Americans, and expected fellow Mountaineers to follow his example.[21]

In the beginning, the Mountaineers was primarily a decorous venture of days spent on mountain hikes and nights filled with dancing, avant-garde plays, poetry readings, and banquets, often followed by Protestant services on Sunday mornings led by Professor Meany. Its social high spot, an annual summer outing, drew scores of participants, men and women, who were instructed to pack a long inventory of equipment and adhere to firm regulations. Strict rules were in place for a reason; inclement weather often repulsed efforts to summit snowy peaks. Without proper gear—warm and windproof clothing, sturdy boots, and reliable backpacks, all purchased from reputable merchants who advertised in the pages of the monthly newsletter—and without appropriate training and safeguards in climbing techniques, club members could be injured or even killed. Not even in the backcountry could the Mountaineers escape their predilection for order in all things. Upon returning to camp after daylong hikes or summit attempts, the men and women split chores: cooking, running the commissary, cleaning, and tending fires. Often, men prepared the meals and women collected firewood, a reversal of the usual roles they played back home in Seattle. The passion for organization continued after the yearly summer outings. Between their hikes up mountainsides, the Mountaineers worked to catalogue all of Washington's backcountry trails while building new ones so "that those who follow may find and enjoy the same beauties."[22]

One group of disaffected wilderness devotees, who found the rules and adherence to decorum too rigorous, formed the Co-Operative Campers in 1916 to "make our mountains accessible through co-operative camps and to encourage a love of simple living in the open air." The "Co-op Campers" organized summer expeditions and local walks, promoted conservation ethics, and taught good camping skills. Unlike the Mountaineers, however, they did not require members to purchase gear or indulge in high-class get-togethers. The primary motivation was to get outside. Members pooled their money to acquire tents, sleeping bags, packs, and food. Membership requirements were minimal and fees were modest, so that everyone who could demonstrate their mettle as able campers could join. Seen one way, the Co-Operative Campers were a mobile, urban-based version of the communitarian settlements that dotted Puget Sound at the turn of the century, including places such as Home, near Tacoma, and Equality in northern Skagit County, both part of a vibrant national cooperative movement. But they were an outing club first. Every summer, the Co-op Campers, like the Mountaineers, organized a summer expedition to Mount Rainier National Park. There, at Summerland, a spacious alpine meadow on the mountain's northeast flank, they built a mini-town subdivided into the "Fire Department, Board of Public Works, Board of Health and Sanitation, Department of Recreation and Commissary Department." Fusing recreation with communitarian ideology to teach blue-collar campers was an extension of how earlier utopian radicals had tried to make consumption a political act. Yet by bringing the urban with them into the countryside, the Co-Operative Campers shared more with the bourgeois Mountaineers than they would have admitted.[23]

Anna Louise Strong, the noted Seattle leftist activist and the Campers' first president, embodied these inconsistencies. On a trip to Mount Rainier in 1916, Strong told her father to warn those without "good mountain boots" to "wear rubbers" instead of regular street shoes so they did not "suffer getting up to camp across the snow." If Meany represented reformist and middle-class Seattle, Strong personified radical and revolutionary Seattle, even though she, like many in the city, had been born in the Midwest, in Nebraska in 1885, and she was not a child of poverty. She acquired her politics from her father, Sydney Strong, who had become an outspoken Congregationalist minister in Seattle's affluent Queen Anne Hill neighborhood and she attended Bryn Mawr, then Oberlin, and earned a Ph.D. in 1908 at the age of twenty-three from the University of Chicago. She worked on child welfare for the United States Education Office, returning to Seattle in 1915 to find a city in the throes of class warfare just as she had declared allegiance to socialism. A year later, Strong ran for the Seattle school board and, with the support of women's groups and organized labor, won a seat as the only female board member. She pushed for programs to aid underprivileged children and did not limit her activities to elected office. She was a reporter for the *New York Evening Post* and covered the Everett Massacre of 1916, a vicious confrontation that pitted an army of private guards hired by lumber mill owners and local police against organizers for the Industrial Workers of the World, in the mill town of Everett just north of Seattle. After a short gunfight, two deputies and five Wobblies lay dead. Such events further radicalized Strong, and although her politics attracted many admirers in Seattle, including fellow radicals among the city women's clubs, incensed fellow school board members orchestrated her recall in 1918. She continued her crusades, writing editorials and poems for the *Seattle Union Record*, the leading labor periodical, and organizing opposition to the American entry into the First World War. Strong wrote in her 1937 autobiography that she "drugged [herself] with forests, cliffs and glaciers" in the summer to escape the stresses of politics. Nor was her affinity for the outdoors at odds with her rebellion against capitalist society; if anything, her time with the Co-op Campers reinforced her working-class loyalties.[24]

Of Strong's many enemies, the ones she feared most in the mountains were the same ones she feared in Seattle: the federal government and the anti-unionists of Seattle's business cliques. By 1917, National Park Service administrators believed the Co-op Campers and fellow travelers were crowding out of the park the largely middle-class auto campers welcomed by the Rainier National Park Company, the exclusive concessionaire for transportation and guiding services. Stephen Mather, director of the National Park Service, and Roger W. Toll, superintendent at Mount Rainier, decided to push the Co-op Campers off the mountain instead. Seattle business leaders applauded. By then

Strong had turned her primary attention to Seattle's burgeoning labor struggle. The earlier concord between local businesses and union leaders at the start of the First World War had all but disappeared, and employment cutbacks during the postwar recession inflamed the mutual distrust. Strong, siding with the unions, took part in the famous general strike of February 8–11, 1919, one of the largest of its kind ever in the United States, but the mass walkout failed and Strong left Seattle for the Soviet Union, inspired by a speech Lincoln Steffens gave on the Russian Revolution. She died in 1970 in mainland China. The Co-Operative Campers and the Park Service continued their struggle until 1922, when the Campers disbanded and disappeared.[25]

Despite divergent political trajectories, the Co-Operative Campers and the Mountaineers had combined forces to open Mount Rainier to public camping. The two associations jointly persuaded the Park Service to expand free public campgrounds in the early 1920s and rein in the Rainier National Park Company's monopoly, which eventually collapsed when investors pulled out during the Great Depression. The broader cooperative movement of which Strong's campers were a part remained robust in Washington State well into the century. Cooperative groceries, established in the aftermath of the 1919 general strike, continued to provide services, if on a limited basis, for union members and the working poor. Unionists and crusading physicians, in 1945, founded the Group Health Cooperative of Puget Sound. For outdoor sports fanatics, however, low fees and availability of recreational equipment came to matter more than an adherence to pure cooperative philosophy or class allegiances. Lloyd Anderson, a city bus driver and Mountaineers member, rebelled at the club's requirements to attend chic parties and, two years after Meany died, he and a group of young supporters took control of the club, ousted the old guard, and installed officials who enjoyed strenuous sports more than elegant social events. Anderson, influenced by his participation in several Seattle cooperatives, did more than remake the Mountaineers: he transformed how its members and other outdoorsy Americans bought their equipment. In 1939, Anderson and his climbing partners organized a cooperative buying organization called Recreational Equipment Incorporated and pooled annual membership fees to purchase high-quality equipment from Europe in bulk and at affordable prices. He quickly gained loyal customers around the country. The new company, soon known by its abbreviation, REI, espoused the ideals of collective action but deployed them to market consumer goods. It was a shadow of Strong's Co-Operative Campers.[26]

Mass consumption, as sociologists Robert and Helen Merrill Lynd found in their 1929 study, *Middletown,* was remaking the contours of American society. As more and more Americans subscribed to glossy magazines, attended motion pictures, and listened to radio shows, eager companies employed mass advertising to elicit the economics of desire. Older forms of recreation faded as the demand for new goods and services—Ford automobiles, Westinghouse radios, and Hollywood movies—compelled manufacturers to make better products for specific groups of consumers. By the late 1920s, the segmentation of the sporting goods market had helped make Seattle a laboratory for sporting goods innovation. Local entrepreneurs like Eddie Bauer, born on Orcas Island in northern Puget Sound in 1899, took advantage of consumer demand for good equipment. Unlike Lloyd Anderson or Anna Louise Strong, Bauer was a capitalist first. He had worked in several Seattle sporting goods stores before opening his own sports shop in 1920, consciously catering to the city's middle- and upper-class outdoorsmen with his hand-tied fishing flies and patented line of goose down sleeping bags and jackets. Bauer advertised fishing and hunting opportunities in his own backyard, working in unison with fishing friends at the *Seattle Times* and the Chamber of Commerce. After weekend trips, Bauer would display his trophies in front of his downtown store—rows of glistening trout and salmon dangling from chains, heads of deer and elk draped over the entryway. When he expanded his operations to include mail-order shopping, his catalogues, rife with images of hunters and anglers pursuing their quarry through tall trees and along swift streams, sold the Northwest to clients across the country.[27]

The outdoors, now big business in Seattle, had become the business of municipal leaders as well, worried that future generations would be deprived of a childhood spent swimming and fishing, climbing trees, and exploring woods. This growing sentiment emerged out of, and fed, the "back to nature" movement sweeping across urban America. Its advocates, like the playground movement reformers who preceded them, warned that city children were at risk of becoming too civilized and soft. With the entire Northwest outdoors a potential classroom, the Seattle public schools, following the lead of educators across the country, launched a gardening program in 1913 to promote a new "vocational [and] recreational outlook" among Japanese and Italian immigrant children while providing "social service to the community."[28]

Seattle's park system, too, began to offer more recreational activities, a move that continued the drift away from the formal landscape aesthetic championed by John C. Olmsted, who had created the city's greenswards as sites of quiet contemplation of beauty, and toward a mission of public health and education. The Board of Park Commissioners' annual report in 1922 stated the case simply: "Seattle makes no mistake in promoting and encouraging all kinds of clean and healthful sports, and for every dollar invested in the common stock of recreation we get dividends of health, strength, alertness, good behavior, obedience, clean habits, or in a word, good citizenship." The Parks Department opened new public beaches at Green Lake, Alki Point, and Lake Washington, and later offered swimming and sailing lessons, rowing competitions, day camping for children, and snow skiing instruction at Snoqualmie Pass. An automobile camp at Woodland Park, opened in 1922, attracted nearly twelve thousand visitors annually within two years of its opening.[29]

Seattleites were now accustomed to seeing outdoor work and outdoor play as allied goals; park officials worked together with local clubs like the Mountaineers and retailers like Eddie Bauer and REI to maintain city facilities and provide skills instruction. As Seattle's leisure needs expanded, the city Parks Department, in 1927, began to push for a new Metropolitan Park District to include most of King County. Meany, the ever-voluble leader of the Mountaineers, joined the campaign and, two years later, in an address on "The Coming City of Puget Sound," invoked the spirit (if not the letter) of the long-dead 1911 Bogue Plan to establish regional parks encircling Seattle. Automobiles and railways had led to "the contraction of space" so that "the whole Puget Sound region" was a "single contact area from the standpoint of work and play." Only by acquiring more "open spaces," said Meany, could Seattle meet future consumer demand and prevent the loss of what had become an economic resource almost as valuable as timber or fisheries or real estate: outdoor recreation.[30]

Putting the Small Men Out of Business

Those who sailed on Lake Washington, hiked on Mount Rainier, picnicked amid the pines of Volunteer Park, or fly-fished on the Duwamish River believed they were living in an Eden designed for their pleasure. Yet if they had considered those who lived on boats instead of sailing on them, who used parks for shelter instead of dining in them, or who caught fish for subsistence or an earning instead of for fun, they would have seen the marvelous landscape of boosters' gazetteers and broadsides in a darker light. By the 1920s, the working poor were not welcome in the city's Arcadian playgrounds. Their eventual expulsion from the recreational garden began along Seattle's waterfront and around the remade watershed encircling the city.

Indeed, some observers had foreseen that changing Seattle's rivers and lakes would transform human behavior as well. The first changes affected how and where Seattleites spent their leisure time. Prior to the construction of the Lake Washington Ship Canal, which utterly transformed the lakes and rivers surrounding Seattle, some boating and fishing enthusiasts disparaged the project as a threat to scenery and recreation. Hiram Chittenden, the Army Corps of Engineers official who directed construction of the canal, had little patience for those who saw beauty and utility at odds. His experience as the primary architect of the road system in Yellowstone National Park convinced him otherwise. The lake, he told critics in a 1910 article, would remain "the chief scenic attraction of the city" because the canal was "a case where utilitarian ends can be accomplished without any sacrifice of sentimental interests." His remarks about Seattle proved prophetic in a city more often attuned to water than to land.[31]

W. S. Phillips, the outdoors author known as "El Comancho," had observed, in 1907, that "in the Northwest coast country the gasoline launch occupies about the same position in the pleasure field that the automobile does in the flat Eastern country." Just as mass-produced automobiles had unlocked national parks and the nearby countryside to city-weary Americans at the beginning of the century, so motorized launches, powered by small and affordable gasoline engines, had opened marine and freshwaters to legions of weekend explorers. Operating motorboats was easier than learning how to sail, and those who favored wind power often looked down on the gasoline mariners. One sailboat enthusiast, in a nod to racial purity as the mark of proper places for play, complained that on summer weekends Puget Sound was "so pregnant with gasoline explosions" that it resembled "an Oriental New Years." But, understandably, boaters of all kinds welcomed the Lake Washington Ship Canal. A reporter for *Pacific Motor Boat,* a monthly magazine, wrote in 1911 that Seattle would soon have "both a land and water system of highways" that would make the city the envy of mariners.[32]

During the canal's inaugural ceremonies in July 1917, the Chamber of Commerce, worried that wartime anxieties might overshadow the celebration, appealed "to the great motor boat and yachting interests of the Northwest" to participate. Boaters eagerly obliged, staging a parade that "amazed thousands of spectators and awakened them to the new realization of immense importance of the motor boating interests of the Puget Sound territory." Recreational boaters soon became the canal's most avid users, and the Seattle Yacht Club built its new headquarters on Lake Union, in 1920, because the man-made waterway had made the lake "the New Center for Seattle's Yachting and Motor Boat Development." In 1924, Seattle hosted an international motorboat race on Lake Washington, prompting one writer to note how "Nature has bestowed upon Seattle every advantage for the holding of such a regatta." Harbor guides and touring maps of this "splendidly situated port" pinpointed clubhouses, moorages, and fuel stations, charting Seattle's new aquatic byways for boating enthusiasts.[33]

Many Seattleites, entranced by the remade waterscape surrounding their city, now made the water itself their home. Those with means built whimsical houseboats: floating bungalows, with covered front porches for summertime sitting and wood stoves for taking the chill off autumn evenings. Some used their houseboats as summer homes; others occupied them year-round. Those who lived on Lake Washington, according to a reporter from the *Seattle Post-Intelligencer,* were not poor "water gypsies" but the "'Poseidons' and the 'Uranians' of the 'Queen City.'" "If the city's soulless populations prefer stuffy apartments to the airy houseboats," the reporter concluded, those who lived in houseboats enjoyed "a taste of liberty and a breath of the free, open, outdoor life."[34]

Yet just as race and class segregated the city's land-dwelling residents, there were also chic and shabby neighborhoods on the water separated by income and ethnicity. The houseboats along the Duwamish Waterway, like the shacks on the shores of Lake Union, were not part of the charmed floating world. They were cheap dwellings clustered next to the shipyards, fishing terminals, and marine shops, often little more than canvas tents nailed to a flatboat deck or simple scaffolds covered with tar paper and scrap shingles. Most housed immigrants from southern and eastern Europe or East Asia; others were homes to Indians, living and working off reservation, or workers thrown out of jobs during the postwar recession. Really water gypsies, many migrated north to fish in Alaska, returning after the season to net fish in Puget Sound and the Duwamish River, where they angered the wealthier sport anglers who were growing concerned over declining salmon runs.[35]

The historical evidence for the salmon fisheries in the Duwamish River and Elliott Bay is incomplete, but most of the available sources suggest a once thriving salmon population that sustained commercial fishers and Indians

Aerial view of the Alaska-Yukon-Pacific Exposition fairgrounds with a retouched image of Mount Rainier in the background, 1909. (Courtesy Special Collections, University of Washington Libraries, Nowell x1040a)

Pleasure houseboats on the western shore of Lake Washington, Seattle, 1912. (Courtesy Special Collections, University of Washington Libraries, Lee 20035)

Edmond Meany, president of the Mountaineers, on a club outing, 1911.
(Courtesy Special Collections, University of Washington Libraries, UW 17998)

Eddie Bauer in front of his retail store in downtown Seattle with deer trophies, c. 1920.
(Courtesy Museum of History and Industry, MOHAI 89.11.1)

Public automobile camp at Mount Rainier National Park, c. 1925.
(Courtesy Special Collections, University of Washington Libraries, WAS 0699)

Bertha Knight Landes, the reform-minded mayor of Seattle who served only one term, with Frank Edwards, her future successor, c. 1928.
(Courtesy Museum of History and Industry, MOHAI 1983.10.4014.1)

Second regrade of Denny Hill, looking south from Fifth Avenue and Battery Street, May 17, 1929. (Courtesy Seattle Municipal Archives, Item 3429)

Aerial view of Hooverville, south of downtown Seattle, 1937. (Courtesy Special Collections, University of Washington Libraries, Lee 20102)

U.S. Navy officers watching tidelands squatter shacks burn at Smith's Cove, Seattle, c. June 1942. (Courtesy Museum of History and Industry, *Seattle Post-Intelligencer* Collection, P-I 2395)

James R. Ellis (center), Mayor Gordon Clinton (far left), and members of the Metro Study Committee, October 3, 1957. (Courtesy Museum of History and Industry, *Seattle Post-Intelligencer* Collection, P-I 86.5.21764)

Pro-Metro campaign poster for the September 1958 election. The photo was later found to have been staged by Dorothy Block, the Metro Action Committee's vice-chair. (Courtesy Washington State Archives, Puget Sound Branch, Metro Photograph Collection)

Metro opening ceremony on the shores of Lake Washington, with "Princess Sparkling Water" banishing "Polly Pollution" and "King Algae," February 19, 1963. (Courtesy Washington State Archives, Puget Sound Branch, Metro Photograph Collection)

Muckleshoot Indians arrested after fishing near the Soos Creek hatchery on the Green River, near Auburn, c. 1962. (Courtesy Washington State Archives, Puget Sound Branch, Metro Photograph Collection)

Muckleshoot Indians on fishing rights march through South Seattle in support of the men arrested for fishing on the Green River, July 13, 1966. (Courtesy Museum of History and Industry, *Seattle Post-Intelligencer* Collection, P-I 86.5.4450)

Real estate developer Kemper Freeman, Sr., with aerial photograph of Bellevue and Lake Washington behind him, 1966. (Courtesy Museum of History and Industry, *Seattle Post-Intelligencer* Collection, P-I 1986.5.25604.1)

Bert the Salmon with Seattle Public Utilities director Diana Gale (center) at "Creekstock" celebration on Longfellow Creek, West Seattle, April 27, 2000. (Courtesy Seattle Municipal Archives, Item 101110)

who supplied the local fresh fish market. Until the 1880s, Seattle stood at the nexus of a productive commercial fishery but, by the end of the century, most commercial operations had moved north to larger runs on the Fraser River in British Columbia or southeastern Alaska. Nonetheless, a survey by the U.S. Fish Commission in 1896 noted that the region still harbored significant stocks of coho (silver), chum, and chinook salmon, plus the nonmigratory kokanee in Lake Washington. By the first decades of the century, observers concurred that the fishery was in decline, with the upper Cedar River, dammed in 1900 to provide water for Seattle, blocking spawning salmon from swimming above Landsburg. Dredging the Duwamish Waterway, which opened in 1917, had removed not only snags and brush but also the sinuous bends where tired migrating salmon could rest in the journey upstream to spawn. The former river delta, replaced by man-made Harbor Island and the hardened banks of the Duwamish Waterway, no longer sustained eelgrass beds to shelter vulnerable juvenile salmon on their way to sea. Diverting Lake Washington to spill through the Ship Canal forced adult fish that spawned in the lower Cedar and Sammamish Rivers to adopt an entirely new and confusing route. Seagulls preyed upon stunned juveniles that spilled over the canal locks on their seaward passage, and sea lions feasted on returning adults that became stuck in the violent eddies. Completing the array of threats facing the beleaguered fish was industrial pollution. Sawdust from lumber and shingle mills turned waters anoxic, robbing salmon and their eggs of vital oxygen; shipyards and marine shops dumped oil, gasoline, and by-products from welding and painting—lead, mercury, and other heavy metals—into the waters; and sewers discharged raw sewage or, when storm drains were overwhelmed by rainwater, untreated runoff from city streets and homes.[36]

The state's waters were still seen as commons, open to all, and, as state fish commissioner Jonathan L. Reisland complained in his biennial report for 1905–6, these industries, "all very essential to the great future of our state," would not "be hampered or held back for the salmon industry." Typical of many conservationists of the era, Reisland wanted "due and proper consideration" for all users, but in the absence of restrictions against industrial use, fisheries managers had to adopt another solution. Reisland's position was the prevailing attitude toward conservation at the time—to balance efficiency, equity, and aesthetics in the protection of natural resources. Without support from the state or the voters, fisheries wardens could do little to prevent mills and factories from dumping effluent, wood chips, and sawdust. Instead, they turned to artificial propagation. In 1900, the State Fish Commission built the Green River salmon hatchery on Soos Creek, just north of the confluence with the Duwamish River, to check habitat loss throughout the entire watershed. Mistakenly referred to at first as the White River Hatchery, for the old arrangement before dredging and diversions changed the Duwamish's drainage and tributaries, it soon became one of the most productive in the state, providing eggs for other hatcheries around western Washington.[37]

By the first decade of the twentieth century, salmon runs in urban Puget Sound, as elsewhere in the Pacific Northwest, were on technological life support. Facing pressure to produce ever more fish to satisfy ever more consumers, state fishing managers turned to hatcheries as a solution to habitat destruction. They also began to demonize the most visible scapegoats, the few remaining commercial fishers who plied the waters of Elliott Bay and the Duwamish River. Most sport anglers were middle-class, native-born, and politically connected. But most local commercial fishers were working-class, often immigrants, who wielded little political sway. Furthermore, rapid industrialization spurred by America's entry into the First World War posed another threat to sport fishing. Shipyards expanded to meet the Allied demand for munitions, flour mills and packing plants grew to feed the soldiers overseas, and shipping traffic increased. Tugboat operators, shipbuilders, and lumber mills complained that commercial fishers' gill nets, made of fine mesh designed to entrap fish by their gills, were jamming their propellers, and fish traps and weirs were damaging barges. Most offenders, said W. T. Isted of the Northwestern Tow Boat Owners' Association, were "ignorant of the English language, or at least they profess to be as they do not pay any attention to the navigation laws of our country." Spotting an opportunity to oust their fishing rivals, sport anglers joined the shippers in the fight. Both groups invoked the mantra of public access to the waterway commons and insisted that it was a right now jeopardized by unchecked commercial fishing.[38]

L. H. Darwin, the new state fish commissioner who was then waging war against Indian fishing throughout the state, blamed aliens, mostly "Austrians, principally from the province of Dalmatia, and other southeastern European peoples," for unsustainable fishing in Puget Sound. Darwin had already overseen the passage of the state's first fisheries code banning certain types of gear, such as spears and snares, and the next to join his list of pernicious technology were purse seines and gill nets, popular among immigrant fishers on Puget Sound for their high degree of effectiveness and low cost. With the most productive hatchery on Puget Sound just twelve miles upstream from Elliott Bay at Soos Creek, Darwin concluded that the very survival of the region's fisheries was at risk. "The greatest mistake of all," he wrote in the biennial report of 1917–19, "is permitting aliens not only to completely monopolize our fishery, but to permit them to destroy it while monopolizing it."[39]

Darwin's anti-alien invective played well against the backdrop of wartime patriotism and postwar apprehension over immigration. Nearly six hundred Seattle residents petitioned Governor Louis F. Hart against commercial net fishing on the Duwamish, which they claimed deprived "thousands of citizens of both healthful recreation and food" and threatened "the natural propagation of salmon" assisted by the "State hatchery located thereon." After seasonal closures from 1918, the State Fish Commission, in February 1920, closed the Duwamish River and Elliott Bay to commercial fishing altogether, as well as Shilshole Bay at the mouth of the Ship Canal. One grateful sportsman thanked the governor and the commission for taking "an interest in our welfare." Commercial fishers did not surrender without a fight and attacked their new nemesis: urban sport anglers. Their laments were a contradictory critique of consumer politics. On one hand, commercial fishermen protested, farmers and city dwellers took too much water from the Green and Duwamish Rivers, leaving too little for the fish at the state salmon hatchery downstream, and, with insufficient water, spawning salmon could not overcome bars, snags, and other obstacles downstream from the hatchery collection racks. On the other hand, immigrant and commercial fishers defended their work as free enterprise and invoked the idiom of producer ideology as their claim to American identity. "Any honest man who wants to work in any lawful pursuit to produce the things that are needed by the people," one group wrote to Governor Hart, should "have an opportunity to do so." Some, like R. Q. Hall, took a cue from their opponents and deflected blame onto others. Closing the Duwamish, Hall testified at a State Fisheries Board hearing in June 1921, was "putting the small men out of business" without doing anything to punish the ultimate culprits for the salmon crisis. But E. A. Sims, the hearings chair, retorted that the Duwamish would never be "opened again in his lifetime."[40]

Sport anglers had won the battle. The war against commercial and Indian fishermen had played out in ways that would have been familiar to poor whites and blacks in the Reconstruction South, or the subsistence hunters in rural England forcibly excluded from Crown reserves by the Black Act of 1723. Backed by the power of state and business, sportsmen then and now had asserted their privilege to hunt and fish as they saw fit. As before, the victors enjoyed their spoils. As Eddie Bauer boasted in 1928, "Seattle fishermen are fortunate to have at their very door the finest salt water fishing in the country." Another local outdoorsman turned retailer, Ben Paris, started the first fishing derbies in the early 1930s to attract more customers to his business on Westlake Avenue, an emporium that combined his hunting and fishing store with a barber shop, cigar stand, and grocery. The work of promoters like Bauer and Paris paid dividends in more sales and more publicity. "Calvin Coolidge would laugh out loud," wrote a reporter for the *Seattle Times,* "if he ever caught a Chinook like they haul in by the boatload in Seattle waters." H. L. Dilaway, another local sport fisher, asked rhetorically if there was another place where someone could "catch salmon off the docks, or under the shadow of a skyscraper?" Meanwhile, former commercial fishermen who once called Elliott Bay and the Duwamish River their workplace hid in the shadows. With their legal access to the fishery blocked, poaching, often under the cover of twilight or night, became the preferred substitute. Fish thievery was a form of what the historian Karl Jacoby has called "environmental banditry." According to the Department of Fisheries report of the state fish commissioner for 1925–27, wardens had arrested fourteen people for net fishing in the Duwamish River. A similar pattern persisted for the rest of the decade. Most of those charged with illegal fishing had Italian, Greek, Slavic, or Japanese surnames. Some were likely Indians.[41]

Contrary to the exhibits and sideshows at the Alaska-Yukon-Pacific Exposition of 1909, the Indians of Puget Sound had not vanished or been assimilated. Although decades of federal assimilation policy and racist exclusion had frayed kinship ties and dispersed families across Puget Sound, they had not sent Natives into oblivion. Wage labor had replaced or augmented older traditions of seasonal subsistence rounds, as it had a century earlier when Indians worked alongside King George's Men and the first Bostons. Natives had to confront conflicting feelings of independence and dependency, pride in being Indian, and shame at what being Indian now entailed. At the end of the Progressive Era, the winds of reform reached Indian country, inspiring a group of government-school-educated Indians, in 1911, to found the Society of American Indians, one of several national pan-Indian organizations. In 1914, Pacific Northwest Indians founded the Northwest Federation of American Indians and, after several years of debate, its leadership decided that the best way for Puget Sound Indians to claim their rights was to look to the unfulfilled treaties negotiated by territorial governor Isaac I. Stevens. In 1923, with the help of sympathetic white attorneys, the federation persuaded Congress and the Bureau of Indian Affairs to let the Indian groups named in all four of the Washington treaties sue for claims arising from unmet treaty obligations.[42]

More than one hundred Indian men and women from around Puget Sound testified in 1927 at a series of depositions later presented to the United States Court of Claims. Because the claimants' attorney, Arthur Griffin, had advised his clients to focus on the value of their lost lands and violations of the treaty clauses, most testimonials told of the changed geography around Puget Sound. The accounts called upon the moral authority of responsibilities ignored and duties imprinted, quite literally, on the landscape. Repeatedly, witnesses talked about hunting and fishing, speaking from personal experience (most of the deposed were in their sixties and older), recalling streams once jammed with salmon, thick forests, and grassy prairies burned intentionally every year to encourage plentiful game forage. In many instances, they complained that whites were expelling Indians from what one Snoqualmie, William Kanim, called their "natural grounds." Sometimes, the evictions had been violent. Joe Bill, a Muckleshoot, complained that "here lately I went down to Green River to try and get fish for my living and I was taken prisoner by a game warden and held in Seattle in jail for two days and two nights." He had to fish for a living, he said, because he had "an allotment of the poorest kind of land" on the reservation. Joseph Swyell, a Puyallup, concurred. "Nowadays, when any of our younger people got caught fishing or hunting, they put them in prison." "The understanding," Swyell continued, "and which we have never been paid, the hunting, the fishing, and the clam diggings [are] all taken away from us . . . and what Mr. Stevens has promised the Indians has never pulled through." Jennie Joe, a Duwamish living on the Tulalip Reservation, north of Seattle, put the controversy in even starker terms: "to-day we haven't got anything and the younger generation has got nothing." Government attorneys pressured those testifying on the accuracy of their memories since no one was old enough to remember the actual treaty ceremonies. Wilfred Steve, another Snoqualmie, conceded that while the lawyers had a point, the Indians' accounts, which had been "handed down . . . from mouth to mouth, through my older people" were the "only way of perpetuating their history."[43]

The Indians on the stand did not speak as one people, but they did speak for one aim: justice for ancestors who had taken the word of Governor Stevens at face value during the treaty negotiations over seventy years before. They wanted their own share of the fish and game, farmlands, and scenery that the original white invaders of Seattle had claimed as their own. In 1934, the United States Court of Claims, while not repudiating the Indians' memories or history, ruled that the federal government owed the plaintiffs nothing. The Indians had their oral histories translated into the written word, but, for the time being, their duly recorded testimony sat on library shelves.

Almost twenty years after the Alaska-Yukon-Pacific Exposition, a magical world's fair that presented Seattle's wonders to the nation and the world, Almira Bailey, an author for the city's Chamber of Commerce, wrote a broadside in 1930 titled *Seattle and the Charmed Land* that proclaimed how, in the Puget Sound country, "our childhood geographies come true." Bailey's remark spoke to an important truth even as it masked a bigger reality. Childhood geographies in the modern world, at least in the twentieth century, are generally places devoid of drudgery. To the legions of fly fishermen and well-appointed hikers and campers, Seattle was a kind of a childhood idyll made real. But to those undeserving of playtime, this "Charmed Land" was neither Oz nor the Big Rock Candy Mountain, a make-believe place celebrated by a hobo dreaming of cigarette trees and lemonade springs. Some of those denied access to Seattle's playground, the itinerant laborers and homeless squatters, probably knew another version of the famous song where the hobo found no candy and no rest at the end of his hard traveling.[44]

Creating spaces for recreation in and around Seattle in fact had yielded both zones of pleasure and zones of combat. Middle-class residents gained beautiful locales wholly dedicated to leisure yet working-class residents, for whom production and consumption could not readily be separated, often had no distinct landscape for play. Indian peoples, shoved aside along with immigrants and alien residents, had no assigned landscape for their sustenance or recreation. Fights continued over who could claim natural amenities and resources. What resulted was not the classic "tragedy of the commons," as articulated by Garrett Hardin, where unrestricted access leads to the destruction of common pool resources, but a "tragedy of the commoners" instead. Those excluded were not without recourse, but they often had limited means of asserting their demands.[45]

When affluent Seattleites embraced outdoor recreation, they were not embracing unproductive and pristine spaces of an ideal nature while rejecting the unsightly landscapes of industrial and commercial production. Sport anglers and commercial fishers, bourgeois mountaineers and socialist campers, picnickers and poachers alike were engaged in both production and consumption. All fought to protect leisure, even if the poor had comparatively less time and resources for play, producing particular spaces and cultivating particular behaviors to promote outdoor recreation, while acknowledging the importance of labor to advance producers' rights and delimit specific locales to create the things necessary for the growing Seattle economy. However, what one side saw as a beautiful playground another saw as the source of their daily food. The problem was that for the new leisure classes, the best places to play were, for others, the best and only places to work. Thus, the allocation of leisure across the changing urban landscape was more than a contest over aesthetics. Campaigns to build more parks or promote sport fishing narrowed the dynamics of consumption to fights between antagonistic consumers, or consumers against producers.[46]

The new outdoor leisure society in Seattle could not have emerged without a new emphasis upon the non-human world, upon nature, as a source of personal regeneration, a conviction that made a fetish out of self-satisfaction and the physical environment. And it could not have emerged without a newly powerful sense of place, one forged through mass consumption. This is not to say the new mass consumer culture was at its core inhumane. The conservation movement was one of the era's greatest political and social triumphs. Americans now realized that in order to consume and enjoy nature, they needed to conserve and protect it. Indeed, many city residents, by reconnecting to Seattle's rivers and lakes, mountains and forests, parks and greenways as consumers, began to realize that decades of unchecked development in the name of progress had imperiled their environmental legacy. Seattle and its region would have been poorer, bleaker places without their efforts. In their drive to protect this heritage, conservationists and outdoor enthusiasts resuscitated something that the Progressive reformers had ignored in their quest to modernize and rationalize urban life: the importance of authentic and organic connections between people and places. What the more prosperous votaries of the backpack and fly rod would not recognize was how their passion for all things wild and natural could be just as self-interested and greedy as that of the most rapacious capitalist or even the most well-meaning reformer. Their ethic of place did not include any awareness of others unlike them, or of the inherent contradictions it embodied. A childhood fantasy come true, Seattle's charmed lands were geographies of arrested development.

CHAPTER 6

Junk-Yard for Human Junk

The Unnatural Ecology of Urban Poverty

In 1915, Department of Health and Sanitation inspectors condemned 547 buildings throughout Seattle, burning or razing 395 of them. Most were typical squatters' shacks, built from lumber mill scraps and splintered wooden boxes, strewn like jetsam across the tide flats at the southern end of Elliott Bay. Inspectors from the Department of Health and Sanitation, visiting the nearby Jackson Street neighborhood the year before, found a similar scene onshore: Japanese, Greeks, Italians, Russians, Chinese, Filipinos, and eastern European Jews crowded above storefronts, crammed into basements, or squatting in decrepit old houses that barely kept "within the limit of the law." The inspectors combing the city in 1915 found "the best of these shacks . . . built on logs for foundation, were floated to places about the bay." Few had toilets or privies, depending on the tides "to carry away the human waste." M. T. Stevens, the chief sanitary engineer, and J. S. McBride, the commissioner of health, explained: "Our city being located on a series of hills and valleys, we find the finer residences on the higher lands and the poor homes in the valleys. These higher districts give us very little trouble, but the lower ones require constant attention."[1]

Those with more power had pushed those with less power to the margins, often unwittingly, sometimes willfully, in the name of civic improvement. To Stevens and McBride, and to city engineer R. H. Thomson, leveling hills and filling tide flats had been a way to provide a "fair foundation for municipal solidarity." But municipal solidarity was not the result. Changing Seattle's physical terrain had reinforced inequality, concentrated it, and made it more visible. Reengineering topography had channeled the city's castoffs, quite literally, to the bottom. Thomson's protégés commanded the city's major agencies, from City Light to the Engineering Department, in the 1920s and 1930s, and their solution to Seattle's new urban ecology of poverty was to remove still more hills, burn more shacks, and expel more squatters. They seemed oblivious, at first, to the association between changing the landscape and generating or reinforcing social inequities. Others, however, were beginning to take notice.[2]

During the interwar years, a growing number of Seattleites began to question the costs of continually reshaping their city to drive growth and progress. Many citizens, radicalized by the tumultuous class conflicts of the previous decade, saw the poverty clustering in the city's poorest neighborhoods as the result of an economic system run amok. Other, more distant observers, including students and faculty at the University of Washington, saw the city's social divides as an organic process, the natural outgrowth of an unstable urban environment. Few made the connection between environmental and social inequality, but in their academic critiques and political reactions to the seemingly perpetual cycle of destruction and rebuilding lay the nascent foundation for a new vision of place grounded in justice.

A Very Cheap and Undesirable Residence Section

George F. Cotterill, a former assistant engineer to Thomson and one-term mayor elected on a reform ticket in 1912, was unrepentant in his faith. He believed that engineering could solve almost any of his city's problems. In 1928, in an address on the role of engineers in building Seattle, Cotterill told his audience "the Engineer was on the watch tower, sighting and striving to prepare for the greater Seattle that he saw upon the horizon. Let it never be forgotten that Seattle is not the child of chance or circumstance, but had its origin in the vision of a surveyor."[3] From homesteads to regrades, the engineer made a city under perpetual contruction.

Thanks to the ceaseless work of the engineers extolled by Cotterill, Seattle was, in many ways, the epitome of what the economist Joseph Schumpeter later called the "perennial gale of creative destruction." This fundamental impulse of modern industrial capitalism, Schumpeter said, devoured everything old—markets, companies, transportation, even places—and continually remade them anew to drive further profits and growth. The genius of the engineer was to make the ceaseless destruction and rebirth of the city an exercise in progress, but progress had its limits. Even Thomson could not annihilate all of Seattle's hills at once. When the hydraulic cannons stopped and the sluices ran dry, over half of Denny Hill lay at the bottom of Elliott Bay, the remainder a steep escarpment more than eighty feet above the busy streets below. Residents had protested high assessment taxes and damage to their property, compelling Thomson, in 1906, to stop the regrades short. Atop the hill, vestiges of the old neighborhood remained and, by the mid-1920s, the aptly named "Regrade District" had become, in the words of V. V. Tarbill, writing in the *Harvard Business Review,* "a very cheap and undesirable residence section" of unpaved streets, unpainted houses, and weedy open lots. The anticipated real estate bonanza never took off. Most there were renters, and their landlords put little energy into upkeep "because of the hope (or fear) that another regrade would soon take place."[4]

Leaving the balance of the hill still standing had created a traffic jam. Thomson had foreseen that automobiles would become inextricable to American city planning. Early automobiles needed gentle grades and, by 1920, nearly all crosstown traffic had to drive around Denny Hill's cliffs. A memo to city engineer James D. Blackwell in 1924 envisaged a Seattle population of over one million residents by 1975, with most commuting to downtown jobs by automobile. To Blackwell, the removal of Denny Hill had become the keystone to a traffic plan for the entire city. The editors of the *Town Crier* concurred with the city engineer, writing, "the business pulse of the city was expanding . . . beating against the walls that enclosed it: More room! was the cry." Eager landowners ready to improve their properties and upgrade their rentals organized the Denny Hill Improvement Club and the Uptown Improvement Club to petition for another regrade and, in 1926, earned the approval of the *Seattle Times,* which had turned against regrading and Thomson during the first round of earthmoving.[5]

As the second regrade of Denny Hill moved ahead, the now renamed Denny Hill Regrade and Improvement Club praised the city's endeavor. City attorneys had prepared the assessment rolls and condemnation awards in advance and assigned work for the entire project to one contractor, hoping to avoid past problems by planning for everything. Their confidence was premature. Denny Hill was the site of Denny Park, the city's oldest public space, "a breathing spot" for downtown workers who ate their lunches amid its tall oaks and pines, enjoying the view of Elliott Bay and the Olympic Mountains. The regrade would lower the park more than eighty feet, obliterating the trees and lawns, and irate citizens from across Seattle mobilized to save the park. The former engineer George F. Cotterill, previously an advocate of regrading, now called the second Denny Hill regrade "an expensive and disastrous blunder," and pleaded for a "remedy" to "protect the park area, with all its beauties of Nature and cultivation." His appeal relied on the language of progress; he had seen the park mature "from the stumps of the primeval forest." But Cotterill failed to persuade his former colleagues, and in 1928 the steam shovels began their dirty work. When they stopped three years later, all that remained of the park was the cupola of Denny Elementary School, itself cut in two, in 1911, by the first Denny Regrade. A new park emerged, but there were no great old trees, no commanding vistas of the city's harbor.[6]

The second regrade of Denny Hill was not as dramatic as the first; a city had grown around and over the hill, making hydraulic excavation impossible. Instead, giant steam shovels, running on special narrow-gauge railroad tracks, dumped excavated earth onto a web of portable conveyor belts running to scows in Elliott Bay. With most of the city's tidelands already paved over and built on, these special self-dumping scows steamed into deep waters, filled their holds with water, rolled over, and emptied their load. A reporter for *Scientific American,* impressed by the engineers' ingenuity, praised the "absence of noise and nuisance." But the shovels split water and sewer lines, spilling the contents onto city streets, while conveyor belts, exposed to sun and rain, abraded by rocks and sand, stretched and snapped. The unique scows often tipped over too easily inside the harbor line and impeded maritime traffic. Onshore, heavy winter rains liquefied the exposed hillside and turned more than nine city blocks into a muddy lagoon. Two workers died on the project, one in a landslide, another crushed by dislodged debris.[7]

Predictably, recriminations, then lawsuits, followed. Laborers forced to work overtime sued for unpaid back wages or quit. Angry at the mounds of debris blocking street traffic, the city sued its contractor; the contractor, hoping to recoup losses, sued the subcontractors responsible for grading, dumping, and street paving. Local residents, besieged by noise and rubble, used the complaints against the contractor to sue the city in turn. Four cases reached the state high court in Olympia, and in three, the court forced the city to recalculate its assessment taxes and damage estimates.[8]

Not until King County proposed building a new public hospital overlooking Skid Row and the late timber baron Henry Yesler's domain did the City Planning Commission, created to coordinate citywide development, understand the full consequences of the earlier regrades. The commissioners found hundreds of Japanese and African American families living in ramshackle houses, or single Filipino and Chinese men renting rooms in decrepit residential hotels, all clustered in the gulches and ravines south of Yesler Way, beset by near-constant mudslides and flooding, or on the steep slopes of Yesler Hill, a section of First Hill commonly known as Profanity Hill. (The name was supposedly inspired by the cursing of attorneys and their clients who had to climb the steep grade to reach the former King County Courthouse.) The commission blamed the cumulative effects of the Jackson Street and Dearborn Street regrades, which ran from 1907 to 1912, for the wretched state of the neighborhood. The Engineering Department reached the opposite conclusion: it had not removed enough hills.[9]

Seattle's municipal engineers were now impelled by a version of what the historian Thomas P. Hughes called "technological momentum," by which a technological system is shaped by society until it shapes society in return. Seattle had already invested heavily in regrading as an instrument for urban redevelopment; municipal engineers now had a stake in expanding their influence within city government; and a civic culture had emerged to justify and promote regrading as the preferred solution to long-standing social and environmental problems. By the 1920s, many businesses and absentee property owners saw regrading as a means of expelling unwanted tenants or neighbors. A group called the Associated Central Business Properties petitioned the city council to dispense with official notification and condemnation proceedings in the proposed Yesler Hill regrade zone. R. H. Thomson, who, at the age of seventy-four had returned to head the Engineering Department temporarily from 1930 to 1931, proposed the most radical regrade, removing more than 2.7 million cubic yards of earth, almost as much as the original Jackson Street project alone.[10]

Thomson's final regrade was to remain forever imaginary. The Yesler Hill project never left the drawing room at the Engineering Department's headquarters. The *Seattle Times,* which had supported the second Denny Hill regrade, came out against it: "Very few people . . . believe that Yesler Hill should be entirely removed or radically regraded." The *Times* had good reason to oppose still more regrades. Read Bain, writing for the *New Republic,* observed that "mastering the natural environment" at the expense of the "defenseless taxpayer" had left Seattle "nearly insolvent." In October 1929, as the nation began to tumble into the Great Depression, the conveyor belts and steam shovels on Denny Hill continued their labors as legal disputes and inclement weather extended the project into late 1931. By then, the depression had killed what remained of the unsteady local commercial real estate market, with property owners fleeing the Regrade District, site of the former Denny Hill, leaving empty lots and abandoned buildings. This time, however, the changing location of Seattle's business and industrial districts stopped regrading permanently. New and less expensive high-rise building technologies employing steel beams and concrete foundations reduced the need for level sites in the hilly city. Residents continued to move into newer bungalows and cottages in Seattle's northerly suburbs, depressing demand for apartments and older houses closer to the city core. Sluggish real estate sales intensified stagnation within the Regrade District itself. Warehouses, automobile lots, and taverns spread haphazardly along the now-flattened streets, and persisted for the next forty years.[11]

Seattle's regrading mania had expired at last.

A Class of People . . . More Vagrant and Radical

Just as the city's lower classes had been stranded in rundown neighborhoods or cornered in landslide-prone valleys, the dispossessed had also collected in Seattle's low-down waterfront places—filled tidelands, shorelines, gullies, and ravines. When the World War I boom collapsed into postwar bust, many fled to what remained of the former shoreline commons. Long before the bottom fell out of the American economy in the 1930s, prosperity had vanished in the Pacific Northwest, especially in rural regions, where sawmills, mines, and farms never fully recovered from the short but severe postwar depression. As the prospects for rural Pacific Northwesterners dimmed, many packed up and moved to the region's cities, which, taken as a whole, had grown an astonishing 254 percent from 1900 to 1920. Urban centers, like Seattle and Portland, fared better than their rural hinterlands, but affluence was relative. Many of the shipyards and factories along the Duwamish Waterway and Elliott Bay had folded after the armistice in 1919. The temporary gains made by organized labor after the general strike that same year evaporated as the more conservative American Federation of Labor unions retreated from radical politics and consolidated power over unruly Puget Sound locals. Seattle's lethargic industrial economy of the 1920s forced many underemployed to move along, looking for seasonal jobs in the region's already hard-pressed agricultural and timber industries.[12]

Decades of unregulated logging had taken their toll on the region, leaving large swaths of cutover land exposed to the elements and elevating the flood risk to those living downhill and downstream. Professional foresters like Hamilton C. Johnson, the ranger hired to supervise the city's Cedar River watershed, appalled at the waste, had warned in 1913 that if the city did not undertake a "systematic or organized effort" to reforest these lands, the watershed would "grow up to weeds and scattering growth of inferior timber having but little value." Regenerating forests could also yield other possible social benefits. Beginning at the turn of the century, back-to-the-land advocates saw the Pacific Northwest's cutover forests as an outlet for the restless and jobless poor and immigrants. Urban professionals, promoters of logged-off lands settlement, and timber companies—among others, Weyerhaeuser Timber Company, the Seattle Chamber of Commerce, and the state Department of Agriculture—saw a merger of profit and reform in their campaign to sell acres of stumps as acres of farmland. They had overestimated the fertility of the logged-over lands, choked with rotting wood and plagued by inadequate drainage. Even if settlers, many with little farming experience, could remove the massive stumps to plant their crops, heavy seasonal rains leached away what little nutrients remained, or washed precious topsoil downstream. By the Great Depression, most observers called the experiment a failure. On the logged-off lands, "social and environmental disaster mingled and became one," according to historian Richard White, and without sufficient financial or political support the poor suffered the costs while promoters reaped the rewards.[13]

Meanwhile, itinerant workers in Seattle began to increase in number by the 1920s, attracting the attention of the sociologist Roderick D. McKenzie, then a young assistant professor at the University of Washington. McKenzie observed what city health officials had noticed in 1915: that segregation was as much spatial as historical, vertical as horizontal. The "settler type of population," he wrote in 1924, who were often "the married couples with children," had fled the city center while "the more mobile and less responsible adults" tended to "herd together in the hotel and apartment regions near the heart." "It is in the Seattle neighborhoods, especially those on the hill-tops," he concluded, "that the conservative, law-abiding, civic-minded population elements dwell. The downtown section and the valleys, which are usually industrial sites, are populated by a class of people who are not only more mobile but whose mores and attitudes, as tested by voting habits, are more vagrant and radical."[14]

McKenzie had earned his doctorate at the University of Chicago, where he was a student of Robert Ezra Park, who, with Ernest W. Burgess, had founded the "Chicago school" of urban sociology. For Park and Burgess, a city was "a laboratory or clinic in which human nature and social processes may be conveniently and profitably studied." They drew from the nascent science of ecology to explain how many of the forces that shaped natural ecosystems, such as Darwinian evolution and competition over limited resources, also governed cities and their populations. As groups fought for access to and control of resources, they occupied distinctive niches in the city, what Park and Burgess called "natural areas," housing people of similar social, racial, or political characteristics. Those with power and money moved away from the city core or up the hillsides; the poor and powerless stayed behind in a process of "succession," an ecological term for structural changes in plant and animal communities through time. In Seattle, McKenzie found a city, like Chicago, full of rivalries: between immigrants and native-born citizens, between seasonal and permanent residents. Successive "invasions" of migrants, usually non-whites, tended "to be reflected in changes in land value." As an invasion took hold, "the value of the land generally advances and the value of the building declines," furnishing "the basis for disorganization": increased vice "under the surveillance of the police" and sometimes "outward clashes." McKenzie left unstated what shape those clashes sometimes took.[15]

One clash was over where people could and could not live because "Seattle, like other cities, has an explicit policy of segregation," wrote Katherine I. Grant Pankey, in 1947, and "a potent weapon with which threatened 'invasions' have been stopped is the use of the restrictive covenant." During the 1920s, racial restrictions were commonplace in real estate deeds and covenants throughout the nation, and Seattle, although still predominantly white, was no exception. Many "married couples with children" that had fled the city core to settle in new housing developments atop Seattle's hills or in neighborhoods north of Lake Union were required by developers to sign deeds that forbade purchasers and their assigns from selling their property to Jews, blacks, Asians, or other non-white buyers. A University of Washington graduate student in sociology discovered that Seattle's central residential district (bounded by Marion Street to the north, Dearborn Street to the south, and Lake Washington and Elliott Bay to the east and west) was "a truly cosmopolitan area" with "minority groups living in a Gentile white setting." The cosmopolitan district had been one of the few areas of the city where minorities could purchase or rent homes without restriction, yet, even here, affluence and whiteness ran uphill. In the valleys and lower slopes were the "Suki Yaki restaurants of the Japanese; the family houses, chop suey restaurants, and language schools of the Chinese; the restricted dance halls, pool halls and rooming houses of the Filipinos; the Jackson and Cherry Street communities of the Negroes, and the cheapest and most overcrowded housing." On the higher elevations, closer to Lake Washington, were "Ashkenazic and Sephardic Jews" with better views and better housing conditions.[16]

A class line in Seattle was harder to define. According to the *Monthly Labor Review* in 1926, the arrival of "poverty-stricken 'tin-Lizzie tourists,'" a euphemism for low-class migrant workers traveling in beat-up old Model T's, posed "a grave social menace" to Seattle because they would "work more cheaply than local wage earners" and their children would put a strain on local schools. John Herbert Geoghegan, a University of Washington sociology student, fused moral outrage and scientific observation in his study of migratory workers in Seattle. He drew upon the earlier work of fellow sociologist Carleton H. Parker, who had investigated the Wheatland Hopfield Riot of 1913, where farmworkers rebelled against brutal working conditions and low pay in one of the most violent events in California's labor history. In Seattle, Geoghegan interviewed nearly four hundred men, although only forty agreed to complete his survey, and found conditions equally dire: migrant workers harassed by employment agencies, inferior housing, and discrimination by race and politics, day laborers with limited savings and almost no access to credit or any social support. About 86 percent of all migratory workers in the city were unskilled and "excellent material for revolutionary propaganda."[17]

Geoghegan might have been describing a man like Fred Strom. Arriving from his native Sweden in 1913 at twenty-one, Strom first lived in a boardinghouse on Profanity Hill and worked as a bricklayer. One day, coming to work late after an all-night binge on Skid Row, he lost his job. He toiled for a time in the forests as a lumberjack and joined the Wobblies. A few years later, as a roving sawmill operator, he aroused the ire of bosses hostile to his union activism. Strom was never a hardcore unionist, but he was a faithful picket-line walker; he participated in the 1919 general strike. By the mid-1920s, Strom moved into a houseboat on the Duwamish River, hopping from job to job, living alongside Indians, Scandinavians, Filipinos, Japanese, Italians, Greeks, and Slavs, who built shacks and houseboats from lumber discarded by nearby mills or flotsam scavenged from the river current. It was a rowdy community, overwhelmingly male, populated by fishermen and loggers, the unemployed and more than a few bootleggers who, during Prohibition, sneaked Canadian liquor into Seattle by night. In 1935, Strom saved enough money to buy a slip on the western bank of the river, built another houseboat, and stayed put despite repeated attempts by the Port of Seattle to expel him. As men like Strom made the waterfront their home, the middle and upper classes that had become literally Seattle's "upper" crust, from atop ridges and hills, looked down at the pollution, ugliness, and disorder beneath them.[18]

Upland property owners had long considered houseboats nuisances, whether they were bohemian retreats or blue-collar shelters, middle-class summer homes or working-class dwellings. Meanwhile, the poor clustered in the empty and damaged places—the city's valleys, ravines, and tidelands that had withstood the brunt of the city's earthmoving and water shifting. Zoning laws and housing preferences reinforced the geography of poverty as commercial and industrial development concentrated what Gordon Beebe kindly termed "blighted residences" in areas with the lowest land values and the most pollution. Although Seattle had diverted much of its waste to Elliott Bay by 1910, several storm sewer outfalls continued to dump effluent directly into Lake Union and Lake Washington. City health inspectors blamed houseboats because they remained unconnected to the sewer system and few, if any, had on-board septic tanks. In 1922, H. M. Read, the city health commissioner, charged the more than a thousand floating homes with turning Lake Union into a "virtual cesspool" and fouling bathing beaches on Lake Washington. He ordered them removed. Houseboaters in Lake Union and Lake Washington formed the Houseboat and Home Protective League and successfully thwarted outright condemnation. Three years later, when landed homeowners renewed their attack, the league outflanked them once again.[19]

During the 1920s and 1930s, conflicts over balancing environmental protection and social advancement puzzled many engineers and planners, including the highly political engineers who longed to build a new and progressive society. Herbert Hoover, the Stanford-educated engineer who had masterminded the feeding of millions of war-ravaged Belgians and would later claim, contrary to the historical evidence, to have rescued the flood-torn Mississippi River valley after the devastating 1927 deluge, was one such idealist. In 1924, discussing plans to build dams on the Colorado River, he suggested that "some waterfalls" were "in the wrong place, where few people can see them," but with a little technological tweaking, better waterfalls "could be constructed with a view to their public availability as scenery." "Human intelligence added to the resources of nature," he said, was an unstoppable force for good and beauty. Elected president four years later, Hoover, the acme of American engineering prowess, argued that technocratic elites could and should merge efficiency and beauty with equity.[20]

Big-scale planning was the order of the day, well before Franklin Delano Roosevelt's famous New Deal projects remade the Far West and the Deep South, and the engineering frame of mind was shared by thinkers on the left as well as on the right. The Left in Seattle was not monolithic; some, active in the public power crusade, which in the Pacific Northwest took on the cast of a secular religion, opposed private utilities like gigantic Stone and Webster, based in Boston, owner of Seattle's municipal electric railway and nemesis of Seattle's publicly owned City Light. Others joined the Technocracy movement, which in 1932 emerged with a plan to solve the depression through radical social engineering. In Seattle, political power slipped away from the clique of businessmen and professionals that had long influenced city affairs and toward reform, focused on what historians have called "municipal housekeeping," an extension of women's domestic sphere into the bastions of male-dominated political life. Washington's women had won the right to vote in 1910, nine years before the Nineteenth Amendment made it the law of the nation, and in Seattle they flexed their muscle to recall, in 1912, the corrupt mayor Hiram Gill. Bertha Knight Landes, a member of the Seattle Women's Century Club, helped to remove Gill from office. With the support of her husband, Henry Landes, a professor of geology at the University of Washington and an ardent suffragist, she won election to the city council in 1922 and, four years later, was elected mayor, the first female chief executive of a major American city. Governing the city, in Landes's words, was akin to caring for "the larger home."[21]

In many ways, Landes's vision of seeing Seattle as "the larger home" was an ecological vision, one that differed from Roderick McKenzie's bleak assessment of the city as combat zone. It had its antecedents in Hull House, where Jane Addams and her settlement house workers fought to improve unsanitary conditions and provide unmet services for Chicago's poor, or the work of Ellen Swallow Richards, the Massachusetts Institute of Technology professor who created the discipline of home economics by merging her research in sanitary engineering with the new science of ecology. How much Swallow and Addams influenced Landes is unclear, but one thing is certain: as mayor, Landes wanted to clean house, literally. Moving beyond the Victorian concept of separate spheres, she did not reject but sought instead to co-opt the engineers and other experts, and then redirect their impulse for reform toward social equity. Landes made the City Planning Commission that had criticized regrading permanent, enacted revisions to building and zoning codes, reformed the civil service system, expanded the number and acreage of city parks, and initiated the first sustained efforts to soften Seattle's persistent cycles of underemployment. She took into account the entire urban environment, from the ground up. But, in 1928, after only a two-year term, Landes lost in a rout to Frank Edwards, a lackluster theater operator cajoled into running by downtown business interests. Despite endorsements from all three of the city's dailies—the *Times*, the *Post-Intelligencer*, and the *Star*—Landes could not outflank Edwards's relentlessly negative campaign mocking her "petticoat politics." But the end of the Landes interregnum did not stop changes already under way in Seattle; it only redirected their effects. The problem for Seattle's less powerful was that without someone as fair as Landes, the fault for environmental degradation and social decay would fall on them.[22]

Some of the club women who had been allies or silent supporters of the Landes administration, like Ella B. Rowntree, a resident of the upscale Mount Baker Park neighborhood, blamed "tin-Lizzie tourists" for destroying the "pure beauty" of the "natural woodland, flowers, and ferns" along Lake Washington Parkway. "We, who own property surrounding this Park must build and maintain homes of a certain standard," she carped, "in order that this district shall be a desirable and attractive place in which to live." Sigrid M. Frisch, who lived atop Queen Anne Hill, said motorists were dumping "mattresses, bedding, clothing; parts of autos, furnaces, heating leads and packing materials . . . by the auto-loadful" in the ravines below Queen Anne Boulevard. The auto camp in Woodland Park, which had opened to applause at the beginning of the 1920s, had become a haven for these automotive vandals. A. S. Kerry, the park board's president, told the city council: "No more disgusting spectacle was ever exhibited in any city in America" than "non-resident campers whose primitive, unsanitary living conditions" rivaled the nation's worst slums. He closed the camp in 1929, citing poor attendance and angry neighbors. Roads were bringing the penniless and the homeless to Seattle, and many of them sought refuge in the city's parks and greenswards. In a 1928 campaign flyer for a proposed Metropolitan Park District, which would expand and merge the city's park system with parks throughout King County, Kerry warned voters to remember the true purpose of scenic greenswards: "We must not have the unsightly back yard features, such as camping places, swamps, tin cans and wash tubs placed in parks in the city or in parks close enough to the city so as to destroy the very thing that we are trying to accomplish."[23]

Many well-off homeowners supported an urban beauty reserved for them, but when changes in the land fouled the waters, sullied the view, and cheapened property, they fought back, and by the interwar decades the contours of class conflict followed the physical lines engineers had inscribed in the landscape. In the principal conservation organizations, the heirs of the municipal housekeeping movement turned to removing the riffraff from the city's public spaces to protect parks and greenswards for flora, fauna, and respectable citizens. Even the most well-intentioned reforms had rebounded to hurt those least able to resist. Discrimination and segregation in Seattle, as in any city, was not a natural process, as Roderick McKenzie had claimed, but an economic and political process. Sometimes, the injuries were intentional; sometimes they boomeranged to hurt those with the power. Non-human nature did not discriminate. Human nature did.

Scrap-heaps of Cast-off Men

During the winter of 1929–30, Allen Roy Potter, a Seattle social worker, spent his time walking up and down Skid Row, standing in the soup lines, huddling in the cold rain, and interviewing men in the flophouses. Gone were the noisy lumber mills and card parlors of the old Skid Road. The renamed Skid Row, stripped of its lumbering past, was now a moribund neighborhood of residential hotels, seedy eateries, relief missions, and pool halls. Most Seattleites blamed the denizens of Skid Row "for not saving enough from their seasonal work in the lumber, fishing and road construction work." Four years later, when the depression worsened, Potter had found that the typical victim was a native white male, forty-two years old, probably single, and unemployed for more than ten months. Skid Row was no longer the home to the chronic down-and-out. It was the changing face of Seattle itself.[24]

The severity of the depression surprised Mayor Frank Edwards, the one-time theater operator, and like so many politicians at the time, he counted on voluntary measures to combat unemployment. Fearful of "Communists" who "seek to embarrass and even destroy our government and its institutions,"

Edwards, in 1931, called on city employees to contribute to "the Star Sunshine Club" and "the Community Fund" to provide relief. He was recalled later that year by a public outraged over his firing of City Light superintendent J. D. Ross in a pique over public utility ownership. The mayor's successors turned to hiring the unemployed on public works projects. The Park Department expanded its long-standing welfare program of using the unemployed to repair park facilities and cut brush and trees. "Seattle's parks have truly been a godsend to the 'Unemployed' during these past extremely trying years," the Park Board stated in its 1934 report, "and have provided practically the only recreation possible to hundreds of people." The Unemployed Citizens League, founded in 1931 by leftist unionists and political activists, protested that city hiring policies were discriminatory, little better than the private employment agencies that had preyed upon migrant workers during the 1920s. In a petition in 1935, the Downtown Local Unemployed Council, frustrated that city officials had coerced the unemployed to work for no more than board and clothes, declared, "we will perform no work for less than Union wages, paid in cash." As the federal government began to assume more responsibility for welfare after the election of Franklin Roosevelt, the federal Civilian Conservation Corps (CCC) and Works Progress Administration (WPA) worked closely with the Park Board and the Washington Emergency Relief Administration, a state agency, to administer relief projects. The CCC maintained camps at Carkeek and Denny Parks to train workers for park maintenance and trail work in the national forests and parks around western Washington.[25]

Don G. Abel, the WPA's state administrator, like the advocates who touted the virtues of poor people settling logged-off lands, believed in the restorative powers of outdoor living and vigorous physical work for human bodies and ravaged landscapes alike. "Old skills have been restored through the opportunity to use them," he wrote to the city's Department of Engineering, and "many workers have returned to private employment, fit again or newly fit for better paid jobs." Park foresters and gardeners realized that stands of aging trees in city parks, the tatters of the original forest, required sustained maintenance to keep them safe and beautiful, but those living next to the parks complained that the jobless were like insects, devouring the woods and leaving their waste behind them. Mrs. Alexander F. McEwan, president of the Washington State Society for the Conservation of Wildflowers and Native Trees, said that Seward Park, with one of Seattle's oldest groves, might have retained "its rural beauty as an example of our native forest" but for logging. In response to such protests, officials limited timber removal and restricted families on relief to one cord each for winter fuel.[26]

Those unable or unwilling to work, unwanted in parks and feared by city leaders, looked to the old watershed commons, home to Indians, migrant workers, and indigent immigrants, as a place to live and survive. On Lake Union, jerry-rigged tankers supplied heating oil to the growing houseboat population, lumbermen sold sawmill debris for stoves, and fishermen in runabouts peddled their catch from slip to slip. But, in 1936, after city engineers diverted the last of the untreated sewage outfalls from Lake Washington, health inspectors tried again to expel houseboats from the city's lakes. By 1938, they had succeeded on Lake Washington, save for one wealthy enclave in Union Bay. Efforts on Lake Union were less successful. Residents in the Montlake neighborhood formed the Portage Bay Improvement Club in 1938 to combat the lake's "houseboat menace," but lake dwellers formed the Waterfront Improvement Club and successfully maintained that sewer outfalls, still not diverted into Elliott Bay, and marine industries were responsible for the bulk of the pollution.[27]

On the tidelands south of downtown, near the earlier battleground of the first white invaders and Native defenders, Seattle's engineered landscape had become a sour joke. "Seattle Tidelands are moving very slowly as industrial property," noted L. C. Gilman, a local agent for the Great Northern Railroad, in 1934. In place of the absent industry, new "shack towns" moved in: clapboard houses on Shilshole Bay, alleys of huts next to the Great Northern yards at Smith's Cove in the Interbay district, and driftwood lean-tos on West Seattle's beaches. Some shack towns adopted names of higher-class subdivisions and apartments on the hills above, or of other cities. Hollywood and Reno, at the foot of Lander Street, were near the switching yards for the Great Northern Railroad; Louisville was squeezed between the quays and wharves of Harbor Island; and Indian Town, a shack village of long standing, was near the base of Yesler Street, adjacent to the remnants of Ballast Island, at the foot of the lumberman's original mill.[28]

But the largest shack town was "Hooverville," built on the site of the old Skinner and Eddy Shipyard along the East Waterway of the Duwamish River. This new "hobo jungle" emerged in late 1929 to house what Donald Francis Roy, a University of Washington graduate student in sociology, called a "scrap-heap of cast-off men" and "a junk-yard for human junk." In December 1930, the chief

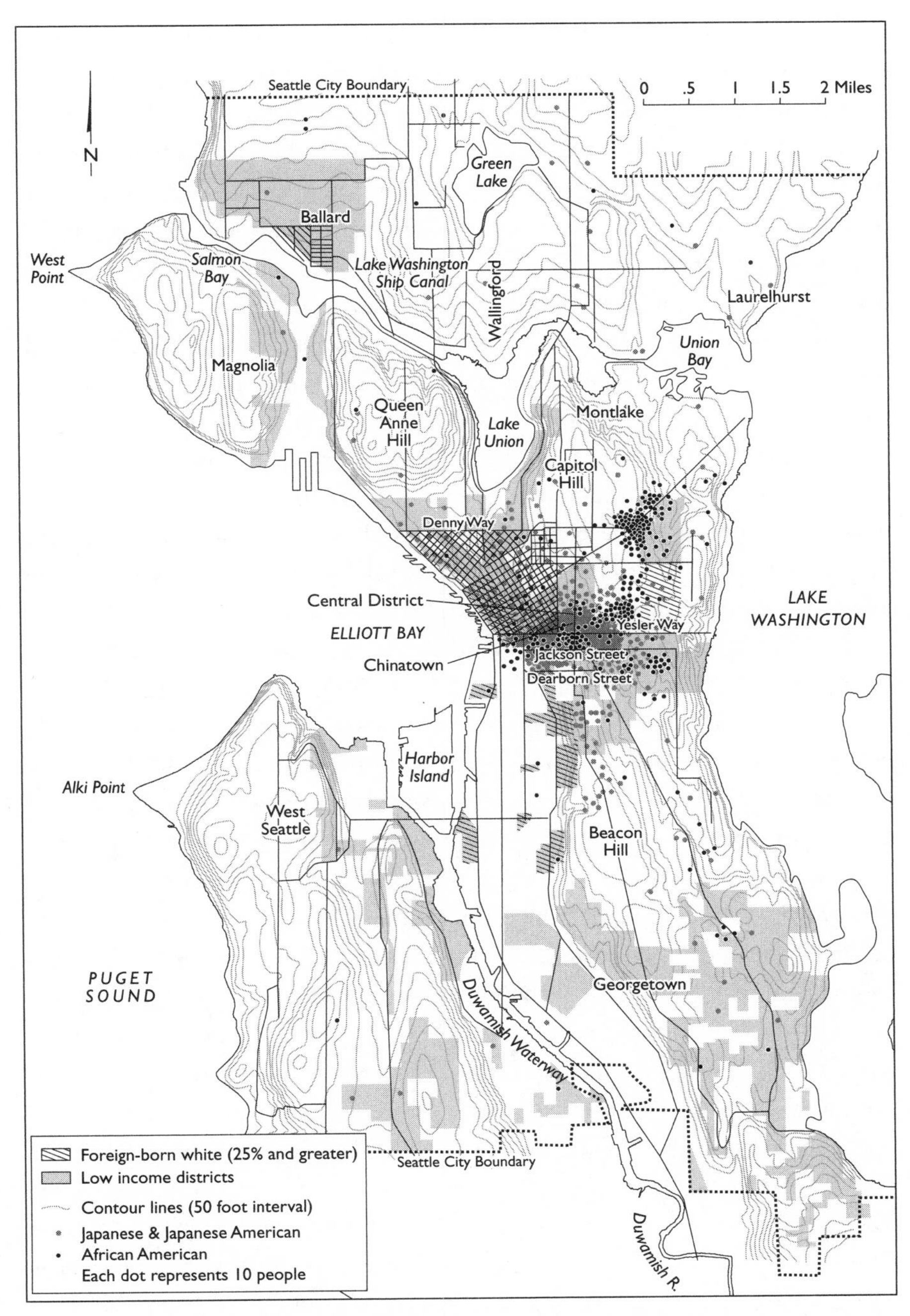

Low-income and minority neighborhoods in Seattle by topography, c. 1940. (Adapted from Calvin Schmid, *Social Trends in Seattle* [Seattle: University of Washington Press, 1944])

of police warned squatters that he would not welcome "drifting criminals who flock to Seattle and commit crime in order to live." Jesse Jackson, an unemployed lumberjack, recalled how, in autumn 1931, "Seattle Health Officials decided our shacks were unfit for human habitation" and gave the squatters seven days to vacate. "At the expiration of seven days notice, at 5 a.m., just as daylight was breaking, in one of the heaviest downpours of rain that fell in Seattle that fall, a regiment of uniformed officers of the law and order swooped down on us with cans of kerosene, and applied the torch." The police and health inspectors set fire to the shacks again the following month; when the squatters returned, they dug in and built fireproof roofs of tin or scrap metal. The next summer, the city relented and let the squatters stay.[29]

"For the first time in many years," the Department of Health and Sanitation reported in 1935, "Seattle has been forced to allow the growth of shack towns." Department inspectors estimated the shanty population between four thousand and five thousand, located "mostly along the waterfront and on our filled-in tidelands." When the downtown local of the National Unemployment Council called on the city to open community bathhouses for shack dwellers, the city health commissioner, Frank M. Carroll, shot back, "Hooverville is all wrong, and should be eliminated," advising the city council to let the "population of Hooverville" apply to the federal government for relief instead.[30]

Roy, the sociology graduate student, visited what he had called a "junk-yard for human junk," living almost two months among the "Hooverites" to collect data for his master's thesis. "The name is deceiving," he wrote; Hooverville was "a segregated residential area of distinct physical structure, population composition, and social behavior." More than five hundred shanties, crisscrossed by paths more akin to "animal trails," clustered on the shores of Elliott Bay in "open, noisy disregard of carefully draughted graphs and diagrams." Yet it was a tangible expression of "the critical state of a bed-ridden economic system," protest made real in tin and tar paper.[31]

Jesse Jackson, charismatic and articulate, became Hooverville's de facto mayor. He and his "town council" suppressed vigilance committees, expelled troublemakers, and arranged for relief aid and sanitary inspections. Hooverville was not an ideal hobo democracy—as in the city that surrounded it, native-born whites were the predominant group—but residents did share power among themselves. The district's ad hoc governance committee seated "two Whites, two Negroes, and two Filipinos" at every meeting and, according to Roy, handled their own affairs in a remarkably egalitarian fashion. Despite their poverty, Hooverites took pride in independence. "Hooverville is a colony of industrious men," Jackson explained in 1935, "the most of whom are busy trying to hold up their heads and be self-supporting and respectable. . . . [T]he lazy man does not tarry long in this place." Most Hooverites were "unemployed lumberjacks, fishermen, miners and seamen," or migrant farmworkers from as far away as Montana and the Dakotas. A former logger, Jackson declared he would "feel much happier if he was out in the woods, standing upon the running board of a donkey engine listening to the signals and working the levers, yarding logs to the landing." "If former President Hoover could walk through the little shanty addition," he added, "he would find that it is not inhabited by a bunch of 'ne'er do wells' but by one thousand men who are bending every effort to beat back and regain their place in our social system that once was theirs."[32]

They built houses from castoffs—discarded lumber, fruit crates, bailing wire, tin sheets. Jackson's desk was an old ice tank. Residents caught fish for nearby markets, hauled junk for salvage, and collected driftwood for fuel companies. But during heavy rains, water pooled in huge puddles, and the high tides of Elliott Bay during winter storms flooded many shacks. "The sea appears to be eternally licking its chops," Roy wrote, "in anticipation of swallowing the entire community in one juicy gulp." The salty soil made gardening unfeasible, and collecting shellfish on the hardened banks of the East Waterway was impossible. Toilets stood at the end of rickety footbridges strung over the water, and ocean currents often returned the waste to the shacks. There were only two taps in the entire district to supply clean water for cooking, cleaning, and drinking. There was no electricity at all. The grounds were, largely, clean and orderly, but Norway rats scurried beneath floorboards and bedbugs hid inside mattresses.[33]

In search of efficiency, Seattle's engineers and developers had helped to create a place without clean water, without sanitation, without steady work, without utilities—in short, without any of the rationalized urban infrastructure that the Progressive project was supposed to secure. Hooverville mocked the engineering hubris that Hoover symbolized and that had created the filled tidelands and the city's waterfront. Engineers like Hoover and Thomson had argued that remaking landscape was an instrument to remake society, and many Seattleites had agreed with them. Hooverville laid bare the poverty of that philosophy.

As the Second World War came, Seattle was not a true technocracy, in the classic formulation of Howard Scott, the California engineer who believed that scientific management could eliminate poverty, but it was a city in which engineers continued to have great influence. They still commanded and staffed the city's major utilities, from City Light to the Department of Water, and the region's largest employers, from the timber and shipbuilding industries, plus now aeronautics, at the Boeing Airplane Company. From the vantage point of the experts and officials who ran the city and its wartime economy, the national emergency provided cover to oust the unwanted Hooverville squatters. Complaints against shack towns had mounted during the depression years. "We recognize the fact," Mary Gamble Young of the North End Improvement Club wrote to the city council, in 1937, "that no funds are available for other housing. That it is unlawful to shoot or drown them. But—we want you to do something about it." By the time of the Pearl Harbor attack in December 1941, plans were already in place to remove the shacks because residential construction in Seattle had slowed during the depression, setting the stage for an acute housing shortage. Desperate to make room for workers and industry, city leaders turned to a 1939–40 WPA survey, which found that out of 86,086 "residential structures" in the city, 28.5 percent were "substandard" and 17,437 had no proper sanitary facilities. A "shack elimination committee," formed in early 1941 and organized by the new Seattle Housing Authority, recommended destroying the shack towns. The Port of Seattle wanted the waterfront cleared for industry, and the United States Coast Guard was eyeing the docks adjacent to Hooverville for a new slip.[34]

In April 1942, the Seattle Housing Authority, citing the wartime emergency, decided to torch the city's shanties. Hooverville went up in flames and its inhabitants were evicted. Some may have moved uphill. The Seattle Housing Authority's mission was to provide safe and sanitary shelter for the poor. Indeed, Seattle was one of few American cities at the time to build high-quality, publicly funded accommodations at a price the poor could afford and without regard for race or national origin. The first director of the housing authority's Board of Governors, Jesse Epstein, had conceived and directed for this purpose construction of Yesler Terrace atop Profanity Hill, backdrop to Hooverville and the muddy tide flats below. The Seattle Housing Authority's first annual report stated that "building public housing on the summit of Seattle's most famous hill, an advantageous position," would convert "a blighted district such as this . . . socially and economically injurious to the community's welfare" into something "desirable from a residential standpoint." With assistance from the United States Housing Authority, "Seattle shall soon see one of its most ugly and disreputable areas transformed into a carefully designed and constructed residential district." "In place of blight," the Seattle Housing Authority explained, would be "substituted 600 homes for low-income families," of varying sizes, from apartments to ranch-style bungalows, with recreational and community facilities, ensuring "a wholesomeness of living." After completing Yesler Terrace in 1941, Epstein secured federal funds to build similar housing projects in other poor districts: Holly Park on South Beacon Hill, and Duwamish Bend in industrial Georgetown. One of the nation's first integrated public housing projects, Yesler Terrace afforded some of the best views of the city, but it was not an amenity. The new housing project, far removed from Seattle's better neighborhoods at the time, was out of sight, just as Hooverville was beginning to fade from public memory.[35]

Roy, the University of Washington sociologist, had anticipated this collective amnesia. In his 1935 master's thesis, he invented an imaginary contest to select America's "Unknown Rugged Individual" for a "future depression memorial." He described the typical Hooverite as a northern European immigrant, likely Scandinavian, in his late forties, a man who worked intermittently in the logging or mining camps: "Jobless, propertyless, familyless, and savings spent, he came to Hooverville in the fall of 1931 to make that community his home." This "Mr. Hooverville" represented a bygone era of fishing dories and lumber mills. Seattle's new working classes would build aircraft and load container ships in spaces where tin-roofed shacks once stood.[36]

Most Seattleites preferred to remember the past in rosier hues. In December 1942, a full-color map and article in the *Seattle Times,* prepared for "the benefit of newcomers who find it difficult to believe that the plants where they are employed now" had once been "out in the bay," detailed how the city had "pushed back the sea." Factories and warehouses lined the Duwamish River valley from Elliott Bay to the end of the waterway near the old Black River confluence, and piers servicing warships and merchant marine freighters covered the unkempt tidelands ringing Elliott Bay. "When the Siwash Indians roamed Seattle," according to a legend sidebar alongside the map, "waves lapped at the foot of Beacon Hill." Even those trained to analyze the hidden dimensions of class and race overlooked how changes in the land connected to and directed changes in Seattle's social fabric. Calvin Schmid, a sociology professor at the University of Washington trained in the Chicago school believed, like his colleague Roderick D. McKenzie, in the concept of human ecology to explain

Seattle's downtown and central business district showing residential and social patterns. (From Calvin Schmid, *Social Trends in Seattle* [Seattle: University of Washington Press, 1944], 295, Special Collections, University of Washington Libraries, UW 23843z)

Seattle's geography as a wholly "natural process." In a study from 1944, *Social Trends in Seattle,* Schmid described the city's socially unstable downtown and waterfront neighborhoods—Skid Row, the Regrade District, Jackson Street, Profanity Hill—as "the result of the natural sifting and sorting processes of competition, conflict, mobility, and differentiation, . . . the pressure of public opinion, internal cohesion, and relatively low economic status." Schmid's map was an atlas of determinative inequality. Absent from his analysis was any historical sense of how this new geography was anything but natural and how different it might have been.[37]

Only Yesler Terrace, the housing project conceived in the spirit of charity, built on top of one of the few hills not occupied by the city's wealthier residents, suggested that McKenzie's "vagrant and radical" classes were not fated by nature to live in the depths and shadows. Meanwhile, many of Seattle's "conservative, law-abiding, civic-minded" citizens, as McKenzie had described them in 1924, had begun to move even farther away from the city and into exclusive neighborhoods like Innis Arden, a subdivision adjacent to the small town of Edmonds, just beyond Seattle's northern city limits. Some followed the failed paths the back-to-the-land movement had blazed decades before, this time with more capital and more support. On the shores of Puget Sound, the Boeing Aircraft Company, which had bought logged-over land from the Puget Mill Company as an investment, cleared away the stumps and, in 1940, had started subdividing land for homes, registering restrictive covenants for prospective homebuyers the following year. According to the standard title document for Innis Arden, "any person or persons not of the White or Caucasian race" could not buy lots or homes, and occupancy was restricted to homeowners and their domestic servants.[38]

Remaking landscapes in Seattle before and during the interwar years had sharpened, not softened, the distinctions between Seattleites. Engineers and planners, politicians and health inspectors, tried to broker the subsequent conflicts over the effects of changing landscapes on the city's social fabric. Usually, they failed to see completely their own complicity in the problems. Power always framed who would benefit, who would enjoy the fruits of the engineers' labors, and who would do neither. The more affluent and more powerful classes and groups—who lived next to the parks and atop the hills, who ran the industries and had the political influence—dictated the shape of the city. Their own concerns, focused on such issues as aesthetics or efficiency, became putatively public and environmental concerns. As a result, some Seattleites began to see their political or economic disputes in terms of humans versus a supplicant or recalcitrant nature and not in terms of humans versus humans.

But not all Seattleites did. Despite the limits of Roderick McKenzie's ecological metaphors or the blind spots on the maps drawn by his colleague Calvin Schmid, the sociologists and social workers who walked the alleys of Skid Row or shared meals with the residents of Hooverville made the correlation. Some, like Donald Roy, infused their scholarly studies with principled positions on the consequences of a political economy based on perpetual urban renewal. Others, who had taken a dim view toward the city's poor and dispossessed, also perceived an association between poor people and poor lands. In their analysis, however, it was the impoverished who had bankrupted Seattle's scenic places, not the other way around. In these conflicts was an ethic of place, nascent but incomplete, that connected the exercise of human power to an unpredictable nature.

Nonetheless, as Seattle's leaders and residents would discover, this new urban environment was a palimpsest of exploitation, conflict, compromise, adaptation, and defeat. Physical forces and creatures beyond human control always pushed back. So, too, did the people who suffered from the changes. The new urban ecology was never the result of purely natural forces but the combination of human power magnified or thwarted by an unpredictable physical environment. The non-human environment that enfolded the city was not predetermined, nor was the poverty that the decades of shaping and reshaping Seattle had aggravated. In the end, the ecology of urban poverty was altogether a human creation.

CHAPTER 7

Death for a Tired Old River

Ecological Restoration and Environmental Inequity in Postwar Seattle

In September 1949, *Seattle Business* asked readers to identify the city's largest river, the Duwamish, from an aerial photograph. The editors assumed that readers could not find the original watercourse once "marked only by the paths made by Indians as they moccasined their ways to fish for salmon." Instead, the former Duwamish was now an industrialized waterway where Boeing Aircraft's enormous Plant Number 2 hatched "mechanical birds" along a "Golden Shore" of related manufacturers, among them Bethlehem Steel, Kenworth Motor Truck, Monsanto, Isaacson Iron Works, and Cascade Gasket. The same year, students at Cleveland High School, in the blue-collar Georgetown neighborhood just down the river from Boeing, traced this transformation in the *Duwamish Diary*. Told from the perspective of the river, the story explained how the Duwamish had been resurrected from a "careless, meandering stream to an important waterway . . . that the gods of progress might be satisfied." The grateful river praised its guardians as "moderns" whose "eyes have ever been on the future."[1]

But only a decade later, Don S. Johnson, president of the Washington Pollution Abatement Council, a conservation group dedicated to sport fishing, protested in a letter to the *Seattle Times* that the Duwamish had become a dumping ground. The untreated sewage flowing into the river had previously been draining into Lake Washington, which was on its way to becoming an ecological wasteland. Alarmed voters, after two contentious elections in 1958, authorized a new regional government agency, the Municipality of Metropolitan Seattle, known as Metro, to protect the endangered lake. Metro's solution was expedient, simplistic, and in tune with the philosophy of the time: remove the waste from Lake Washington and put it into the Duwamish. Johnson, horrified, had predicted nothing but "death for a tired old river." Metro's executive director and chief engineer, Harold E. Miller, replied that the Duwamish was the perfect outlet for sewage; once treated, the water would be "clear, well oxidized, stable liquid" with enough dissolved oxygen "to maintain fish life." But the sacrifice of the river had begun and the choice was for the lake that supplied leisure and scenery to many comfortable Seattleites and suburban neighbors. It had not occurred to Seattle, its affluent suburbs, or Metro to save both at first and thus serve larger principles of environmental and social equity.[2]

In the years after World War II, many Seattleites, like their counterparts across the nation, came to fear that city and nature could no longer peaceably coexist. The stage was thus set for the modern environmental movement, a fundamental shift in thinking and policy. Before the war, Americans had tended to consider an unimproved environment threatening to social order, physical health, and economic development. After the war, in a reversal of this equation, environmental "improvement" usually meant degradation. Killer smog, mountains of garbage, and burning rivers, the new symbols of urban postwar America, drove more and more people to the suburbs in search of an Arcadian refuge. Modern environmentalism's start in America's cities was no accident. Some historians, like Andrew Hurley, have argued that the quest to preserve amenities like clean air and water had alienated some urbanites from others, and that race and class trumped any common effort to protect the environment for everyone. It was the middle class and the better-off working class, usually white, who decamped for the evergreen suburbs, and the less powerful, often poor, and non-white who were left behind to face industrial pollution and urban blight. Other scholars, like Samuel P. Hays and, more recently, Adam Rome, have suggested that suburban living exposed the newly affluent middle class itself to environmental decline. As tract housing gobbled up open space, septic tanks spilled effluents into groundwater, sullying the fresh promise of suburbia and arousing political activism. These two explanations are not at odds with each other but, like highways at the edge of town, they briefly intersect before parting.[3]

In Seattle, a small group of influential middle-class citizens banded together in the late 1950s with like-minded professionals from the largest King County suburbs to save and restore Lake Washington—and, in the process, they almost lost a river. Blaming unchecked urban growth for the pollution, their solution was regional government, a seeming heresy in a city addicted to the fast buck and political independence. But Metro's founders were pro- not anti-progress. They fused the old faith in engineering and reform with a new faith in ecology and environmentalism. Quality of life need not mean the sacrifice of affluence or recreational amenities. They failed to anticipate not only the consequences of ruining a river but also the virulent opposition to anti-pollution efforts in the smaller, less-affluent suburbs around Lake Washington, where people feared government intrusion and resented urban intellectual elitism more than dirty water. In the clash between city and suburb, blue-collar and white-collar, the Duwamish River became the region's open drain, and its salmon began to die. The Indian peoples still fishing along its shores would be blamed for the death of the river.

The City of Tomorrow, a Cure for Suburbanitis

On April 21, 1950, with the opening of the Northgate Shopping Mall, one of the nation's first suburban shopping centers, the 2,500-space parking lot was filled to capacity, shoppers lured by the Bon Marché, Nordstrom, an Ernst Hardware, plus an A&P, Newberry's, a National Bank of Commerce branch, a medical-dental center, and a supervised play area. Rex Allison and Ben Erlichman, the two enterprising real estate developers who planned Northgate, and the mall's first president, James B. Douglas, erected a 52-foot faux totem pole at the mall's entrance, sponsored a Cadillac giveaway, and made the cover of *Life* magazine with a 212-foot Christmas tree trucked from the Cascade Mountains. As Douglas later claimed, "downtown [Seattle], with its inadequate parking and difficult traffic congestion, was no match for Northgate." By 1965, when Interstate 5 reached Northgate, Douglas had added twenty-five new stores, a movie theater, and a hospital, while more than doubling the size of the parking lot. Three years later, more than 50,000 cars drove every day to Northgate, now a bedroom community to Seattle with its own surrounding apartment buildings and offices. Across Lake Washington, another developer, even earlier, in 1946, had opened the Bellevue Square shopping center. Good jobs at Boeing plants in Renton, at the south end of Lake Washington, and Everett, some twenty miles north of Seattle, had lured suburban home buyers, but the largest growth was in unincorporated King County, and the most dramatic gains were in the ring of smaller towns and new subdivisions surrounding Lake Washington.[4]

In 1942, four incorporated towns, including Seattle, bordered the lake.

Twenty years later, that number had more than doubled on the eastern shore alone. In 1950, the Eastside population stood at nearly 33,000. By 1960, it had doubled. Developers like James Douglas and Kemper Freeman, Jr., owner of Bellevue Square, had converted acres and acres of former farmland and empty lots into row upon row of new homes, following the pattern pioneered by William Levitt, architect of the "Levittowns" of the postwar 1940s. The National Housing Act of 1934 provided low-interest loans and the G.I. Bill put cash in the hands of home-hungry veterans. In western Washington, where wartime shortages had compelled defense workers to live in houseboats, apartments, and hastily built public housing, those who stayed now sought permanent homes. The anticipated postwar recession quickly evaporated with Cold War defense spending. Heralded by Boeing's success, Seattle became the nexus of the Pacific Northwest's "military-metropolitan-industrial" complex, mirroring the Los Angeles–San Diego and San Francisco Bay areas. Homebuilders ran for office, and politician-developers worked with Boeing executives to create new lakeside subdivisions adjoining the towns of Renton and Bellevue. New roads, followed by the interstate highway system, generated still more development. The Mercer Island Floating Bridge, opened in 1940, connected Eastside towns to Seattle and did away with the slow and expensive commuter ferries.[5]

The new bridge across Lake Washington united Seattle and suburbs as never before, and, in August 1950, a group of business leaders joined to create Greater Seattle, Inc., a boosters' group. It invented a Seafair to celebrate Seattle's centennial in 1953, and the event soon became an annual salute to Seattle's maritime heritage, drawing tens of thousands to Lake Washington for hydroplane races, waterskiing championships, Boeing-sponsored aircraft shows, fishing contests, and water sports. Summer outdoor galas had been a Seattle tradition ever since an earlier generation of boosters first hosted the annual Golden Potlatch in 1911, named after the ceremony indigenous to the Native peoples of coastal British Columbia and southern Alaska. The longtime link to leisure denoted the good life. Boeing machinists and real estate agents trolled for fish in an annual Teamsters' Salmon Derby. Dockworkers, suburban mothers, and corporate attorneys set up summer clambakes to watch the boat races at Seafair. Gatherings like these gave the illusion of city and countryside blending, thus proving the theories of such urban thinkers as Lewis Mumford, who saw metropolitan growth emanating from the city center in concentric rings. Yet, as Mumford himself realized, cities never grew naturally, like ripples in a pond. Even planning alone was insufficient. Politics and economics, plus the physical environment itself, dictated where the waves would flow.[6]

Seattle's centennial in 1953 stirred a growing number of critics of the region's rapid growth. Among them, the most vocal and the most powerful was James R. Ellis, an idealistic attorney and native of Seattle. As a student at John Muir Elementary, he and his classmates had celebrated their school's namesake with a yearly festival, dressing up as wild animals, planting groves of trees in nearby parks, and holding debates on the merits of conservation. Ellis's father, an import-export broker to China, took the family on summer fishing trips to the Cascade Mountains and camping on the Pacific coast. One summer, he drove James and his brother, Robert, to the end of a logging road in the Cascades, left them alone with a knapsack of food and tools, and instructed them not to return until they had built a log house from scratch. The two teenage boys emerged at the end of the summer, their task completed, and the cabin became the family's vacation home. Ellis left Seattle for Yale University, but after Pearl Harbor, he enlisted in the army and served stateside. His brother, another army enlistee, served in Europe and later died at the Battle of the Bulge. In grief, Ellis and his new wife, Mary Lou Earling, moved into the cabin he had built with Robert and holed up like Henry David Thoreau, a childhood idol. There he found his path back into society: committed to serve the public good and protect the Pacific Northwest's open spaces.

After law school at the University of Washington, Ellis joined the venerable Seattle firm of Preston, Thorgrimson, and Horwitz and became active in politics, with the Municipal League of Seattle and King County; his wife, Mary Lou, was involved in the League of Women Voters. He helped lead the Municipal League's push to reform the King County charter, in 1952, and create "home rule" with an elected county commission. When the measure went down at the polls by nearly two to one, the league narrowed its campaign for regional government to essential services and targeted Seattle exclusively. In a speech to the Municipal League in 1953, Ellis outlined the agenda of regional or municipal government: to stem decay in the downtown business core, rein in suburban growth, cut through traffic jams, and streamline utilities and waste disposal. The league drafted a $5 million bond levy for the 1954 election to upgrade an overtaxed sewage system, but voters again defeated it.[7]

A greater Seattle had arrived, but it was not the "Greater Seattle" envisioned by Virgil Bogue, author of the failed 1911 *Plan of Seattle.* Residents in the Eastside towns, including many one-time inhabitants of Seattle, distrusted what they saw in the big city: political radicalism and corruption, high taxes, and social unrest. Many suburbanites worried that regional government would boost their property taxes. Seattleites, for their part, saw the suburbs as siphoning off taxpayers while failing to contribute to services used by commuters. Overlapping agencies responsible for sewage, transportation, and crime prevention in the suburbs created a jurisdictional jungle of services that city dwellers had come to take for granted. It was a political balkanization endemic to suburban America. One study claimed that King County had as many as 180 separate agencies, excluding school districts—more than any other county in the United States except Cook County, Illinois.[8]

Ellis urged the Municipal League of Seattle and King County to convene a special Metropolitan Problems Advisory Council, and two years later, in 1956, the council published its first report—*Seattle: The Shape We're In!* The report recounted the region's physical and social decline, fusing moral indignation with technological optimism in language reminiscent of the boldest Progressive Era decrees issued years before by the city engineer, R. H. Thomson. The league cast the region's problems in principled terms that reversed the image of the city as the center of evil and the suburbs as the wellspring of virtue. Growth had afflicted King County with, in the words of the report, an acute case of "suburbanitis," its symptoms being unchecked sprawl and inner-city rot. Seattle and its environs, the nation's seventeenth largest metropolis, was a "single metropolitan community" and required a metropolitan government, like Toronto's pioneering system. A follow-up study in 1956—*Seattle: The Shape . . . Of Things to Come!*—spelled it out: "The city of tomorrow can open new doors to a better way of life, or it can strangle in its own haphazard growth." Suburbs and outlying areas were growing at a rate twenty times greater than Seattle, and the new super-city faced common problems of unclean water and air, congested traffic, and insufficient parks. If Seattle and its suburbs made common cause, the region would be a source of pride, but if it "kept in the track of Topsy-like local government," its natural advantages would disappear forever.[9]

This "Topsy-like" growth had yielded, by 1955, some twenty-one independent sewer districts, representing almost forty communities, all clustered around Lake Washington with ten new disposal systems since 1941, dumping treated and untreated waste into the lake. Beyond the reach of municipal systems, real estate developers, eager to sell lots and homes, had built septic tanks to attract prospective buyers. Even lawns and household pets, two ubiquitous features of the suburb, were sources of pollution; phosphorous and nitrogen from fertilizers and pet droppings seeped into storm drains and aquifers. Yet pollution in the suburbs remained unnoticed because the Duwamish River, now the region's industrial sink, attracted attention first. The sport anglers who, in the 1910s and 1920s, had fought so viciously to evict commercial and subsistence fishing from the Duwamish had a new enemy, pollution. The state laws, passed at the beginning of the century, gave the Department of Fisheries and the Department of Game (the latter regulated inland fisheries) the power to approve all industrial permits, but prohibited them from regulating waste. But even if fish wardens could have levied fines and written citations, pollution had gone beyond discrete offenders.[10]

Seattle's combined sewer system, designed by city engineer Thomson in 1894, with storm water and sanitary lines running together, was now antiquated. The city had diverted the last of its thirty-three raw sewage outfalls from Lake Washington into Puget Sound by 1936, but the combined discharge still spewed filth whenever heavy rains inundated the system. The Duwamish River itself received no relief at all, and angry sportsmen pressured Governor Arthur Langlie to create the Washington State Pollution Control Commission. Understaffed and overworked, its investigators quickly discovered that aquatic pollution surrounded Seattle on all sides, especially along the now heavily industrialized Duwamish River, where high concentrations of creosote, oil, gasoline, calcium hydroxide, muriatic acid, chromates, cyanides, lead, mercury, and other heavy metals turned up. In West Seattle, after every downpour, storm surges gushed into the Duwamish and raw sewage spilled onto the beaches, forcing health officers in the late 1940s to ban swimming at Alki Point.[11]

When the Washington State Pollution Control Commission pleaded for a comprehensive resolution, Seattle asked a prominent sanitary engineer, Abel Wolman of the Johns Hopkins University, to study the city's sewage crisis. Wolman's findings in 1948 restated what his predecessor, George Waring, had said almost sixty years earlier—separate the storm water from the waste lines. E. F. Eldridge, the commission's director, considered the plan inadequate. Eldridge was a limnologist, a specialist in the study of lakes, trained by the pioneering biologists Arthur D. Hasler and Chauncey Juday at the University of Wisconsin, and he knew that, left unchecked, sewage could ruin Lake Washington. Nearly fifty years of changes to the lake and its watershed, from the opening of the Ship Canal in 1917 to the redirection of the Duwamish River, made it more susceptible to pollution. The postwar building boom had put more pressure on the lake. Bulkheads, piers, roads, and marinas had reduced its surface area by nearly three thousand acres and opened new pathways for pollution. Constraining the lake's drainage to the Ship Canal reduced its flushing abilities, and the Army Corps of Engineers, which operated the locks used daily by hundreds of commercial and recreational watercraft, needed to impound the lake to keep Puget Sound's salty waters at bay. The canal locks, built for transportation and flood control, could not cleanse the lake and, instead, trapped the dirty waters.[12]

To Eldridge, the only answer was to remove pollution from both the lake and its former tributaries entirely. He and his commission colleague, Wallace Bergerson, decided to concentrate first on Lake Washington and recommended building a massive network of interceptor sewers and treatment plants around its shores, a system they hoped would be applied to the Duwamish River and Seattle proper as well. But the city's leaders pleaded poverty and stuck with the Wolman report. (Even then, Seattle followed Wolman's suggestions piecemeal and continued to rely on the combined system.) Not until feces and foam washed up on the beaches of West Seattle did the city council recognize the growing magnitude of the pollution problem and agree to an additional interceptor system and treatment plant at Alki Point. And as Eldridge and Bergerson had predicted, Lake Washington soon began to suffer pollution problems of its own. Beginning in the late spring of 1955, swimmers and boaters noticed thick mats of algae coating the lake's surface in a foul, green muck that stuck to arms and legs, oars and propellers, water skis and fishing lines. When the cool and rainy spring turned to the warm and sunny days of summer, the blue-green algae browned and putrefied in the heat. Public health authorities repeatedly shut down popular swimming spots along the lake's shores for days at a time.[13]

Eldridge had cautioned Governor Langlie directly, in 1949, that "the time [when] treatment will be needed" to address unchecked water pollution in metropolitan Seattle "is much nearer than indicated . . . and in the case of some of the areas the time is now here." Langlie had refused to listen, Seattle's political leaders had refused to listen, and Eldridge's prediction was now coming true. For James Ellis, looking for a solution to the region's "Topsy-like" approach to growth and regional governance, spoiled beaches and filthy waters made for compelling evidence. Yet even Ellis did not grasp the gravity of the pollution crisis at first. It would fall to a respected scientist, unknown outside of a small circle of fellow zoologists, to give Ellis the data he needed.[14]

An Over-exaggerated Pollution Problem, or a Good Place to Live Imperiled?

Wallis T. Edmondson, better known to his friends and colleagues by his nickname, Tommy, was an unlikely activist. He was a scientist's scientist. Born in Milwaukee, Edmondson had been enchanted as a boy with the lakes, ponds, and vernal pools, the ephemeral puddles of snowmelt dotting the glaciated hills of Wisconsin every spring. He had his first microscope at twelve, and when he was a high school student in New Haven, Connecticut, the eminent Yale zoologist G. Evelyn Hutchinson gave him access to his laboratory and plankton samples from around the world. When Edmondson entered college, he was already a published scholar, an acknowledged expert on the phylum Rotifera, the microscopic animals found in freshwater environments and moist soils. He earned his Ph.D. under Hutchinson, and then spent a year at Wisconsin with the noted limnologists Chauncey Juday and Edward A. Birge. In Madison, he met his future wife, Yvette Hardman, a Ph.D. in microbial ecology, who became his lifelong collaborator. They moved to Seattle in 1949, and he joined the zoology department at the University of Washington. The Pacific Northwest was perfect for Edmondson, with its mosaic of deep saltwater inlets and abundant lakes, and no locale was more fruitful for his studies or training graduate students than Lake Washington. Two of his doctoral candidates, George Anderson and Gabriel Comita, made the lake their laboratory, Anderson focusing on phytoplankton (microscopic plants), and Comita on copepods, tiny crustaceans that fed on phytoplankton.[15]

On a June day in 1955, Anderson brought back a beer bottle full of water, telling his mentor, "That is not the lake I knew." Swimmers and boaters had already noticed a dramatic increase of the blue-green algae now identified by Anderson as *Oscillatoria rubescens*. "This, of course," Edmondson later recalled in his memoir, "was wildly exciting to us." The algae species was a sign of human-induced eutrophication—a condition whereby excess nutrients such as phosphates and nitrates, abundant in untreated sewage, hastened algal growth, which, in turn, stripped the waters of dissolved oxygen, killing all but the hardiest flora and fauna. Sewage effluent in particular was an extremely effective fertilizer, so nutrients liberated by dying algae could continue the cycle unabated. In subsequent experiments, Edmondson and his students found that visibility in the lake's waters, which in 1939 had exceeded three meters in depth, was now less than one meter.[16]

During the warm summer months the water now looked like pea soup. With the help of Donald R. Peterson, a scientist for the Washington State Pollution Control Commission, Edmondson and his students rushed to publish their findings in leading scientific journals, and the commission released its own report in 1955 based on the research. Edmondson began to move from neutral scientist to concerned citizen, probably with James Ellis's help. Soon, the *Seattle Times* had published an editorial entitled "Lake's Play Use Periled by Pollution," citing research done at the University of Washington. In turn, the Metropolitan Problems Advisory Council, chaired by Ellis, capitalized on the publicity in its report from 1956, *Seattle: The Shape . . . Of Things to Come!* Ellis enlisted Edmondson to advise the council on the grim state of Lake Washington and on the possibilities for its revival. The scientist's conclusions were sobering. Combined effluent dumped into the lake by the ten suburban sewage treatment plants topped twenty million gallons daily, with the rate of algal growth directly proportional to the increase in sewage. Poisoning the algae or scooping it from the lake for removal would not solve the problem, either. "The effect is no different in principle from increasing growth of grass by fertilizing lawns," Edmondson explained, and if fertilization continued, "within a few short years we can expect to have serious scum and odor nuisances." "Definite action must be taken if the lake is not to deteriorate," he warned.[17]

In the past, Seattle residents had turned to engineers for answers; now, they consulted another authority, the ecologist. A relatively new term to the layperson, "ecology" stood for the notion that human actions disrupted the balance of natural systems. The public embraced the term just at the time that the scientific discipline itself was in intellectual ferment, incorporating theoretical advances from engineering, cybernetics, and systems theory. Many ecologists, steeped in the older tradition of field observation and laboratory work, were resisting the new approaches. They were also wary of applying pure science to solve political or technical problems. Edmondson was an agnostic on such matters. He spoke to those who were best able to take action, the professional residents of Seattle and King County: lawyers, doctors, engineers, labor leaders, homemakers, and businesspeople—almost all college-educated. Active in such venerable reform groups as the Municipal League and the League of Women

Voters, they were what historian Carl Abbott has called "neo-progressives," sequel to the earlier Progressives. Seattle's neo-progressives took cues from their counterparts in Washington, D.C., and other major cities, but politics then and now remained local. So, too, did the expertise that informed local politics. Ties between academia and the federal government, which had drawn closer in the name of national security during and after the Second World War, now drew closer on the regional and local level as well. The University of Washington was among the leaders in the establishment of the Washington State Pollution Control Commission; the National Science Foundation funded Edmondson's Lake Washington research; and the Pollution Control Commission and the Municipal League drew on University of Washington graduate students and faculty for their reports.[18]

Edmondson would later defend his role in the Lake Washington cleanup as not "for the purpose of pollution control" alone. This was only half true. Accused of research that was no more than applied science, he defended his Lake Washington studies as "an example in which the results of basic research could be put to immediate application." Ellis, a moderate Republican, took the lead in using Edmondson's findings for political goals that, he hoped, would transcend ideological differences. Governor Arthur Langlie, a former Seattle mayor and conservative Republican, appointed a special state committee in early 1956 to study solutions to aquatic pollution and urban overgrowth. Mayor Gordon S. Clinton turned to Seattle's Metropolitan Problems Advisory Council, which, in turn, hired a San Francisco engineering firm, Brown and Caldwell, to reexamine the sewerage question. As the engineers began their study, Ellis and the council worked closely with Seattle city hall to draft legislation to create a new political entity, the Municipality of Metropolitan Seattle, popularly known thereafter as Metro.[19]

Inspired by Toronto's example, Metro was to be a federated municipal corporation, insulated from the political pressures of direct elections. It was more like a super public utility but one with taxation powers to build regional sewerage and mass transit, regulate zoning, and create parks. The Seattle delegation, which introduced the bill to the state legislature in autumn 1957, faced fierce opposition from legislators representing suburban King and Snohomish Counties. Support for Metro cut across party lines, but so did opposition. The fight was entirely over local control, pitting small towns and suburbs against Seattle and the larger cities on Puget Sound. State Representative Daniel J. Evans, a former civil engineer and freshman Republican from Seattle, tried in vain to assuage fears over losing local sovereignty, and only the intercession of House Speaker John L. O'Brien ensured passage in the lower chamber, by a single vote, and by a whisker in the Senate. The bill became law after opponents had won a key concession: Metro would have to pass by a two-thirds majority in Seattle and King County elections the following March.[20]

Both Ellis and Edmondson forgot or overlooked that Lake Washington, which united city and suburbs in play, was dividing the two in politics. Nowhere was resistance to regional government stronger than in the small towns and working-class suburbs around the lake. Nicholas A. Maffeo, an attorney from Renton, president of the King County Taxpayers' League against Metro and the Metropolitan League of Voters, was an ardent advocate for local autonomy, limited government, lower taxes, and private property rights. He claimed that the proposed measure "would destroy our historical form of local self-government" and "simultaneously place a staggering financial and tax burden upon us." "The proponents of 'Metro,'" he exclaimed, "in their haste to solve a grossly over-exaggerated pollution problem . . . would remove the safeguards of freedom placed by our ancestors at the local level." Maffeo stated his opposition in stark terms: "I wish to say, my friends, in all earnestness that if it came to a choice between 'loss' of Lake Washington or loss of any part of our great American heritage of freedom, I would personally choose to lose Lake Washington." It was Seattle, he said, that was ultimately responsible for most of the raw sewage "contaminating and polluting the beaches."[21]

Maffeo's broadside was only the beginning of a bitter political campaign in the early spring of 1958. He proposed draining the lake, then filling it in or using atomic energy, somehow, to remove contaminants, and he called for an "independent investigation" because of Edmondson's "suspected collusion with Metro supporters." An anti-Metro advertisement, running in the *Bellevue American* and *Highline Times* on the eve of the March 1958 election, caricatured Metro as a snarling octopus, a "Metro monster," strangling a hapless taxpayer in its tentacles while picking his pockets clean. Another ad in the *Washington Sentinel* of Kirkland and the *Seattle Times* was even more blatant; again, Metro was an octopus, but this time with the face of Adolf Hitler. Maffeo harangued Edmonson by phone and mail, prompting the scientist to warn the Metro Action Committee, a group of supporters working for the measure, that continued public disinformation might lead to electoral defeat. Edmondson later wrote that the behavior of "the man from Renton" and his allies "made a weird sort of sense" to those who knew Washington politics.[22]

METRO MEANS...

1. Seattle rule and domination.
2. Higher taxes.
3. Arbitrary property assessments, even if 100 per cent of property owners object.
4. Garbage collection, park purchasing and water installation can be added to sewers, planning and transportation without your vote.
5. Unlimited demands on the cities, towns and county for "supplemental income."
6. Tax-supported transportation system.
7. Granting unlimited authority to Seattle officials elected by and answerable only to Seattle.
8. Once Metro is voted in, there is no "getting out" or turning back. Seattle rules forevermore!
9. If Metro passes, the many constitutional and legal defects will result in years of expensive litigation, rather than any constructive action.
10. It is a grab for power by Seattle officials in the fields of transportation, planning, park purchasing, water and garbage collection in addition to sewers, clothed in the emotional appeal of "save Lake Washington now." Once this power is voted in the property owner is subject to unlimited taxes for any of the above purposes, at a time over which he has no further control or voice but solely at the pleasure of Arbitrary Seattle officials "who know what's best for their 'country cousins'." The second Lake bridge is a graphic example.

Anti-Metro advertisement, published shortly before the first Metro election, in the *Highline Times,* March 5, 1958. (James R. Ellis Papers; courtesy Special Collections, University of Washington Libraries, used with permission of Robinson Newspapers, Seattle)

The actual context of opposition to Metro was far more complex. Although the enabling legislation for Metro limited its jurisdiction to only three functions—sewage, mass transit, and zoning—many suburbanites saw it not only as a spendthrift bureaucracy, but as something more ominous. Conservatives in postwar suburban King County, like conservatives from Orange County, California, to Maricopa County, Arizona, deeply resented the federal government and urban elites. They were ardent supporters of growth and free enterprise, but racially homogeneous. (Of course, the same was true of Metro supporters.) Many were first-time homeowners, newly arrived in the middle class thanks to high-paying defense jobs. Military and defense-related industries had helped to shape their self-image as rough-hewn individualists suspicious of state authority who had cut loose from the fetters of a corrupt and radical society back East or in town.[23]

Not surprisingly, suburban conservatives hated all things Soviet and leftist, and after the Second World War, Washington State Republicans made anticommunism a central plank in their platform, sweeping into office in the 1946 elections. One prominent Republican was a lifelong Communist hunter, Representative Albert Canwell of Spokane, who quickly made good on his campaign promise to chair a commission to investigate the infiltration of state agencies by Communists. The new commission began a series of aggressive public investigations into the activities of the Washington Pension Union and the University of Washington, forcing the university to fire three tenured professors for alleged "un-American" activities and to put three others on probation for charges of lying to administrators about their political beliefs. Although the 1948 election erased the Republican majority in both houses, and the Canwell Committee disbanded three years later, electoral defeat did not temper the rhetoric. Conservatives on the far right still tended to label all adversaries as "Reds" and found a new following in the mid-1950s in the smaller blue-collar and lower-middle-class towns ringing Lake Washington. Several organizations, such as the John Birch Society, beginning during the Second World War, had injected themselves into the local politics of Seattle's suburbs, as in Lake Hills and Bellevue. One of their tactics was to play upon suburban homeowners' fears of rising taxes. They found a ready audience. Eastside and south King County homeowners, with the help of the Birchers or other right-wing activists, soon equated regional government with a collectivist state dominated by an elitist oligarchy. They planted fears that high-minded scientists and Seattle politicians had concocted an ecological crisis to mask a Communist coup.[24]

Metro proponents, meanwhile, organized the Metro Action Committee to get out the vote, targeting fellow professionals likely to support the measure's aims not only on water pollution but also on mass transit, coordination of planning functions, and economizing of government services. The deteriorating state of Lake Washington was the emotional core of their pitch. Five days before the election in March 1958, Boeing president William M. Allen made his peace with Metro in a letter to his employees. "I have put aside any such fears about Metro," he said, in hopes that it would help Seattle "to continue to be a good place to live." Although a big, tax-hungry, city-dominated agency might siphon off corporate profits and impose environmental regulations, Allen worried that without some form of planning, unrestrained growth would ruin Seattle for future employees. His support for Metro was by this time a safe political bet; both of Seattle's major dailies had come out in favor of the proposal. Fred Marshall, an editorial cartoonist for the *Seattle Post-Intelligencer,* illustrated his own rendition of the "Metro monster"—an octopus labeled "Pollution Problem" swimming in Lake Washington, with residents standing on the shoreline and pointing at one another to do something. Other local newspapers, civic clubs, and town councils throughout King County, plus the sewer districts in the suburbs of Hunt's Point, Beaux Arts, Mercer Island, Bellevue, and Medina, the most upscale environs fronting Lake Washington, all called for a yes vote. Yet Metro went down to defeat. Maffeo and his angry suburbanites had outflanked Ellis, the Metro Action Committee, and the most powerful of Seattle's leaders and institutions. In the smaller, less prosperous communities of Renton, Kirkland, Kenmore, Bothell, Redmond, and Issaquah, organized right-wing opposition had undermined support.[25]

A counter-strategy was needed, and John F. Henry, director of Bellevue's sewer district, took the lead. He had listened to E. F. Eldridge's original pollution warnings and proposed, in 1956, his own plan to divert sewage from Eastside suburbs into the Duwamish River. He joined the Metro Action Committee and, in the words of its leader, Richard H. Riddell, was "one of the most important factors . . . in swinging the reaction of Bellevue voters." In an open letter to Eastside residents before the March 1958 election, Henry said no other agency would be able "to remove pollution from Lake Washington on an area-wide basis" and opponents were "either insincere or misinformed." Bellevue voted more than two to one in favor. Riddell had advised Henry after the election of the strategy for a new campaign. One tactic was little more than simple gerrymandering. Ellis, Henry, and other committee members pored over vot-

This Monster Is EVERYBODY'S Problem!

Pro-Metro editorial cartoon by Fred Marshall, published before the first Metro election in the *Seattle Post-Intelligencer,* February 18, 1958. (Used with permission of *Seattle Post-Intelligencer*)

ing tallies and precinct maps during the summer of 1958, correlating districts where opposition ran highest, notably at the southern end of the lake basin, just south of Renton, with those towns and suburbs where the sewerage needs were less acute. They shrank and reshaped Metro's service area to fit the desired electoral map.[26]

Meanwhile, Henry asked the Bellevue city council to back another Metro vote that September. The Bellevue councilors conceded on one condition: that Metro's power should extend over sewers only. The revamped plan also stipulated three other concessions at the request of Eastside town managers: Metro would compensate communities for existing sewage plants made obsolete by new facilities; a two-thirds vote of Metro council members would be required to authorize any new bonds; and the redesigned system would follow Henry's earlier plan to use the Duwamish as the region's toilet. The restructured Metro had become an agent for promoting further growth, not for constraining it. The suburbs had flexed their muscle again. With pollution identified as a major impediment to further development, new and better sewers would guarantee continued progress without ruining what, in the view of boosters, made that progress possible: the region's beautiful waters. To bring the suburbs to Metro's side, Seattle reformers had to accept a scheme that hid the underlying cause of pollution: urban sprawl. With the most rabidly anti-Metro locales removed from the new service area, thus unable to vote, proponents launched another offensive. The long-awaited Brown and Caldwell engineering survey, a 600-page tome, urged the establishment of a regional sewer utility that neatly coincided with Metro's redrawn boundaries. Harold E. Miller, the lead author of the Brown and Caldwell report who would later go on to become Metro's first chief engineer, said how "sewerage and drainage deficiencies," left unsolved, would soon become "a major obstacle to continued growth and development." Miller did not have to wait long to see if his forecast held true.[27]

During an unusually hot, dry summer in 1958, fewer than two inches of rain fell from May to August, less than half of normal. City and county health officials again closed local swimming beaches, caked in noxious algae, around the lake just as residents were seeking relief from the heat. The editors of the *Seattle Times* declared contaminated beaches and spoiled water anathema to a city whose aquatic recreation was "one of its most characteristic and most prized assets." In response, Governor Albert Rosellini announced that all new subdivisions in King and Snohomish Counties would be required to spray their wastewater and sewage over land rather than disposing it into the lake. But Metro's antagonists were not dissuaded. The Washington State chapter of the Daughters of the American Revolution and the John Birch Society co-published a pamphlet, "The 'Metro' Monster," raising the alarm that the "traditional form of American government is under attack by forces here on home ground." The "Metro movement," a nationwide scourge of fifth columnists, was an attempt "to separate citizens from their property" and thereby aid "an old enemy—totalitarian dictatorship." A sympathizer scrawled a warning in the margins of one pamphlet: "Metro is *not only* sewers!"[28]

This time, the Metro Action Committee was ready. It rallied women and children to march with placards through Seattle and suburban neighborhoods, ringing doorbells and passing out pro-Metro literature in the days leading up to the September election. Letters and editorials in local papers cited responsibilities to present-day families and future generations. The committee took out a full-page newspaper advertisement the day before the September 1958 election showing five children in bathing suits on a Lake Washington beach closed to swimming, with the warning "EMERGENCY! CLEAN UP OUR FILTHY WATERS—VOTE METRO." (Dorothy Block, the Metro Action Committee's vice-chair, later admitted she had staged the photograph using her children in front of a planted sign, with an old tire thrown in for added effect.) The ploy worked, and voters approved Metro by a comfortable margin.[29]

Metro engineers and scientists now fanned out across the Lake Washington basin, measuring water quality and mapping routes for sewer lines even as the fight against Metro continued, its opponents not fazed by electoral defeat. In 1960, the Metro council adopted a ten-year plan and issued $125 million in construction bonds, to be repaid by an annual sewerage charge of $2 per household. In a last-ditch effort to slay the "Metro Monster," Nicholas Maffeo joined a lawsuit against Seattle, one filed by Metro itself to compel the city to surrender its disposal facilities for the new regional sewers. Metro, the attorney from Renton said, had violated the Fourteenth Amendment by taxing citizens without representation. The state Supreme Court waved off Maffeo's argument in a unanimous decision upholding the election results. The court also ruled that although Seattle still owned its own sewage facilities, it had also agreed, through the democratic process, to make them available to Metro. Two years later, with all legal challenges behind it, Metro reached agreements with fifteen municipal agencies and sewage districts across King County to guarantee construction of the giant interceptor sewers to funnel the region's sewage to treatment plants before discharge into Puget Sound. It took the next two decades to gain control of all necessary plants, major trunk sewers, and pumping stations.[30]

All that sewage and wastewater had to go somewhere, and that somewhere was the Duwamish River, reconnected to its historical headwaters by a new artificial drainage system of concrete and steel pipelines. Metro "groundhogs" often found themselves digging, quite literally, backward into time, and in 1962 they unearthed the old channel of the long-absent Black River that once drained Lake Washington into the Duwamish. A large pocket of water, still embedded in the river's aquifer deep beneath the surface, gushed out. When workers tried to block up the breach with sand, the waterlogged embankment collapsed. It took more than six hundred bags of cement to seal the break. "The men digging out the tunnel one shovel full at a time," reported the *Renton Record-Chronicle*, "were spooning out a river. And in a way they were right." To assemble the large sections of concrete pipe and build the gravity-driven sewers, the "groundhogs" had to excavate beneath streets and subdivisions in suburban King County. The city of Renton lost its major sewer line for days at a time, and an accident severed Seattle's sixty-inch water main. Robert Hillis,

Metro's public information director, later quipped without irony that when the sewerage charges went into effect, Metro was "second only to the Communist Party as the most unpopular organization in town."[31]

Cold War suburbanites were cool to big government and Communists, but most were enthusiastic about big science and technology. And, in April 1962, on the site where the former Denny Hill once loomed, Seattle opened its Century 21 Exposition, a fantasy world of the future in the Far West where the old frontier met John F. Kennedy's New Frontier. The gleaming United States Science Center designed by Minoru Yamasaki looked like spun sugar, and the Space Needle pierced the sky. A sleek Alweg monorail zipped around the exhibits of thirty-five governments, General Motors, and Boeing, now the nation's largest airline manufacturer. This "American Temple of Science," built in the rundown Regrade District, an attempt at urban renewal, was a symbol of how wonderfully science and technology could solve almost any problem. Metro officials capitalized on this enthusiasm and held a festive ceremony on the shores of Lake Washington the following February to herald the official completion of the first sewer line. On a cold, clear morning before a bank of reporters and photographers, an actor in a shiny flowing dress played "Princess Sparkling Water" and, with a magical wave of her hands, banished scum-encrusted "King Algae" and "Polly Pollution" from the lake. In this humorous morality play, nature typecast as feminine and regal, though younger than Mother Nature herself, was redeemed by engineers and ecologists, and made ready to serve the needs of man.[32]

Edmondson, the University of Washington ecologist for whom Lake Washington seemed a kind of giant scientific experiment, bet a skeptical Swiss colleague a bottle of whiskey that, after sewage diversion, the lake would rebound and water quality recover. But until the first diversion and treatment plant went into operation, the lake continued to decline. The *Seattle Post-Intelligencer* ran an editorial, in October 1963, titled "Lake Stinko." Edmondson had predicted this and urged patience. His optimism was well-founded; with each branch of the Metro system completed, water quality in the lake improved, steadily and dramatically. Dissolved phosphorous inputs declined from more than a hundred metric tons per year in 1961 to less than twenty metric tons in 1969; raw sewage effluent, thanks to new treatment plants, all but disappeared. Edmondson found that water transparency, a key indicator, increased from less than one meter in 1963 to more than three meters in 1969—the highest in almost fifteen years. Indeed, as Edmondson said later, Lake Washington was now one of the best examples of lake restoration in the entire world. *Harper's Magazine* called Metro's creators "modern-day vigilantes," and Metro itself won an All-American City award—even though it was not a city—for setting an example of good government and community building. The rescue of Lake Washington became a staple of how "bioremediation" could save aquatic ecosystems. Scientists and politicians, too, from Ontario to Switzerland would later apply and extend Edmondson's findings to resuscitate Lake Erie and Lake Zurich. The National Research Council later concluded that Lake Washington's revival, "a unique blend of scientific judgment and public action," abetted effective restoration of countless aquatic environments around the world.[33]

Success came at a cost, even if that cost was at first unseen, and the first to complain about Metro's meddling were sportsmen on the Duwamish River. They had sounded alarms over pollution years before, in 1959, when Don S. Johnson, the angler and conservationist, had first predicted "death for a tired old river." Sport anglers soon had additional evidence. In 1960, as the first Metro sewers neared completion, the King County Outdoors Sportsmen's Council and the Washington State Sportsmen's Council publicized scores of fish kills along the Green and Duwamish Rivers over the previous two years: rafts of dead salmon, trout, sculpin, and other fish pooling in the river's eddies. Angry fishermen blamed industries along the Duwamish, particularly Boeing, and predicted that Metro's sewers would kill even more. After several more die-offs that spring, biologists from the state Department of Fisheries and Department of Game joined in protest. Metro officials instead blamed meat-packing plants, foundries, marine yards, and metal-plating shops along the river for the fish kills, and mentioned that the Washington State Pollution Control Commission had endorsed the Duwamish outfalls. Building the sewers to reach all the way to Puget Sound would have added $18 million to the project and killed off any chance of saving Lake Washington itself. By 1964, Ken McLeod, president of the Washington State Sportsmen's Council, was lashing out against Metro and the state pollution commission for a "buck-passing and stalling attitude" that had put the Duwamish on the brink of death.[34]

A Town Needs a River to Forgive the Town

The salmon fisheries in urban Puget Sound, too, had been pushed to the verge of collapse. Sportsmen, reluctant to impose seasonal restrictions or equipment prohibitions on themselves, had become more and more dependent on hatch-

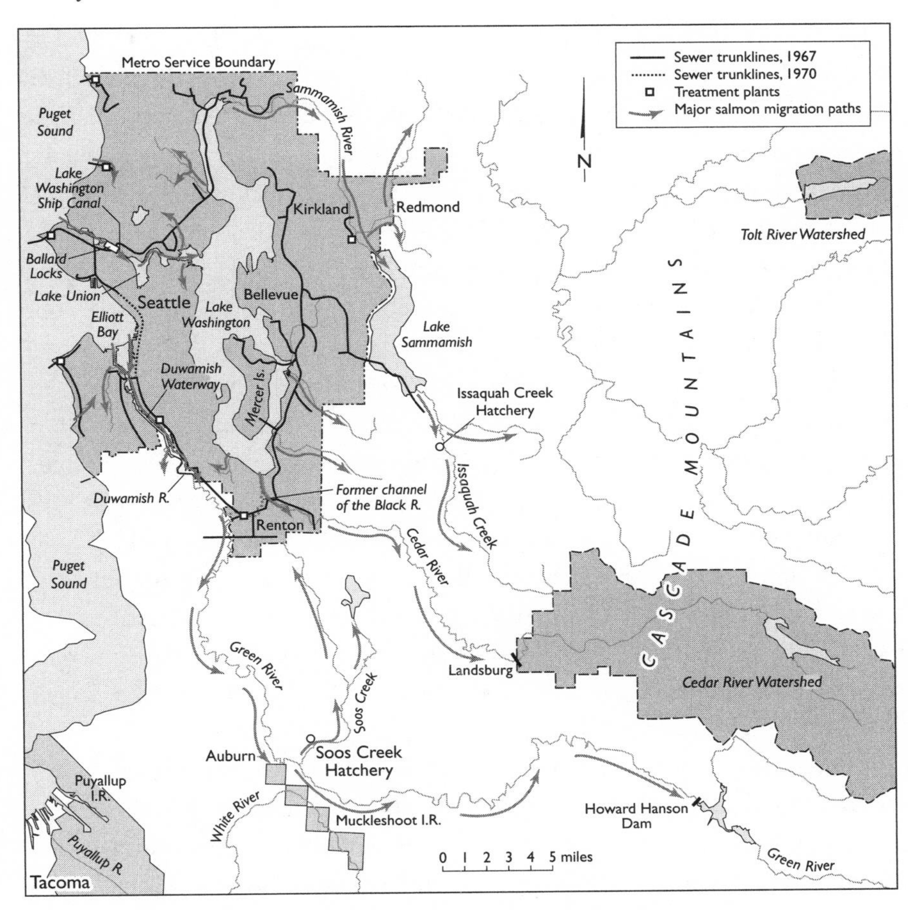

Metro regional sewers and major salmon migration paths, c. 1970.

eries. The state-run Issaquah Creek hatchery, built in 1936, did seem to bolster the runs of steelhead, an anadromous type of rainbow trout prized for its beauty and feistiness. But the Soos Creek hatchery, which had been operating since the turn of the century on the edge of the Muckleshoot Indian Reservation just above the confluence of the Green and Duwamish Rivers, had achieved only uneven success in breeding chinook and coho salmon. As native species waned, sportsmen persuaded state fisheries' biologists to plant rainbow and eastern brook trout, but both were voracious predators that gobbled young salmon fry before they migrated to the sea.[35]

A few anglers, realizing that the hatcheries, no matter how many fish they produced, were not enough to revive the Duwamish fishery, called for the restoration of the Black River, eliminated and dried up by the completion of the Lake Washington Ship Canal in 1916–17. The state's fisheries director, B. M. Brennan, had called such solutions impractical. Rebuilding the Duwamish, he had written way back in 1939, "could not be divorced from the problems facing other streams" in the Lake Washington drainage or "the lack of adequate water" needed for fish ladders at the Ship Canal locks. In 1901, Seattle had closed the upper Cedar River, the lake's largest remaining tributary, to spawning salmon out of fears that decomposing fish would contaminate drinking water supplies. The Sammamish River, outlet for Issaquah Creek and the hatchery on its banks, was falling to dangerously low levels every summer after irrigators turned its current onto their fields and orchards. Decades of digging, dredging, and draining had made the Green and Duwamish Rivers "one and the same." Absent hatcheries, Brennan intimated, there would be no salmon at all.[36]

Native-born white sportsmen had earlier expelled immigrants and Indians, considering them responsible for the fishing crisis of the 1920s. But Indians had never gone away, and they continued to assert their treaty fishing rights on the Duwamish and Green Rivers despite setbacks in state and federal courts. When the Great Depression drove Natives elsewhere into deeper poverty, many in the Puget Sound area had survived by catching shellfish and salmon, fishing legally on their reservations, sometimes sneaking off to fish on private and state property. During the Second World War, when they could find work in labor-starved Seattle, Portland, and other northwestern cities, fishing conflicts subsided, but when white soldiers came home to take back the jobs, confrontations intensified. Seattle residents who wanted unimpeded access to the fisheries they had enjoyed before the war blamed Indians for ruining their sport, and fish wardens, angered over Indian intransigence, began to harass and prosecute those caught fishing off reservation. Although local agents for the Bureau of Indian Affairs tried to protect their wards, more often they tacitly supported fishing restrictions as a way of controlling tribal politics themselves. Dismissed from wartime jobs and sinking into indigence, Indians slowly came to see fishing as survival plus something more. By the late 1960s and early 1970s, treaty fishing rights would become "an emblem unfurled," as historian Alexandra Harmon put it, for unity.[37]

Even as Natives were becoming ever more politicized, state fisheries officials were also coming to see Indians as a threat. In an address to the Washington State Sportsmen's Council in 1962, Edward M. Mains, the assistant director of the state's Department of Fisheries, compared the salmon fisheries of Puget Sound to "an endless chain" in which "unregulated Indian fisheries" were the weakest links and hatcheries the strongest. An official with the state Department of Game, the agency responsible for all inland fisheries, compared sport fishing and hunting "to a manufacturing process of raw materials converted through the labor of a skilled organization." Hatcheries and game farms, administered by trained wardens and scientists, were the skilled organizers; happy sportsmen were the customers. The easiest solution was to make more fish—and keep the Indians out. Unless spawning salmon could return to the hatchery unimpeded so wardens could harvest their eggs, the system would break down.[38]

In February 1962, fisheries wardens arrested three Muckleshoot Indians on the Green River, just downstream from the Soos Creek hatchery, and, the following September, another group of Muckleshoot began fishing again near the hatchery, upstream from the new interceptor sewers on the Duwamish. Sportsmen and commercial fish operators accused the state of doing nothing to protect the salmon. One enraged angler, Jack Schwabland, wrote to Governor Albert Rosellini in September 1963. "Can it be true that you actually plan to do nothing while a resource of this magnitude is wiped out by 37 indigent bums? Everyone knows that the treaty fishing rights are being interpreted far too broadly today, that it never was contemplated by the formulators that Indians would mass gill-nets bank to bank on spawning streams and sell their catch commercially." "Time is running out at Soos Creek," he warned. "For Heaven's sake get those fish into the hatchery, by whatever means are necessary!"[39]

In a letter to the state assistant attorney general that same year, Mains from the Department of Fisheries contended that without the hatchery at Soos Creek or proscriptions against Indian fishing the "successful run of chinook salmon" on the Green and Duwamish Rivers "would not have existed." It was a proposition that Indians tested in court. During the trial in May 1963 of the three Muckleshoot Indians first charged with illegal fishing on Soos Creek—James Starr, Louie Starr, and Leonard Wayne—defense attorney Frederick W. Post said that the tribe had suffered because of what state and county engineers had done to the rivers decades before. In 1914, "the white men changed the course of the [White] River" with levees and dams "so instead of flowing down the Duwamish it goes out the Puyallup River." Now, at "the mouth of that river, the Puyallup tribe have the right to take fish," yet the Muckleshoot, who depended on the Green and Duwamish Rivers, were denied the same rights. The state's attorney general inquired whether the three men on trial really were "constructive Muckleshoot"—meaning recognized and authentic Indians by blood and ancestry—eligible for treaty rights. If not, they enjoyed no special protection; if they were, their actions nonetheless endangered the fragile fishery. The presiding judge sided with the Muckleshoot, ruling that the state had failed to demonstrate that protecting salmon required restricting Indians' treaty rights.[40]

Later that year, in September, state and Metro representatives held "a lengthy discussion" in a specially called meeting on Indian fishing activism. Ralph Anderson, a Fisheries Department administrator, maintained that "eventually Federal action would be required if the fishery was to be saved," since "the Indians apparently had a legal right under their treaties to fish when and where they pleased." Letters from Metro officials "to the Congressional delegation and interested state and Federal agencies" would be more effective "in the long run than any stop-gap measures taken to harass the Indians." Fred E. Lange, Metro's technical service director, urged making the public more "aware of the problem" of Indian fishing "so that more pressure could be exerted on the Federal government." Anderson added that the Department of Fisheries planned to "place an article in the local papers at least once a week" during the chinook salmon migration that autumn "for just this purpose."[41]

For their part, Metro officers promised to play up the threat to the fisheries from renegade Indians as further proof that the sewers would save the salmon. Metro Council chairman C. Carey Donworth, responding to the Muckleshoots' growing militancy, wrote to the Department of Fisheries that October to lend further support in the anti-Indian fishing campaign. It was "ironic," he said, that salmon runs on the brink of extinction for years from pollution "should be ruined by uncontrolled Indian fishing on the very eve of the end of the pollution threat." He insinuated that Indians were preventing control measures from taking effect despite early evidence from Alfred T. Neale, a scientist for the Washington State Pollution Control Commission, that the Duwamish and Green Rivers, polluted beyond their capacity to flush the waste to sea, could not carry any more effluent and still support a viable fishery.[42]

Two Muckleshoot elders, Bertha McJoe and Olive Hungary, maintained that fishing below the Soos Creek hatchery made historical sense regardless of technical opinions about sewage disposal. The state salmon hatchery stood on the site of an old Indian fishing village, a place long revered for its abundant runs, and the hatchery was so productive that it shipped salmon eggs and fry for planting in streams around Puget Sound. In lieu of deserved compensation, McJoe and Hungary suggested, the state should share the wealth it generated at the hatchery with the tribe. In a letter to Governor Rosellini in 1964 they warned the state to compromise now because the tribe would "be able to win important rights" in court later. Rossellini had no incentive to cooperate. The year before, the state Supreme Court, in a split decision, had ruled that modern Indians had no privileged rights to fish in waters regulated by state law.[43]

One former Seattle resident, the poet Richard Hugo, remembered the past that the Muckleshoot referred to, or at least a particular version of it, and the history of the river that now divided Indian and white, sportsmen and engineers, liberal environmentalists and conservative homeowners. Born in 1923 in White Center, a working-class community southeast of West Seattle, Hugo had avoided violent beatings at home by fleeing to the brush-choked banks of the Duwamish River. After serving in the Army Air Corps in the Second World War, he returned to study under the poet Theodore Roethke at the University of Washington and to work as a technical writer at Boeing before finding literary fame. As he reflected in his autobiography many years later, Hugo often returned in his thoughts to the river he loved as a youth, a river once home to houseboats and dories of Indians and Greek immigrants, to tough-talking bootleggers and former union radicals, a river that nursed flocks of waterfowl and schools of fish. It was this river lost that he reawakened in his poem "Duwamish Head," from 1965, an elegy to his memories of place.

> With salmon gone and industry moved in
> birds don't bite the water. Once this river
> brought a cascade color to the sea.
> Now the clouds are cod, crossing on the prowl
> beneath the dredge that heaps a hundred tons
> of crud on barges for the dumping ground.

Instead of spinning the story of progress told by the flashy exhibits at the Century 21 Exposition or the slick Metro campaign flyers, Hugo plotted Seattle's fall from grace. As he wrote later in another poem, "The Towns We Know and Leave Behind, the Rivers We Carry with Us," Hugo's Duwamish was the place where he and his hometown might find redemption.

> I forget the names of towns without rivers
> A town needs a river to forgive the town
> Whatever river, whatever town—
> it is much the same.
> The cruel things I did I took to the river.
> I begged the current: make me better.[44]

The ultimate fate of the Duwamish River, like the ultimate outcome of the Indian fish fights, was still unclear in the early 1960s when Hugo published his first poems and the Muckleshoots were arrested and tried. The contests over saving Lake Washington from pollution exemplified how ecological restoration in one context could fail to translate to environmental justice in another. Each effort to protect environmental amenities for some was a political act that endangered or abolished rights and privileges for others. The suburbs had co-opted the city, placing limits on regional government. The conflict was between the haves and the have-mores, between the middle-class professionals and reformers from Seattle and the larger suburbs and the newly middle-class homeowners in the smaller towns. Those who lived closer to Lake Washington tended to have finer homes and higher incomes. They enjoyed the lake's amenities and pushed for immediate solutions to the pollution crisis. But those who lived farther from its shores doubted that the cleanup costs were worth the price. To them, the pollution crisis was a stalking horse for an intrusive supergovernment. As angry homeowners fought back, so, too, did Indians who had never accepted their role as scapegoat for the salmon crisis. The Muckleshoot who cast their nets into the Green River and the right-wing activists who accused Metro of a left-wing conspiracy were not in common cause, but both groups did react against a political solution that did not account for their interests. Seen one way, the story of Metro confirmed what the historian Andrew Hurley found in his study of pollution and politics in Gary, Indiana: that the rise of modern environmentalism often went hand in hand with environmental inequality. Lake Washington rebounded, the Duwamish River deteriorated, and, in the end, the poor and minorities suffered.

But even this answer is too pat. Long before the pollution crisis in Lake Washington, patterns of social inequality and environmental injury were already innate features of the region's geography. It was easy for Metro's engineers and scientists, for political leaders and voters in Seattle and its suburbs, to justify polluting the Duwamish. It had become the natural place to waste. Technically, Metro was a success, its architects demonstrating well before the more celebrated case of Lake Erie that endemic aquatic pollution was reversible and that regional government could unite people behind environmental protection. Its supporters had cast a harsh light on the pro-growth orthodoxy prevalent in so many booming metropolitan areas across the nation. And like similar initiatives in metropolitan New York City or Virginia's Fairfax County, Metro's success ensured that the politics of suburban growth would never be the same. In this sense, Metro's unheralded feat was to forge Seattle's new millennial identity as an ecologically conscious city by inventing an environmentally benign past and a new sense of environmental citizenship.[45]

Yet Metro's triumph came at a cost. As with other attempts at regional planning, like those along the Northeastern Corridor from Boston to Washington, the optimism of reform did not translate into effective government. The public did not always know what it wanted, local governments remained splintered and at odds with one another, and the energies for reining in development that had united suburbanites and urbanites soon dissipated as the economic costs of growth management came due. Suburban growth continued even if environmentalists now cast themselves as the loyal opposition. But Metro's success obscured another important consequence often ignored by historians: that ecological restoration, like the earlier gospel of improvement and progress preached as far back as the Progressive Era, tended to benefit those who had power. Environmental citizenship could be an exclusive club, something that Richard Hugo and the Muckleshoot both knew, and therein lay a hard lesson: if some places were deemed worthy of protection, someone else or somewhere else, or both, would pay the price for an exclusive beauty.[46]

CHAPTER 8

Masses of Self-Centered People

Salmon and the Limits of Ecotopia in Emerald City

In 1975, Seattleites found an owners' manual for how to build a city worthy of its scenery. It was Ernest Callenbach's million-copy bestseller, *Ecotopia,* a utopian novel set two decades in the future. *Ecotopia* told of an armed secession by Northern California, Oregon, and Washington, fortified by nuclear weapons, to create an independent nation founded on ecological principles. A fictional New York City journalist, William Weston, was the first American sent to report on life in this supposed paradise. His initial impressions confirmed his stereotypes. Ecotopians were lazy and carefree, contemptuous of the Protestant work ethic and dressed in handspun clothing. Environmentalism was akin to a state religion, and its rituals included gardening, handicrafts, and mandatory recycling. Bicycles and high-speed trains replaced internal combustion engines. People in Ecotopia's major cities had high-paying jobs in high-tech enterprises, or they worked instead on nearby organic farms. Its citizens enjoyed universal health care, a temperate climate, and an outdoorsy lifestyle. "Unusually well versed in nature and conservation lore, and experienced in camping and survival skills," Weston reported, Ecotopians were "sentimental about Indians" and their "lost natural place in the American wilderness." Most lived in communal homes filled with cutting-edge technology, sharing computers and picture telephones connecting them to friends and family. It was the best mix of two contrary worlds, the modern and the rustic.[1]

When Weston asked his hosts why they had renounced the United States, Ecotopians said "things were not getting any better . . . something had to be done." After winning independence through atomic blackmail, the secessionists set about undoing the damages of the past. They dynamited dams and opened the liberated rivers to recreational boating and salmon runs, which had been "reestablished with great effort and . . . much public support." Tree-lined walkways flanked by burbling streams replaced concrete sidewalks and car-choked freeways. Legions of volunteers replanted logged-off lands with loving care, trying to coax nature back to life. Ecotopia's charms eventually beguiled the jaded journalist. Weston ceased to be an objective reporter; he fell in love with an Ecotopian woman, joined the revolution, and abandoned the United States. "There are a lot more things about Ecotopia that the world needs badly to know," he told his editors. "Maybe I can help in that."[2]

At the end of the twentieth century, some twenty years after Ecotopia's imagined war for independence, Seattle had become an ecological utopia of sorts. It had a fleet of electric buses, miles of bicycle lanes, and one of the nation's largest curbside recycling programs. Bands of volunteers pulled up invasive weeds—Himalayan blackberry, Scotch broom, English ivy—in city parks and planted native species. Schoolchildren raised salmon hatchlings in their classrooms and released them into nearby streams, restored to entice the fish to spawn. When the clouds parted and the sun shined (and just as often when the rains fell), Seattleites took to the streets and trails by foot or bicycle, or to the lakes and streams by kayak or sailboat. For those in the tree-lined neighborhoods north of Lake Union, or atop Queen Anne Hill, or in its posh Lake Washington suburbs of Medina or Mercer Island, this city, Seattle, was an emerald city.

The Emerald City resembled Ecotopia in ways that many wanted to ignore or forget as well. In the imaginary country, the reporter William Weston had encountered "surprisingly few dark-skinned faces" after he left the streets of "Soul City," the former real-life city of Oakland. Blacks had fled Ecotopian society's restrictive conventions. Many owned automobiles and conducted a "lively trade" in liquor and other banned luxury goods imported from the United States and Europe. The same was true for Chinatown, which also broke away from San Francisco to become its own city-state. Racial disharmony, Weston said, was "surely one of the most disheartening developments in all of Ecotopia." Likewise, those who lived in the Central District or Rainier Valley, South Park or Georgetown, largely non-white and poor, did not always enjoy Seattle's many outdoor pleasures: its urban parks, surrounding wilderness, or its once-abundant fishing grounds. Seattle did not have a walled-off "Soul City," but it did have its vanishing salmon, their numbers plummeting as asphalt and concrete spread across Puget Sound. The salmon's plight seemed to connect city and suburb, rich and poor, minority and majority, yet the fish, as before, suffered the most in those places where people suffered the most, too—degraded, forgotten, and impoverished neighborhoods.[3]

Green-minded Seattleites were reluctant to make the connection between salmon and inequality, their aversion dating back to the 1950s and the pollution menace that gave voice to a nascent political movement, environmentalism, and the group that fought most aggressively to protect the environment, the middle class. Concerned reformers like James Ellis, the architect of the Municipality of Metropolitan Seattle, rallied voters to reverse Lake Washington's decline. Metro's sewers spared the lake but did not rein in unbridled growth. They merely pushed the waste elsewhere and let the building continue. By the 1960s, when suburban sprawl had become one of the region's most pressing concerns, Seattle's freeways and subdivisions had already paved over or choked many of its salmon streams. The consensus around managing growth and, by extension, protecting salmon would prove to be short-lived, because stopping growth required confronting what had driven environmental concerns since the end of the last century—protecting consumer privilege. In frustration, some citizens tried to forge a shared identity, based on Seattle's unique geography and ecology. Salmon was the symbol, and saving salmon would put an ethic of place into practice. The problem with this strategy was that salmon meant different things to different people. Espousing clean water and tall trees and wild salmon did not always account for the poor and dispossessed, or those landscapes most degraded through time. Ultimately, the plight of the salmon demonstrated why the politics of using place to champion environmental utopianism could fall short.

How Man Tried to Do Nature a Favor

Miller Freeman, publisher of *Pacific Fisherman,* the largest and most important trade journal of its kind, had already warned that civilization and salmon were potentially incompatible. After attending a 1934 meeting of the Washington State Planning Council, founded to guide the state's long-term economic development, Freeman sent an angry letter to a colleague. "If you had attended," he wrote, "you would have had graphically presented to you that every salmon stream from the Sacramento to Puget Sound is being most seriously affected by power projects, irrigation, denuding of the forests, pollution, etc., which if continued cannot fail to materially reduce if not exterminate the salmon fisheries originating in those streams." Freeman, long a tireless crusader for fish, especially salmon, had pushed the University of Washington to create its now-famous College of Fisheries in 1919, then to hire John N. Cobb, an editor at *Pacific Fisherman,* as the first dean. Freeman served on the first international Pacific halibut commission, and later pushed to ratify a treaty with Canada in 1937 to protect sockeye salmon in Puget Sound. The publisher now wondered what was to become of his efforts.[4]

Freeman had grown up in Puget Sound after his parents moved there in 1889. His father ran a print shop, but the Panic of 1893 ruined his business. Undaunted, the young Miller, armed with some cash, started *Pacific Fisherman* in 1903, and then branched into local newspapers and trade magazines. By the 1930s, he was one of the nation's largest trade publishers and a real estate magnate, buying and selling Seattle's improved tidelands to the railroads. And like William Randolph Hearst, whom he admired, Freeman was an unabashed nationalist. He was critical of the fishing habits of Austrians, Finns, and immigrants from southern Europe, but he reserved his most potent venom for the Japanese. He inveighed in his papers and magazines against Japanese immigrants, like many whites on the West Coast at the time, seeing Japan as a threat to America's hegemony in the Pacific, including its mastery over the fisheries. In a 1921 interview with the *Great Northern Daily News,* a Seattle-based Japanese American newspaper, he called them "a wonderfully bright people, frugal and industrious . . . but they are Orientals. We are Caucasians. Oil and water do not mix."[5]

The Issei, the first generation of Japanese immigrants, and their American-born children, the Nisei, ran some of the most productive farms in the region, growing fresh berries and vegetables for local markets in Seattle and other Puget Sound cities. Japanese agricultural success aroused white envy, and Freeman exploited that frustration. He supported the Alien Land Law, passed in 1921, to prohibit residents of Japanese descent from owning or leasing land in the state of Washington. After decisions in as many as one hundred separate cases in Washington courts stripped the Issei and Nisei of their property, Freeman and other white speculators snapped up their farmland on the eastern shores of Lake Washington. He also backed the Mercer Island Floating Bridge, opened in 1940, which connected the Eastside to Seattle, in the name of economic progress, and moved his family to the suburbs so they would "have space around them and the opportunity of seeing nature and beauty and life beyond the city streets.... [T]he city could be the place to work, but the country was the place to live." During the Second World War, Freeman testified before Congress in favor of forced internment and led a special committee of Bellevue residents to investigate disloyal Japanese. He then acquired properties from the land they were compelled to evacuate.[6]

After the war, as returning Japanese tried, often unsuccessfully, to recover their former homes and businesses, the elder Freeman purchased an old strawberry farm north of downtown Bellevue as a gift for his son. Kemper Freeman turned the land into Bellevue Square, the region's first suburban mall, and soon became a real estate magnate like his father before him. The younger Freeman's new shopping center was a savvy move because the customers were coming to him. Population in King and Snohomish Counties swelled from 844,572 in 1950 to 1,107,213 by 1960. In 1962, the Evergreen Point Floating Bridge opened, giving the Eastside its second link to Seattle. Within a year, it was jammed with commuters. Three years later, Bellevue surpassed Seattle to become the state's fastest-growing city. The earlier battle to build Metro's sewers and revive polluted Lake Washington signaled that not everyone agreed more suburbs were good for the region, and, in May 1962, the reliably pro-business and Republican *Seattle Times* ran a special issue of its Sunday magazine devoted to sprawl. A local writer, Rillmond Schear, predicted in 1965 that if growth continued unchecked, Seattle could "become the Northwest version of Los Angeles, only without the saving grace of constant sunshine to make the nightmare barely bearable." Citing local demographers and planners, Schear predicted that the lands east of Bellevue, all the way to the Cascade foothills, would be one uninterrupted suburb before the end of the century. When that happened, a "crusty old settler in busy mid-Bellevue" would "stare out toward the mountains and take revenge on his old tormentors."[7]

James Ellis, the founder of Metro and longtime critic of urban sprawl, warned in a 1968 speech that residents of what some were calling "Pugetopolis" had become "lemming-like... masses of self-centered people" marching "sightless into a sea of blinking signs and tin cans." That February, voters in Seattle and King County had turned down major components of an ambitious proposal designed by Ellis, "Forward Thrust," a $2 billon omnibus of twelve bond initiatives, including funds for regional mass transit. Ellis had faced electoral defeat in the Metro campaign ten years earlier, but he remained undaunted. Despite opposition from the state highway department, the Automobile Club of Washington, and suburban developers, including Kemper Freeman, Ellis revamped

Forward Thrust and, with the help of local civic and business leaders, put it before voters again in 1970. By that time, an even larger number of Seattleites wanted to apply the brakes to growth. Interstate 5, completed three years earlier, had forced the condemnation of 4,500 individual properties and hacked a huge asphalt and concrete trench through downtown over the protests of outraged residents. One group, over a hundred strong, mostly women from the city's various social clubs, had picketed before construction started, waving "sprigs of colorful greenery in their hands." Highway protestors in Seattle, like their compatriots from San Francisco to New Orleans, worried that freeways would beget more freeways. When highway engineers and downtown businesses unveiled plans to build several belt routes to encircle Seattle, a loose coalition of clubwomen, elderly pensioners, young environmentalists, and African American activists in Seattle's Central District raised the alarm. The Central District contingent condemned the proposed Interstate 5–Interstate 90 interchange plan, centered in their largely black neighborhood, as serving "racist welfare and pleasure." Emmett Watson, an irascible *Seattle Post-Intelligencer* columnist, sympathized with the protestors and launched a "Lesser Seattle" campaign in mocking contempt of Greater Seattle, Inc., the local booster group, and coined its slogan "Keep the bastards out!"[8]

With typical aplomb, Ellis maneuvered to shore up support for the second round of Forward Thrust, cajoling Senator Warren Magnuson to find almost a billion dollars in federal funds to help pay for mass transit, the proposal's centerpiece, plus new bonds for storm water sewers, community centers, and improved public health services. Of the total $615.5 million in bonds, $440 million would be for buses and light rail. This time, Ellis faced an older, more formidable opponent than real estate developers and state highway engineers: recession. The "Jet City," home to Boeing, was in a tailspin, with congressional cancellation of a planned supersonic transport and declining commercial airline sales. As Boeing slashed its Puget Sound workforce from 100,800 in 1967 to fewer than 40,000 in early 1971, panicked voters defeated the second version of Forward Thrust in 1970. Unemployment skyrocketed to 12 percent, and *Newsweek* asked if Appalachia had come to Seattle. In the 1974 election, Seattleites rejected building more urban arterials, leaving the city with a financial mess: over $2 million in debts, hundreds of condemned properties to sell, and unfinished freeway ramps suspended in midair. In a commonly told anecdote, two real estate agents paid for a billboard that read, "Will the last person leaving SEATTLE—Turn out the lights?"[9]

The decades of urbanization and suburban sprawl, temporarily halted thanks to the Boeing bust, had already attracted the attention of fisheries scientists who had warned, in 1948, that "the effects of civilization and real estate development" were ruining the salmon habitat around Seattle. The philosophy of perpetual growth collided with the limits of nature. Private dams and obstructions had impeded migration, and road culverts had spread stream flow in sheets too shallow for fish to navigate. Homeowners diverted small brooks to water gardens or fill swimming pools, while others straightened creeks and removed riffles, logs, and overhanging vegetation that sheltered nesting salmon and their young. Even larger changes, like flood control dams and agricultural irrigation, were even more worrisome. State officials wondered what effect "continued consumptive diversions" would have on the fisheries.[10]

When floods inundated the Duwamish River valley in the winter of 1946, residents thought little of the salmon and cried out for permanent protection. That same year, a report by the Seattle Planning Commission said it was "inevitable that a portion of these valley lands will be required and used as sites for expanding industry." Citing the deluge, Seattle and King County politicians, backed by affected industries, lobbied for a massive concrete dam at the headwaters of the Green River. Sport anglers and fisheries biologists objected, but the Army Corps of Engineers, which had first considered the plan in the 1920s, started construction on the Eagle Gorge Dam in 1958. It was the era of big dams and aggressive flood control, so when repeated flooding along the lower Cedar River threatened the growing town of Renton and the massive Boeing aircraft plant located near the channel of the former Black River, county engineers dredged the Cedar's channel and built levees. The debris spilled into the river and smothered thousands of salmon redds, or nests. On a float trip to count salmon at the height of the autumn spawning season, in 1964, the state fisheries biologists gave up in frustration because the turbidity made their work impossible.[11]

"We have witnessed over the years dramatic changes in the environment of the entire basin," state fisheries biologists announced in a memo to the Army Corps of Engineers in 1967, "from that of an undisturbed natural watershed to the modern highly urbanized and industrialized community we have today." Salmon had fewer and fewer places left to reproduce. In 1967, the corps proposed enlarging the Lake Washington Ship Canal locks to accommodate more traffic, the bulk of which was recreational boaters. Fisheries managers were deeply concerned. The locks were now the lake's sole outlet, and although improved fish ladders might allow more mature salmon from the sea to return and spawn, without sufficient water juvenile salmon might not survive their return to the ocean. Salt water continued to push its way beneath the locks, industries along the canal route spilled pollutants into the water, and recreational boaters spewed oil and gas behind them with every trip. The city of Seattle every year sucked more water from the Cedar River, the lake's largest tributary, and was proposing to take still more.[12]

Fisheries managers like Thor Tollefson, director of the state Fisheries Department, took heart that the growing environmental movement had inspired local residents to take a greater interest in what was happening to their local streams and rivers. But Tollefson and other fisheries managers were unwilling and unable to confront the enormity of habitat destruction, continuing to rely on hatcheries as a convenient cure-all, a position supported by most sportsmen and commercial operators, who blamed one another for diminishing stocks, especially king (or chinook) salmon, prized for its firm flesh and fighting abilities when hooked. "It has become increasingly apparent," wrote Ed Sierer, a local television journalist and avid sport angler, "that the salmon, for all his worth and majesty, cannot go on serving both camps." Now, another camp was demanding salmon solutions: people living next to disappearing urban creeks and streams. When developers associated with Northgate Shopping Mall began construction on new apartments in 1969, homeowners living next to Thornton Creek in northeast Seattle raised the alarm. William H. Rodgers, Jr., a law professor at the University of Washington and a member of Citizens for the Preservation of Thornton Creek, warned the state fisheries officials of their legal obligation to protect spawning grounds. The mall parking lot already encased the creek's headwaters, and bulldozers had damaged the creek bed. "The use of culverts to direct the flow of water is anticipated," wrote Rodgers, and "heavy flooding will invariably result in erosion" that could threaten the "fish life in the creek," notably silver or coho salmon and cutthroat trout.[13]

Yet, surprisingly, "in a world full of horror stories about man's destruction of the environment," the *Seattle Times* reported in November 1970, the salmon fisheries of Lake Washington had become an unexpected success. The story began decades earlier, in 1917, just after the Ship Canal's completion, when state fish wardens imported sockeye salmon eggs from Baker Lake in northern Washington and Cultus Lake in British Columbia, reared them, and planted the fry in Lake Washington and the Cedar River. Additional periodic plantings followed, in 1935 or 1937, in the hope of establishing a new fishery by artificial means. There was good reason for optimism. Like all anadromous salmon, sockeye spawn in streambeds, but unlike other species, they spend up to two years in lakes as juveniles growing to maturity before heading out to sea. Sockeye also require large rivers for spawning, and almost all of the major waterways in Puget Sound that had once supported runs were, by the early 1900s, dammed or developed. The Cedar River, free of dams in its lower stretch, was one of the best remaining sites. Modest numbers returned to spawn there until 1967, when the number jumped from 45,400 to 190,000, astounding biologists. The following year, the state opened the first-ever commercial fishing season on Lake Washington, followed by a recreational season three years later.[14]

The reasons for the sockeye's success were complex but, ironically, connected to the sewage problems that had frightened scientists and lakeside residents a decade earlier. As the concentration of blue-green algae (*Oscillatoria rubescens*) increased with pollution, the algae suppressed the population of *Daphnia,* a microscopic freshwater crustacean that could not survive in the oxygen-depleted waters or swim in the thick strands that, in turn, affected the vitality of *Daphnia*'s main predator, a tiny crustacean called *Neomysis.* Once Metro's sewage diversions reduced the phosphorous levels, the algae disappeared and *Daphnia* returned. As its numbers soared, long-fin smelt and, possibly, juvenile sockeye, devoured *Neomysis,* which had also jumped in number, helping to boost both the fish and *Daphnia.* The completion of flood control work along the lower Cedar River in the late 1960s may have enhanced sockeye populations as well. Clean, fast-flowing water and new gravel beds were the unintentional gift of the dredges, and once the digging ceased the fish were able to establish a new run.[15]

The *Seattle Times* reporter noted the apparent irony that "man tried to do nature a favor and perhaps got himself in a bind as a result." The Army Corps of Engineers needed more water for the enlarged locks at the Ship Canal; the city wanted more from the Cedar River for its citizens; and state fisheries scientists needed more to sustain the sockeye. Kenneth Lowthian, Seattle's water department chief, complained that the city supply supported both the Metro sewer system and the sockeye plantings. He could not balance the fisheries with "the basic needs of the people and industry of King County—water supply for one-third of the state's population." "Maybe we've cut our own throat," he said. The *Times* reporter framed the contest in even starker terms: "Next fall, one million sockeye will be competing with one million human beings for the pure, clean water of the Cedar River—and there isn't enough for both."[16]

Some of those one million had an even older claim to the water and the salmon, too. The sporadic fishing protests by the Muckleshoot Indians, starting in the early 1960s, that had earned the scorn of Metro officials had become a national issue by the end of the decade. Indians were fishing throughout Washington and Oregon, flouting in public and before the national media regulations many had refused to recognize privately for decades. The actor Marlon Brando, comedian Dick Gregory, and folksinger Buffy Saint-Marie joined the Indian fish-ins. The American Friends Service Committee provided logistical and legal support, just as it did for civil rights protestors in the American South. In response, state authorities used tactics worthy of Bull Connor and those opposed to the civil rights marchers in the South to push Indians back onto their reservations. On September 9, 1970, a company of police and fish wardens invaded an encampment on the Puyallup River near Tacoma, firing tear gas and beating the sixty-odd nonviolent protestors in full view of news reporters before arresting them. Nine days later, the federal government filed a lawsuit against the state of Washington on behalf of the treaty tribes. As Valerie Bridges, a Nisqually Indian and niece of Billy Frank, Jr., a leader in the Indians' resistance, explained: "It's a treaty we're fighting for."[17]

After a long and contentious trial, Judge George H. Boldt ruled, in 1974, in favor of the tribes, a stunning decision from a conservative Republican appointed to the bench by President Dwight D. Eisenhower. In a sweeping 203-page opinion that traversed the legal history of white-Indian relations, the judge reviewed territorial governor Isaac I. Stevens's treaties. Stevens had assured northwestern Indians that they could continue to fish in perpetuity. Treaty signatories, according to Boldt, had retained the right to fish in all of their usual and accustomed places. They were also entitled to the opportunity to take up to one-half of the entire catch. Boldt rejected arguments made by various lawyers for state departments that modern-day Indians were neither fully authentic nor entitled to obligations made more than a hundred years earlier. "The mere passage of time has not eroded, and cannot erode," he emphasized, "the rights guaranteed by solemn treaties that both sides pledged on their honor to uphold." When state fish and game officials continued to harass Indian fishers after his decision, Boldt ordered them to adopt a new management regime, which effectively made Indians co-managers of the region's fisheries. The verdict, later known to many simply as the Boldt Decision, enraged angry white commercial anglers and sportsmen. Some hanged or burned the judge in effigy, while others held their own fish-ins, or blockaded ferry terminals during rush hour.[18]

As the Boldt Decision wended its way through the federal appellate courts, Hazel Wolf, at the age of eighty, convened a panel of Indian activists at the annual convention of the Federation of Western Outdoor Clubs in 1978. Wolf was a fiery former union organizer who had narrowly avoided deportation for her leftist politics. She had helped Seattle's unemployed during the depression and, during the McCarthy era, spoke out against the hearings to expose alleged Communists convened by state legislator Albert Canwell. She became an environmentalist late in life but soon emerged as a major figure, serving as president of the federation, an organization that included the Sierra Club among its members. At the 1978 meeting, the editor of *Outdoors West*, the official publication of the Western Outdoor Clubs, opposed the Boldt ruling and sponsored five anti-Indian resolutions. Wolf took the floor and condemned the resolutions as racist. A majority of delegates, swayed by her arguments, defeated the motions and the editor quit in protest.[19]

The following year, Wolf stepped down as president after pushing the federation to file an *amicus curiae* brief for Phase II of *U.S. v. Washington*. In a letter to the membership, she noted how "much blame for the decline in the fishery resource has been laid to the aggressiveness of Indians, with little attention to major sources of degradation," including logging, road building, dams, and stream alterations. Her hope was to change "the widespread misconception of this as a racial conflict" and "provide a new source of environmental control of value to the entire community." But Wolf was in the minority. Sport anglers, who considered themselves the original environmentalists, opposed sharing salmon, and commercial fishermen cited lost jobs. Both made sure the local press covered their side. Front-page stories of Indian rejoicing following the Boldt Decision, according to some observers, fanned feelings of resentment. A reporter for the *Seattle Post-Intelligencer* described the mood at Fisherman's Terminal in Ballard as similar to "a slow tide bringing in crud to foul the gear."[20]

In an interview with a local television station in 1979, Boldt, then retired, admitted to being "quite elated . . . so much so that I was quivering a little" after the United States Supreme Court upheld his decision that year. He told the *Seattle Times* the ruling was "a victory for justice." Billy Frank, Jr., a Nisqually leader at the center of the controversy, later praised Boldt for letting the Indians "tell our stories, right there in federal court." "He made a decision, he interpreted the treaty, and he gave us a tool to help save the salmon," Frank said later, remembering the ruling and the man who wrote it. Frank's generosity could not paper over what had become a nasty, sometimes violent struggle. For several more years, the continuing furor bared the divisions between Indian and white, rich and poor, city and suburb, in ever more complicated ways, testing the limits of any optimism.[21]

At first, the Boldt Decision seemed simply to be a fight between Indian and white anglers over who had the right to catch fish, where, and in what numbers. Those charged with protecting the city's most precious utilities—City Light and the Cedar River—knew better. The dams on the Skagit River, which provided inexpensive power, were salmon killing machines, blocking fish from reaching their historic spawning grounds or altering the course of the river downstream with every water release. Frank knew this and said that the Northwest's cheap electricity had "all been paid for by the salmon." "When these lights come on," he explained, "a salmon comes flying out." The same was true of the water spigots in every Seattle home. City officials had blocked spawning salmon from swimming above the intake at Landsburg since 1901 out of fear that rotting carcasses would contaminate supplies. Farther upstream, the Water Department continued logging to pay for watershed maintenance and land acquisition, dumping silt and debris into the river where the currents washed it downstream, depositing the mess on top of the spawning grounds. The fish were always a secondary concern, but that position was about to be challenged.[22]

Save Your Watershed for One Latté a Year

After the "Boxley Burst" of December 1918, when heavy rains undermined the wings of the Cedar Lake reservoir and unleashed the torrent that swept away the mill town of Edgewick, city engineers and foresters had pursued an aggressive policy of selective logging and reforestation to fund watershed maintenance and prevent further flooding. The Water Department negotiated an exclusive arrangement with a private contractor, the Pacific States Lumber Company, to log city-owned lands as well as those owned by the Northern Pacific Railroad, which, when logged, would be sold to the city at a reduced rate. Seattle city attorneys, at the urging of the health commissioner, condemned seventeen-odd logging camps and towns scattered throughout the forest in the name of public safety. The local lumber companies, who hired Japanese and Scandinavian immigrants to live in the towns of Barneston and Taylor, as well as the smaller lumber camps, were compelled to sell out under eminent domain and evict their employees. Unfortunately, the acres of stumps and mud that remained, according to city health commissioner Dr. E. T. Hanley, did not provide "a very pleasant sight for train passengers" passing through "our highly recommended watershed." Downstream, consumers complained that their tap water had "a slight tang and discoloration" that marked it as "a different 'brand' from the famous Cedar River water of . . . earlier boasting." Despite reassurances from Water Department officials that all was well with the city's supply, a twelve-part series in the *Seattle Times* in 1930, titled "The Forests a City Forgot," exposed what the paper called "mismanagement, arising from a combination of politics, ignorance, and willfulness." Goaded by the *Times* reports, the city blamed the Pacific States Lumber Company and went to court to break its contract. The city lost in litigation, and the controversy subsided when the depression all but ruined the local timber economy.[23]

By the early 1940s, with wartime demand for Northwest timber soaring, the Northern Pacific Railroad, looking to divest itself of its Cedar River holdings, offered to donate its remaining lands to Seattle if it could sell the timber first. Once the offer became public, opinion split between those supporting William C. Morse, the Water Department superintendent, and those backing James Scavotto, a populist on the city council who decried further cutting. Morse explained that water customers would benefit by taking "two crops from this watershed—a timber crop and a water crop." Scavotto accused Morse of brazenly taking a "gamble with public health" and a watershed "generally recognized as the best in the country." A two-year battle ensued and, predictably, the city council convened an outside panel, chaired by the sanitary engineer Abel Wolman, to settle the dispute. The report, released in February 1944, determined that logging presented no problem so long as the watershed was in the hands of professional foresters and engineers to assure "some guarantee of permanence and of stability." Citing Wolman's findings, the city council signed an agreement with the U.S. Forest Service and the Northern Pacific to put the watershed's forests under sustained yield management, a policy developed earlier that year by Congress to promote timber production. After the war, the Water Department had continued to log aggressively and often below its quota. Some of the revenues were used to acquire remaining watershed lands until 1996, when the Forest Service ceded its holdings and left the city with full ownership of the watershed's approximately 91,400 acres, just in time to face the full brunt of the salmon crisis.[24]

The completion of the city's second water supply at the headwaters of the Tolt River, north of the Cedar River, in 1964, had alleviated pressure on the Cedar but did not diminish its importance as Seattle's primary supply. Facing potential lawsuits from environmentalists, sport anglers, treaty Indians, and the federal government to protect the salmon, the city of Seattle, in 1995, created a habitat conservation plan for its municipal watershed. Habitat conservation plans, designed by Congress that same year as an exception to the Endangered Species Act of 1973, permitted non-federal property owners who had sufficient cause to receive "incidental take" permits indemnifying them if they killed or injured federally protected flora or fauna. In the plan's first draft, the Seattle water department proposed limiting logging to approximately one-third of the watershed forests and building a new $25 million sockeye hatchery to mitigate destroyed habitat. Department scientists and engineers drafted, as federal law required, an environmental impact statement detailing the plan's costs and benefits, then met with the Muckleshoot Indian Tribe, King County, the Municipality of Metropolitan Seattle, and the National Marine Fisheries Service to ensure approval. Diana Gale, the water department head, expressed hope that participants would "transcend jurisdictional boundaries" and help construct "one of the nation's premiere environmental restoration success stories."[25]

Gale's optimism was quickly put to the test. Under federal and state guidelines established by the National Environmental Policy Act, city officials had to open the process to public comment. They also had to open the river above the city's water intake at Landsburg to spawning salmon, a major change in policy that came after public health officers said the detrimental effects of dead salmon on water quality were negligible. But turning the Cedar River into both source for municipal water and habitat for salmon would be expensive. Seattle Public Utilities, the agency that subsumed the former Engineering and Water Departments in 1997, wanted to stick to the almost ninety-year-old policy of harvesting timber to pay for operations in the habitat conservation plan. Several major environmental organizations, including the local branch of the Sierra Club, appalled at what they saw as political double-dealing, lined up in opposition, joining ecological concerns with a reverential affirmation of the wilderness ideal. Yet the Cedar River had never been an uninhabited wilderness; archaeologists had found evidence of human activity as far back as 9,400 years, the oldest known human presence at that elevation in the Cascade

Mountains. Various Native groups had gathered to trade, hunt, fashion projectile points, and collect berries, nuts, and roots. Human-set fires were commonplace, a traditional method to improve conditions for game and edible plants, like camas.[26]

For environmentalists, however, the Cedar River was a wilderness in spite of its history, because of its "old growth" or "ancient" forest, the mature stands of Douglas fir, red cedar, and hemlock that once blanketed the Pacific Northwest. Wilderness, as applied to the Cedar River, was more of a political canard than an ecological fact. Only about 14,000 acres, mostly concentrated in the upper reaches of the watershed, about 15 percent of the total, contained trees older than two hundred years. The majority of acreage, covered with second-growth forest ranging from thirty to eighty years old, had come back thanks to earlier reforestation efforts or natural processes after the loggers had left, but with relatively homogeneous coniferous species mixed in with deciduous trees.[27]

The ecological or historical complexities did not matter to old-growth advocates. At the end of the 1980s, they were locked in a bitter struggle over the fate of logging ancient forests on public lands throughout the Pacific Northwest, in the United States as well as Canada. The conflict pitted small rural timber towns from Fort Nelson, British Columbia, to Forks, Washington, to Coos Bay, Oregon, against national environmental groups like the Sierra Club. In the northwestern United States, the symbol of this war was the northern spotted owl, an endangered species that nested only in mature conifer forests, and because some biologists had found evidence of the owl's nesting in the Cedar River watershed, it was swept up in the larger struggle over ancient forests. But the real icon for this particular skirmish over the Cedar River and its trees would be the various species of Pacific salmon.[28]

For wilderness campaigners, saving the native salmon was, like saving the spotted owl, the way to stop the cutting of old-growth timber, and the proposed hatchery was as abhorrent as chainsaws and logging roads. Since the 1970s, hatcheries had been anathema to many environmentalists and sport anglers as mounting scientific evidence pointed to the detrimental effects of artificial propagation, ranging from diluted genetic diversity of wild stock to increased susceptibility to disease. Yet hatcheries did provide, in a way, a regular supply of fish. Greens and Indians could agree on protecting habitat to save salmon; they could agree on little else. Focused on their treaty rights, Indians wanted to ensure a reliable supply of fish. They backed the proposed sockeye hatchery in addition to supporting habitat preservation. Major environmental groups disagreed. In a letter to Stan Moses, chair of the Muckleshoot tribal fish committee, the Sierra Club's Charlie Raines cited his organization's first priority, "maintaining and restoring ecosystem function (wild salmon and habitat) with artificial measures as a last resort." He sympathized with the Indians, but his understanding ended at hatcheries.[29]

The Muckleshoot were adamant in defending their treaty rights by almost any means necessary, and hatcheries were one of many mechanisms for ensuring compliance. Indians, however, were not inflexible in their approach toward salmon management, and they supported protecting and restoring salmon habitat in addition to artificial propagation. Indeed, the Muckleshoot, like other Native peoples, had often exploited environmentalist sentiments to their political advantage. They had collaborated successfully with local environmentalist groups, in 1982, to stop a multimillion-dollar marina project near the mouth of the Duwamish River in the name of protecting salmon habitat, and five years later they created a tribal fishing reserve in Elliott Bay, a move that earned the tribe high praise from the *Seattle Times* editorial board. But the Muckleshoot wanted more than kudos from environmentalists and the press. As Stan Moses explained to the Washington Environmental Council, neither habitat restoration nor hatcheries entirely satisfied treaty obligations. As he saw it, the Boldt Decision reaffirmed that the watershed was Indian land; thus, the river's water belonged to the Indians as well.[30]

Following this logic, the Muckleshoot did not trust Seattle Public Utilities to manage the river with their interests in mind because, like the Sierra Club, they realized that the problem facing salmon was not saving the trees alone. The Cedar River was Seattle's faucet and the city's utility managers turned it on and off depending on consumers' needs. As a result, when demand was at its highest in summer and fall, the waters of the river ran low and warm. It was a deadly combination for salmon returning to spawn. The fish required cool temperatures and ample volume, and the Cedar was now an unreliable river for the fish, which made it an unreliable river for Indians, too. The Muckleshoot wanted full access to the watershed as co-managers, and as tribal chairwoman Virginia Cross explained, by "locking gates on roads that lead into the watershed with no consideration of the Muckleshoot's subsistence," the city violated federal law. The Muckleshoot realized that without their support, Seattle's plan would fail. They used their advantage to push hard for concessions from the city. In exchange for backing the salmon hatchery, the Muckleshoot expected to receive guarantees in the habitat conservation plan for hunting, fishing, and ceremonial visits to the watershed in perpetuity.[31]

Curiously, by pushing for the salmon hatchery, the Muckleshoot allied themselves with their longtime adversaries, recreational anglers, many of whom wanted the hatchery as well. Sport fishermen, while understanding that hatcheries were no longer the solution they had once embraced, were also unwilling to concede that they might have to live with a diminished resource. Frank Urabeck of Trout Unlimited argued that "habitat restoration projects were very much experimental with little documented success to date," whereas removing blockages faced by spawning salmon, such as the Ship Canal locks, or using hatcheries selectively were more likely to succeed. Other sport anglers claimed that environmentalists were out to destroy their avocation. John C. Evensen, writing to the Seattle city council, said that artificial propagation would benefit everyone and "not just the rich elitists," and Mark Scalzo concurred, insisting that hatcheries would ensure "an urban fishing experience unlike any other, . . . a calm Lake Washington on a clear morning, Mt. Rainier on one end, the downtown skyline to the west, no cars, trucks, TV to get in the way yet minutes from home."[32]

Parties on all sides of the watershed debate invoked their own entitlements and, beginning in late 1997, wilderness advocates joined the chorus, banding together as the Protect Our Watershed Alliance to drown city hall in letters, phone calls, and electronic mail. Their position was simple: stop the logging. Some offered creative ways to raise money to protect the forests, with one scheme calling for bottling and selling Cedar River water under the label "Seattle Rain." Another idea was to charge an annual fee of $3–4 per customer to cover the estimated $83 million cost of the habitat conservation plan over its fifty-year span. Still another suggestion, titled "Save Your Watershed for One Latté a Year," exhorted Seattleites to "give up that one latté . . . or make it one beer, or a pack of cigarettes or maybe forgo that cheap video" to gain "a green watershed, one with fewer roads and more trees." The new mayor, Paul Schell, elected in the autumn of 1997, had tested the political wind and announced in early 1998 that watershed management should not be subsidized by logging what he called a pristine area.[33]

Schell was the face of a new kind of environmentalism in Seattle, a mutant version of James Ellis's Eisenhower-era civic altruism that promoted ostensibly ecological principles in the name of urban growth. A wealthy real estate developer and former dean of the College of Architecture and Urban Design at the University of Washington, Schell, with a snow-white head of hair and a personality that even supporters found detached and academic, was a fervent cheerleader for "Cascadia," an invented region running from the Oregon-California border to southern British Columbia. He explained, at a 1992 conference on urban planning and development, that Cascadians were "united by a love of the outdoors and reverence for the environment passed to us from the native people." Cascadia, based in part on Ernest Callenbach's novel *Ecotopia,* embraced by free-trade advocates and hard-core greens alike, was a version of what some historians and geographers have called bioregionalism—the notion that regions can be defined by common environmental characteristics. Seen another way, it was the old mantra of Seattle's boosters now wrapped in flashy and progressive green packaging.[34]

Schell's idealism had captured the attention of the voters when he won election as mayor, and, sensing that popular opinion was turning against logging, he played the environment card to claim the zero-logging stance as his own. The editorial boards of the *Seattle Times* and the *Post-Intelligencer,* and the city's two iconoclastic free newsstand weeklies, *The Stranger* and the *Seattle Weekly,* published enthusiastic endorsements. Ed Zuckerman of Washington Conservation Voters said Schell "surprised us in the strength of his words and his language," his Cedar River plan the "administration's first major environmental test." But, if so, Schell soon failed in the eyes of environmental voters, when, with his blessing, Seattle Public Utilities suspended discussions on maintaining minimum stream flow and refused to take the hatchery off the table. Muckleshoot tribal council chairman John Daniels, Jr., angered over the city's hydraulic parsimony, threatened to "exhaust all avenues necessary to ensure recovery and protection of salmon stocks." In an open letter to Schell and the city council, eleven prominent fisheries' biologists and environmental engineers belittled the city's reliance on hatcheries for fostering "a false sense that 'everything is OK'" by "diverting limited financial resources from broader salmon and river recovery goals." Another correspondent asked Schell and the city council if they wanted to be "remembered for having made our salmon an extinct species."[35]

In the barrage of charges and countercharges, the federal Departments of Interior and Commerce handed down, in 1999, what fisheries' scientists had anticipated for years: the announcement of the endangered status of several species of Pacific salmon, including the Puget Sound chinook. Schell, in a *Seattle Times* op-ed essay, asserted there was "no one villain in this drama," and blamed instead "the growth and development that surrounds us." Schell's assessment was historically, if narrowly, correct but politically naive. His city was now the villain, a role reversal that American Rivers, a national environmental group, threw back in the mayor's face when it named the Cedar River as one of the nation's ten most endangered waterways. Schell and Margaret Pageler, a council member who supported logging and hatcheries, lambasted American Rivers for slandering "any organization inclined to pursue collaborative, pro-environment approaches to natural resources management." Environmentalists were not the only ones enraged by the city's vacillation. Utilities districts throughout King County, major consumers of Seattle water, were now worried that protecting salmon could portend higher rates and limited supplies. "We should not forget that people are as much a part of the environment," warned the head of one suburban water district, "as are the spotted owl, salmon, and trees."[36]

As in the 1958 campaigns against Metro, the power of the tax dollar was strongest in the suburbs and small towns that purchased Seattle's surplus water, and one Renton man wondered if Schell was "getting bad advice from his Sierra Club friends." If environmentalists could lay low the powerful city, suburbanites and rural residents wondered, what would become of small property owners? Maxine Keesling, an "affected rural landowner" in Woodinville and a member of several property-rights groups, said, "When state agencies, working with environmental-lockup groups, gain their desired regulatory authority over local land use, the public and landowners will be out-of-pocket and out-of-use." The real threat to salmon, these critics maintained, was not logging, which prevented devastating fires by removing excess fuel, but animal and human predators, especially Indians and foreign commercial fishers at sea. Scientific studies suggested that the most significant hazards facing salmon were not natural predation or overfishing but habitat destruction. Expert consensus mattered little, however, in the ongoing drama of salmon politics.[37]

After five years of meetings and studies, hearings and panels, Seattle Public Utilities issued its final habitat conservation plan in April 2000, a staggering thousand-plus-page document detailing the watershed's geology and hydrology, complete with management plans for fourteen endangered bird, fish, and mammal species, including the chinook salmon and bull trout. (It omitted plans for another sixty-nine "species of concern.") The city council, at Schell's request, demanded an end to all commercial logging and the removal of 38 percent of existing logging roads. Later modifications permitted salmon and trout to spawn above the intake at Landsburg for the first time in a century, since the city seized the river as its own. Seattle Public Utilities pledged to help restore and protect fish habitat in the lower Cedar River, improve fish passage through the Ship Canal locks, and fund further studies of chinook and sockeye populations in Lake Washington. The total cost of the plan was almost $78 million. The sockeye hatchery remained and the city retained the right to an incidental "take" of threatened and endangered species to ensure sufficient water supplies for Seattle customers. The Muckleshoot Indian Tribe had the access it wanted for hunting and ceremonial purposes, but tribal leaders were concerned that the annual minimum river flow was inadequate, and prepared to mount a lawsuit against the city. Sport anglers split; some supported the hatchery, as did their Indian foes, while others condemned the city for ruining native fish populations.[38]

In the end, as one disgruntled citizen remarked, the habitat conservation plan was "a gutless, political choice following the path of least resistance," one designed to empower "a bunch of biologists measuring and monitoring ad infinitum." "Nature is not static and will not be preserved," he concluded. Such comments were oddly prescient. From the crest of the Cascades to the shores of Puget Sound, the waters and lands that encompassed the region were neither fully natural nor completely artificial, but a fusion of the two. In this complex ecosystem, separating the human from the non-human was impossible; the attempts to use salmon to unite a real-life Ecotopia called Cascadia pointed to this shortcoming. "The fish that might save Seattle," as Paul Schell had confidently predicted, had once again divided it.[39]

Re-creating Nature as Best We Can

Seattle seemed ready to afford such an ambitious plan for its municipal watershed because the boom times had come back. Emerald City was the poster child of the high-tech nineties. At the end of the previous century, gold seekers had passed through Seattle en route to the Klondike. Now, at the end of the twentieth century, Seattle itself was the destination and the gold fields were computer terminals made of silicon, powered by cheap electricity generated by the Northwest's dammed rivers. At the close of the 1980s, as Seattle emerged belatedly from the nationwide recession, a local son and Harvard dropout, William Gates, Jr., joined his high school pal, Paul Allen, to found a computer software company whose business practices would have impressed the Empire

Builder himself, the railroad tycoon James J. Hill. Another entrepreneur started selling books on the Internet; still another dreamed of putting gourmet coffee in the hands of every tired commuter or computer-bound office worker.

Even the venerable Recreational Equipment Incorporated, or REI, the plucky cooperative founded by Lloyd Anderson and his fellow Mountaineers during the Great Depression to provide inexpensive equipment for hard-core campers, had transformed in a way that made it, too, a symbol of this new Seattle. It was now a multimillion-dollar corporation with more than fifty stores from California to Massachusetts, selling imported equipment and clothing manufactured in China or Mexico. In 1996, at the height of the dot-com boom, the company built a 98,000-square-foot flagship store at the edge of the old Denny Hill regrade. There, customers could test Gore-Tex jackets in a rain simulator, rock climb on a 65-foot freestanding pinnacle weighing 110 tons, or hike on trails amid native northwestern plants, a cascading waterfall, and views of the Olympic Mountains and Puget Sound. With the bicycle department's cache of 350 bicycles and an equal number of shoe and boot models, all housed in a $30 million edifice of recycled wood and concrete, the store was the apotheosis of hip, outdoorsy consumerism. Timothy Egan, a *New York Times* correspondent, said that some people were calling it "Hike-town," a sarcastic reference to the new Niketown, built by the Oregon-based sport shoe manufacturer, in downtown Seattle. When pressed by Egan to say if the cooperative had strayed from its founding principles, the company president retorted, "This is our roots." Eddie Bauer had already gone national by marketing its rustic Northwest pedigree, and Seattle's beloved homegrown cooperative followed suit. Indeed, Seattle had become a brand name. It even had its own soundtrack in "grunge," the proto-punk and heavy metal brew first popularized by local bands like Green River and Mudhoney. By the end of the decade, at any given moment, someone on the planet was probably buying a book from Amazon.com using a computer running Microsoft Windows and Internet Explorer while sipping a Starbucks Grande latté, wearing an Eddie Bauer or REI polyester fleece jacket, and listening to a song by Soundgarden or Nirvana or Pearl Jam.[40]

R. W. Apple, Jr., of the *New York Times* proposed that, with the exception of New York and Los Angeles, no other city had "shaped modern living in the United States" as much as Seattle. From aircraft to software, Seattle was American ingenuity made real in your kitchen or office cubicle. Writer and British émigré Jonathan Raban, lured by the energy and beauty, had moved to this "new last frontier" and this city "built in a wilderness and designed to dazzle." The Seattle–King County Convention and Visitors Bureau capitalized on sentiments like Raban's in its new marketing slogan, "Seattle: The Emerald City," submitted by a California writer as part of a nationwide contest in 1982, meant to evoke the imaginary Land of Oz. The accompanying logo set Seattle's skyline against the backdrop of Mount Rainier in an emerald-cut gemstone, a romantic blend of city and sea and mountains and green that made it easy to tout the city's outdoorsy yet sophisticated lifestyle. Hiking alpine trails in the morning, kayaking in the shadow of gleaming new skyscrapers, grilling fresh salmon for dinner, and attending opera at night—Seattle was an ideal blend of the urbane and the wild. A *Seattle Times* poll in 2000 found that almost two-thirds of all respondents would live nowhere else. Almost four out of ten listed the environment as the reason.[41]

Once a provincial backwater best known for rain showers and loggers and faded flannel shirts, Seattle now faced, in the words of Emily Baillargeon Russin, a native who had gone back east for school and returned at the height of the high-tech bubble, "an emerging sense of the city as a vast personal playground." Russin's contempt was for the newly wealthy, but many middle-class Seattleites, influenced by the lingering elements of 1960s counterculture, had embraced an upscale environmentalism as a lifestyle choice. They shopped at Puget Consumers Cooperative, founded in 1961 to sell low-cost natural food, which has morphed into a high-end grocery store. They joined organizations like the Mountaineers and supported the creation of the Alpine Lakes Wilderness Area, just north of Interstate 90 in the Cascade Mountains, to enjoy the great outdoors close by while fighting to set aside even more wilderness. They moved into old houses in previously blue-collar neighborhoods like Fremont and turned home improvement into a political fashion statement. They were skeptical of the push for more freeways, scornful of the suburbs that had backed more roads, and suspicious of outsiders, Californians above all, whom they blamed for sullying their paradise. Many of the most strident were themselves closeted former Californians, the "one group of immigrants who could be openly abused in bars and public places, in the newspapers and on television," in polite and progressive Seattle according to Raban. The latest arrivals, too, wanted to seal off their city of green.[42]

Urban ecological chic soon became big business when, in 1995, Microsoft co-founder Paul Allen backed a plan to turn the south end of Lake Union, including the still-defunct Denny Regrade district, into a new city park, dubbed the Seattle Commons. Its supporters imagined a public green, the rival of New York's Central Park or San Francisco's Golden Gate Park, surrounded by trendy, high-density housing, high-tech employers, and even restored streams filled with salmon. The $111 million plan, supported by a consortium of real estate developers and urban planners and sold as a latter-day version of Virgil Bogue's 1911 *Plan of Seattle,* twice failed to entice voters, after which Allen began to acquire the properties on his own, achieving with his checkbook what he could not win at the ballot box.[43]

Fred Moody, a former *Seattle Weekly* editor, looked on his hometown and wryly wrote, "Seattle, long a haven for dropouts and rebels, had turned into a high-tech Rome, begging to be sacked." That comeuppance came in late 1999 when raucous street protests over the World Trade Organization (WTO) ministerial meetings convulsed the city and shocked the nation. Mayor Paul Schell, a longtime proponent of free trade, had wanted to show off Northwest industry and trade facilities to the world, but he and his collaborators from the Port of Seattle and the Seattle–King County Chamber of Commerce did not anticipate that the world, angry over free trade, would also come to the WTO meeting. For five days, from November 29 to December 3, shipyard steelworkers and Boeing machinists agitated for trade tariffs, marching alongside environmentalists and animal rights activists dressed as sea turtles and salmon. The protests started as a peaceful mass action but soon turned violent when a small band of self-proclaimed anarchists, fortified by restless hangers-on, looted and vandalized major retailers at the height of the holiday shopping season. Besieged police officers reacted, at times brutally, by firing tear gas and rubber bullets into the throng. They arrested thousands; fifty-six police officers and almost one hundred protestors were injured. In the aftermath, Norm Stamper, hired as police chief for his progressive policing philosophy, resigned, and Schell lost his reelection bid. A city review committee concluded that Seattle's leaders had failed their citizens "through careless and naive planning." The editors of the *Seattle Weekly,* echoing Fred Moody, puckishly called the WTO protests "the Seattle World's fair on acid . . . only better."[44]

For those on the edges in nouveau riche Seattle, the WTO protests, hailed as a triumph by many mainstream environmentalists, were another example of their exclusion from a movement fixated on distant causes and faraway places. Kristine Wong, an organizer for the Community Coalition for Environmental Justice, had successfully shut down two medical waste incinerators the year before at the Veterans Administration Hospital and Northwest Hospital that were spewing hydrogen chloride, mercury, cadmium, and dioxins across Beacon Hill neighborhoods populated by poor African Americans, Asian Americans, and Pacific Islanders. Buoyed by the victories and in anticipation of the WTO meeting, she tried to convince protest organizers that poor Seattleites living next door to polluting industries were victims of burgeoning global trade and the flourishing high-tech economy. Few in the "overwhelmingly white mainstream non-governmental organizations" wanted to hear her opinion. After all, she explained, most Seattleites defined "the environment as forests, trees, [and] salmon" without considering how recent immigrants or Native Americans depended on "contaminated fish" for sustenance.[45]

Perhaps no place symbolized Wong's frustration better than South Park, an enclave of homes sandwiched between Interstate 5, Seattle-Tacoma International Airport, and the Duwamish Waterway. In 1990, the South Park district had one of the city's lowest median household incomes, about $20,000 per household, and the city's largest concentration of non-white residents: 15 percent Hispanic, 13 percent Asian–Pacific Islander, 9 percent African American, and 3 percent Native American. One South Park resident called his neighborhood the city's "dumping ground." Life expectancy there was almost nine years less than for other Seattleites; a local doctor reported higher than normal rates of respiratory ailments from his Latino patients; the air quality was routinely below federal standards. Another resident claimed he had found crawfish with extra legs crawling in nearby streams. Metro had tried previously to address the pollution problem in the Duwamish River with new treatment facilities and interceptor sewers. To the agency's credit, the improvements worked quite well, yet they could not remove the decades of contaminants already lodged in the riverbed. Heavy metals like mercury, polychlorinated biphenyls (commonly known as PCBs), and dioxin, to name a few, were entering the food chain. Metro scientists found increasing numbers of tumors, fin erosion, and liver failure in bottom-dwelling rock sole and sculpin. Based on these and other findings, the U.S. Environmental Protection Agency designated the lower Duwamish as a Superfund cleanup site in 2000 in advance of dredging by the Port of Seattle to open slips for more large container ships. Scientists had found high concentrations of PCBs in 91 percent of three-hundred-plus sediment samples, and later studies found the same toxin in Puget Sound chinook salmon, raising alarms that Indians dependent on fish for food could be at risk for certain types of cancer. The next year, the state Board of Health approved a report citing environmental justice as an important public health priority and identified the Duwamish River corridor as a location of particular concern. For the first time, state health officials warned children and pregnant women to limit their consumption of Puget Sound fish, particularly salmon.[46]

Salmon and other aquatic creatures carried the problems of South Park and the lower Duwamish with them into almost every neighborhood in Seattle and King County, stored in the adipose or fatty tissue that fueled their epic spawning runs. The 1999 endangered species listing drove home an important biological point: almost everywhere that water flowed to the sea was potential salmon habitat. As federal and local agencies scrambled to come up with new regulations after the endangered species listings, they imposed tight restrictions to err on the side of caution. Federally funded low-income housing developments were held up until contractors assessed the effects of each project on nearby salmon habitat. Costs for other construction projects, from highway expansion on the car-clogged Eastside to new light-rail train tracks planned to ease gridlock, increased exponentially the closer the construction was to wetlands or spawning streams. Dogs were prevented from playing in the Sammamish River in Marymoor Park in Redmond, near Microsoft's plush corporate headquarters, when chinook salmon spawned. One reporter summed up the cumulative effects of the salmon listing: "Think of the spotted owl on a much grander scale, touching on every industry instead of just one."[47]

Historian Joseph Taylor has called the battle over salmon "a durable crisis," but until the end of the twentieth century the costs of this crisis were largely borne by rural, working-class, or Native American residents. Following the endangered species listings in 1999, newly rich and environmentally conscious urbanites began to pay the price for sustaining salmon, a shift in politics that failed to explain the resolution, passed by the Seattle city council in autumn 2000, calling for the removal of four dams on the lower Snake River to save salmon in the Columbia River. The caustic replies from state representatives and citizens living in rural eastern Washington, whose livelihoods depended on irrigated agriculture, took many liberal Seattleites by surprise. One resident from Wenatchee, the apple-growing capital of the Northwest, called Seattleites sanctimonious hypocrites and said they should demolish the Ship Canal, since "there was no salmon run through the locks before the locks were built."[48]

There may have been merit to the rejoinder. Even if salmon survived their trip through the fish ladder at the locks, they faced a new threat. California sea lions were timing their migration to match the seasonal sockeye and steelhead migration, stopping at the locks and each eating upward of fifteen fish per

Editorial cartoon by David Horsey depicting Seattle's possible future after the Endangered Species Act listing of Puget Sound chinook salmon, *Seattle Post-Intelligencer,* March 17, 1999. (Used with permission of *Seattle Post-Intelligencer*)

day. The locks were a man-made seal buffet. Federal fisheries biologists and state game wardens threw firecrackers and fired rubber bullets. The sea lions ignored them. One animal named Hondo, relocated to California, returned nine months later in time for his meal. Frustrated sportsmen petitioned Congress to amend the Marine Mammal Protection Act so they could kill Hondo, Thumper, and the most famous sea lion, Herschel. The Muckleshoot even offered to hunt the seals, earning rare praise from white sport anglers.[49]

The cumulative effect of almost a century of changes to Seattle and its environs cut across a dizzying range of space and time, confounding easy attempts to save salmon. Hatcheries provided no sure relief to the limitations of diminished habitat, and even protecting habitat could not ensure success because salmon at sea faced other threats: commercial harvesting, sport fishing, El Niños, and the rise of salmon aquaculture, an industry imported from Norway and the Canadian Maritimes to meet consumer demand for fresh fish. Salmon farms in coastal Washington and British Columbia that raised genetically altered Atlantic salmon increased local concentrations of sea lice, potentially devastating parasites that attached themselves to passing wild salmon. Equally worrisome were the escapees that could spread dangerous epizootics or infectious anemia viruses among Pacific salmon or, possibly, colonize streams and crowd out native populations. By the late 1990s, scientists had found evidence of Atlantic salmon spawning on their own in coastal British Columbia.[50]

The editorial board of the *Seattle Post-Intelligencer,* responding to the endangered species listing, said sarcastically that Seattleites had it coming to them since they had "long been the state's most vocal proponents of saving the environment in other people's back yards." It was "poetic justice" for Seattleites "to show their willingness to do the same." Seattle political leaders tried to meet the challenge, if partially and with an eye on political expediency. Mayor Paul Schell had inaugurated the Urban Creeks Legacy project in 1999, before the WTO debacle ruined his reelection, to restore four of Seattle's largest remaining streams: Pipers Creek, Taylor Creek, Longfellow Creek, and Thornton Creek, the latter paved over at its headwaters in the 1940s to build the Northgate Mall. Denise Andrews, urban creeks coordinator for Seattle Public Utilities, explained that restoring the city's streams centered on "recreating nature as best we can." As Schell had written in an essay promoting the effort, "people and fish can live together, even in an urban setting." But Schell, a real estate developer, miscalculated the cost. A geology and engineering professor at the University of Washington, Derek Booth, put the total price for bringing back salmon habitat at more than $1 billion for urban Puget Sound alone, and this estimate was likely modest, Booth added, because "the science of salmon living in disturbed watersheds is pretty young."[51]

James Fallows, writing in the *Atlantic Monthly,* rightly noted that the choice was about saving not the salmon but a way of life. The recent high-tech boom had put a strain on the region's capacity to absorb growth. The newly wealthy and newly arrived wanted to live close to the water and trees—the very places that wild salmon needed most. When Seattleites talked about saving salmon, they often failed to separate three intertwined but contradictory goals: protecting the salmon as a charismatic species, maintaining fisheries to sustain Indians and commercial fishermen as well as consumer demand, and preserving the habitat necessary for salmon to thrive. Each group had its own agenda and its own solution—remove dams, or limit new construction, or restore streams, or boost hatchery production, or restrict or ban fishing, or protect treaty rights—but no one group was willing to curtail its own behavior. The century-plus of environmental changes that made the good life in the Northwest possible were at the root of the conundrum. "The debate about whether this transformation should be undone," Fallows said, "is worth carrying out on its own terms—not on the backs of the fish."[52]

To Fallows, the Indians of Puget Sound might have replied that undoing the changes to their homeland should not be carried out on their backs, since, as they had argued for decades, Indians suffered the most when salmon began to disappear. The cooperative arrangements negotiated through the Boldt Decision had worked reasonably well until treaty Indians pushed to include shellfish as part of their treaty rights. The problem was that tidelands in Washington, unlike in most of the nation, were often in private hands. Negotiations between state officials and treaty Indians in the late 1980s quickly broke down over the question of access to public and private tidelands. Tribal officials stressed that they did not want to drive private shellfish growers out of business or usurp waterfront owners' rights. Opposing the tribes was the United Property Owners of Washington, a property-rights group backed by powerful title companies, such as Chicago Title and Transamerica, and the state of Washington.

The tribes' motivation was simple: healthy shellfish habitat was fast disappearing thanks to decades of shoreline encroachment and pollution, ruining beds in all but the most remote reaches of Puget Sound. Private shellfish growers, too, worried about pollution, but were more concerned that Indians might lay claim to half of their oysters and clams, driving them out of business. Shellfish growers considered treaty rights as trespassing, especially on lands improved specifically for shellfish farming. Property owners ignored the fact that the treaties were specific in their intent. Government agents and Indians had not created special rights for Indians but had reached a contractual agreement reserving some of the Indians' preexisting rights. In late 1994, Judge Edward Rafeedie sided with the treaty tribes, except for finding that Indians could not harvest from private commercial beds. A later ruling protected private waterfront owners who did not operate shellfish farms by requiring that treaty Indians give ample notice before harvesting.[53]

Natives had won another significant victory, yet it would be hollow if there were no salmon to catch and all the clams were poisoned by heavy metals and sewage. Earlier court rulings had given the treaty tribes power over fishery management, hatchery production, and catch quotas. Now they wanted to sit at the table to discuss how to bring salmon and shellfish back. Some legal experts and some environmentalists had already seen this coming in the courtroom maneuverings that followed the original 1974 case, *U.S. v. Washington*, when Indians claimed that the treaty fishing rights included the right to protect the habitat salmon needed to survive. These arguments were set aside at the time for later litigation even as the United States Supreme Court upheld the main thrust of the Boldt Decision. Eventually, the claims made their way back to trial, and, in 1980, Judge William Orrick, Jr., ruled in favor of the right to habitat protection. Orrick's decision, later vacated on appeal to the Ninth Circuit on procedural grounds, led some scholars to conclude that the treaty fishing right was indeed "made manifest by three other rights: the right of access, the right of equitable apportionment, and the habitat right." In early 2001, eleven western Washington treaty tribes, citing the Orrick decision, filed a lawsuit against the state of Washington to repair shoddy road culverts, contending that the tribes should have a say in fixing them—and, by extension, the affected salmon fisheries. Phil Katzen, an attorney representing the tribes, said that if his clients were successful they would "try to use this in other arenas to protect habitat."[54]

Some environmentalists hailed the treaty Indians' campaign without realizing that the Indians were acting in their own self-interest and not as environmental saviors. Their first goal was to protect their treaty rights. The treaty tribes generally supported hatcheries, like the facility proposed by Seattle Public Utilities on the Cedar River, as a way to enhance flagging salmon runs. And the city realized that it had to compromise. In November 2006, the Muckleshoot Indian Tribe and the City of Seattle reached a $42 million accord that included the controversial sockeye hatchery, now deemed critical given the uncertain effects of global warming on salmon runs, and gave the tribe access to watershed lands for hunting and gathering. The city also agreed to pay the tribe $250,000 per year for the next decade to study the effects of Indian hunting in the watershed. Environmentalists, sport hunters, and several state agencies, notably the Department of Fish and Wildlife, which was not a party to the agreement, were incensed. Ratepayers were angry, too, since the settlement threatened to boost annual residential and business water fees. P. M. Morris, a guest columnist for the *Seattle Post-Intelligencer,* warned that the settlement would prove disastrous: "Wake up! They are giving away your state while you're 'Sleeping in Seattle.'"[55]

To the officials responsible for the city's watershed, reaching an understanding with the Muckleshoot was the easiest way out of a political quagmire. Rand Little, a senior fish biologist for the agency, had explained this position three years earlier, in 2003, saying that "we need the full meal deal: the habitat and the hatchery," since Seattle was "not in the fisheries business" but only "trying to do what everybody wants." It was a compromise position that, if read in a different and cynical light, bolstered farming, construction, and other business interests to avoid making even tougher choices. As one reporter explained, if "salmon can be raised in hatcheries, why bother protecting their wild cousins?"[56]

Whether Rand wanted to admit it, Seattle was in the fisheries business now and could not buy itself out. In 2001, city agencies ranging from the Department of Parks and Recreation to Seattle City Light and Seattle Public Utilities began to craft a recovery plan that went beyond the Urban Creeks Legacy. The final document, released in 2003, focused on the habitat needs for chinook salmon in five distinct urban environments: Lake Washington, Lake Union and the Ship Canal, the Ballard Locks, the Duwamish Waterway, and the Puget Sound shoreline. It called for a multifaceted approach. Builders, developers, and city agencies would replant destroyed eelgrass beds at the mouths of city creeks and streams, replace antiquated storm sewers, acquire and safeguard undeveloped shorelines, remove seawalls and restore formerly entombed beaches, and blast away concrete and asphalt to "daylight" still more creeks beneath Seattle's streets. "When we see salmon swimming upstream to spawn," said the new mayor, Greg Nickels, "we can't help but be moved by their struggle to continue the cycle of life."[57]

That struggle made for even stranger, more convoluted politics because, as Seattle city council member Richard Conlin complained, "ecosystems are complicated and developments are complicated." In order to make the life of salmon easier, city leaders had to weigh the costs of making Seattleites' lives harder. Proposed new fish-friendly building codes, introduced in early 2006, were even more stringent than before. Property owners would have to preserve native plants and gravel, and would be restricted from home improvements within one hundred feet of the water's edge. Public parks and golf courses faced limits on applying fertilizer or pesticides within fifty feet of wetlands or salmon streams. Developers would be obligated to restore creeks previously stuffed into pipes and culverts. As angry homeowners and contractors, still riding the housing boom, vowed "to race to beat the change," lost in their complaints and speechifying was the fact that these squabbles were largely among those who had done well in the booming nineties. Seattle had sold its charms too well. Middle-income families were finding it difficult to purchase a home in Seattle, thanks to what Knute Berger, an editor for the *Seattle Weekly* called "the new meth"—the region's addiction to growth for its own sake. He posed a difficult question: was Seattle "strictly a commercial zone where everything is for sale to the highest bidder? Or are there other community values that need to be asserted?"[58]

Not unexpectedly, it was the well-off who asserted their values most vocally. In affluent Montlake, a formerly middle-class district adjacent to the Olmsted-designed Washington Park Arboretum where home prices had pushed past the million-dollar mark, residents united to repulse a proposed $4-billion-plus expansion to the Evergreen Point Floating Bridge. If approved, the new super span would slice through the neighborhood. It was a threat many had foreseen almost thirty years earlier when city and state engineers, pressed by downtown retailers, tried to build another multilane highway, the R. H. Thomson Expressway. Residents had thwarted the expressway in 1972, leaving on-ramps to nowhere hanging in midair above Lake Washington. Since then, traffic to and from the mushrooming Eastside suburbs—commuters traveled to the huge Microsoft corporate headquarters in Redmond, and others shopped at Bellevue Square—had grown nightmarish. Motorists on the fourteen-mile stretch of bridge and highway, squeezed down to four lanes across the lake, faced hours of stop-and-go traffic daily. By 2005, the commute from Seattle to the Eastside was worse than the reverse. Worried that a major earthquake could ruin the vital commuter link, state engineers wanted another floating bridge. The Montlake Community Club countered with its own proposal for a soaring suspension bridge; the University of Washington, north of the viaduct, protested that the club's idea would only exacerbate congestion on its car-choked campus. Environmentalists warned that any expanded bridge would destroy the arboretum. Planners and highway engineers answered that the existing bridge was dangerous and getting more so every year. Doris Burns, a seventy-nine-year-old activist who fought against the R. H. Thomson Expressway, had her own solution: "Let the freeway sink." But why was another bridge necessary, anyway?[59]

One reason was the gridlock that choked the bridge every weekday, but another was that Bellevue Square, the suburban shopping mall opened by the Freeman family in 1946, had helped convert Bellevue into an urban center of its own. Kemper Freeman, Jr., who had inherited the family real estate business from his father, renovated the aging mall, which had gone from retail pacesetter to commercial has-been, into a premier destination in the early 1990s. After giving Bellevue Square a pricey makeover, he turned his attention to transforming Bellevue into a full-blown city, complete with high-rises and a proposed $70 million performing-arts center. Bellevue was hot again, and so were Freeman's real estate investments. "I am obsessed with making things work in general," he said. It was ambition he "learned on the farm," the home his grandfather Miller Freeman had built back in the 1920s in the middle of the now-vanished strawberry fields once tilled by Japanese farmers. To guarantee that Bellevue thrived, Freeman called for 1,400 miles of additional lanes to existing highways alone and, like his father, criticized plans for mass transit. When his plans for freeway expansion were blocked, he bankrolled a citizen's group called Let's Get Washington Moving to put pro-highway initiatives before state voters. "I have a greater vision for Bellevue," Freeman told a reporter in 2005, describing the city as "in the adolescent stage of its urban growth."[60]

Freeman and others pushed for an expanded Evergreen Point bridge to help Bellevue and the region through their teenage years, even as studies found that the span's more than 350 drains, designed to remove storm water for motorists' safety, emptied a mixture of mercury, lead, copper, and other pollutants into Lake Washington every day. Tattered brake pads, leaky gaskets, worn transmission lines, and overfilled gas tanks all spattered their refuse onto the bridge deck in concentrations that, at their highest during rush hour, exceeded state standards for acute or episodic levels. "I was really surprised to see how many metals there were," said Sue Joerger, director of the Puget Soundkeeper Alliance, one of several organizations that had conducted the studies in 2003. She realized, "Wow, that's all coming off our cars." In fact, subsequent research revealed that the sources and composition of the pollutants were even more complex and potentially damaging than previously believed. A "toxic stew" of antidepressants, fire retardants, caffeine, and birth control hormones, to name just a few ingredients, was seeping into nearby streams and eventually into the ocean. Scientists could now measure what Seattleites consumed, from what they purchased to what they ate, using sophisticated instruments. One team of University of Washington scientists found an upsurge in vanilla and cinnamon over the Thanksgiving weekend. Holiday spices were likely benign, but as Rick Keil, a chemical oceanographer, explained, even baking and eating had environmental implications. "To me it shows the connectedness," he concluded.[61]

The discovery of still more pollutants entering the watery mazes of urban Puget Sound was an ominous finding for restoration advocates like Judy Pickens, who had worked for over a decade to bring back West Seattle's Fauntleroy Creek. "We've done everything, and it's not right. . . . What more do we need to do to bring this creek into a healthy habitat for the fish?" Volunteers like Pickens had spent thousands of hours removing garbage, planting native trees, replacing logs and gravel torn from original streambeds, and unclogging culverts to entice salmon to spawn. The problem was not entirely man-made. Each individual salmon species had evolved to take advantage of particular environmental conditions, regardless of human changes to the original watershed. Unlike chinook, coho favor the smallest coastal creeks and tributaries, often waiting until late in the fall when there is sufficient water to swim upstream. This trait that enabled coho to colonize a niche environment now placed them at greater risk for poisoning when autumn rains flush pollutants deposited on city streets into the small streams. The toxins stunned the fish, causing them to flail about on the surface in what Nat Scholz, a biologist with the National Marine Fisheries Service in Seattle, called "the Jesus walk" before dying. In one study, 88 percent of the coho observed in Fauntleroy Creek died this way. Yet coho were not an endangered species; federal authorities had listed them only as a "species of concern" for Puget Sound. Biologists cautioned that other salmon species could still benefit from restoration even if the coho did not, but the scientific omens were unpromising.[62]

Replanting native vegetation, putting logs into streambeds, and other cosmetic changes made for good politics but were not always enough for effective restoration. Most scientists now conceded that the best that Seattleites could hope for was partial rehabilitation. The weight of history was too great. The cumulative effects of the past continued to pile up like cars on the Evergreen Point Floating Bridge. At the beginning of the new century, the threat of global climate change, stoked by more than two centuries of industrialization and consumption of fossil fuels, was no longer idle doomsday speculation. Warming posed potentially devastating effects for Puget Sound in the form of increased air and water temperature, decreased stream flows in the summer, increased flooding in the winter, shrinking glaciers, and rising sea levels. "The evolutionary environment in which Puget Sound's many species of plant and animals have developed," noted one report by the Puget Sound Action Team, an environmental task force commissioned by the state governor's office, "is undergoing significant changes." Some changes proved positive for some salmon. Chum salmon, which were more tolerant of temperature swings, were now the region's most abundant species. Most changes, though, placed all salmon species in far greater jeopardy. Calling salmon "wild" in this environment strained the definition of the word, yet restoration advocates remained hopeful, if skeptical. As Judy Pickens concluded, watching the coho dying in Fauntleroy Creek every autumn, "if we get [salmon] back and they don't spawn, then all we have is one long controlled aquarium."[63]

Walking along the beach at Lincoln Park in West Seattle, taking in the majestic views of the Olympic Mountains and Puget Sound, you may come across a sign warning visitors not to catch and eat crabs, clams, oysters, or fish, especially after a heavy rain. Nature here is poisonous. Tellingly, the sign posts this dire message in seven languages: English, Spanish, Chinese, Vietnamese, Laotian, Khmer, and Tagalog. Similar signs appear along the banks of the Duwamish River and all around Elliott Bay, semaphores of the history of salmon in the Emerald City.

The signs point to an important historical truth. Salmon are no longer the fetish of white sportsmen or Indian activists alone. The fish are, in a very literal sense, the city, and fights over salmon say less about the fish and more about the human combatants. For the treaty Indians, success in the courts may deliver salvation to the salmon, yet the potential outcomes might not please those who call themselves environmentalists or the sport anglers who first laid claim to that title. The invocation of protected rights by treaty Indians so far has focused narrowly on the right to subsistence, a right that has empowered some Native peoples with new cultural vitality. Those same rights could potentially undermine rights cherished by other Seattleites, especially those who have identified as environmentalists—the right to inexpensive water and electricity, the right to play in the outdoors at whim, and the right to live in a beautiful city with efficient transportation and a vibrant economy. An environmental prerogative for one group could become an environmental wrong for another. Complicating this new political reality was the dynamic structure of the physical world, which could constrain some choices and open other avenues for political action. Individuals and groups who operated as independent actors remained, as always, part of a larger and more entangled and interdependent history.

The consequences of that history were on full display in Seattle by the century's end. An earlier generation of residents had built a great city and congratulated themselves, appropriately, for their efforts. Yet their ideas were not separate from the physical world they had tried to master, and their actions set into motion a series of unanticipated events that led to fights over salmon and riots over trade. Each group played to type, and all had their own vision of an ideal community where spectacular scenery and temperate climate would enhance their own lives. Their sense of place was utopian and each group framed its vision in light of its own narrow vision of the past, often in blatantly self-serving terms: Indians as the deserving dispossessed, environmentalists as custodians of an unsullied nature, taxpayers as virtuous proprietors defending the free market, and so on. History as refracted by political power determined whose position had the greatest credence. The problem with utopias, however, from the original imaginary island devised by Sir Thomas More down to the present, is that their creators erected gates and built walls to keep out anyone who contaminated the ideal. Utopias could not exist without an enemy, without someone or something to vilify.

Toward the end of Ernest Callenbach's novel, the journalist William Weston, in assessing the benefits and costs of the Ecotopian experiment, explained that his hosts believed secession from the United States was "desirable on ecological as well as cultural grounds," because "small regional societies . . . more subtly and richly" exploit their environments. "It seems likely that different ways of life always involve losses that balance the gains," he mused, "and gains that balance the losses. Perhaps it is only that Ecotopians are happy, and miserable, in different ways from ourselves." At the end of the millennium, Seattle really was like the fabled Emerald City of Oz, beautiful and alluring yet also flawed and not quite what it seemed. Behind the shimmering fish swimming through the locks of the Lake Washington Ship Canal every year, or thrown, freshly killed, by fishmongers to the cheers of camera-toting tourists at Pike Place Market, lay the power of the non-human world, manifest in the lives of the Emerald City's inhabitants, in their streets and buildings and homes. Emerald City was built on the backs of the salmon, but also on the backs the less fortunate humans who still shared that city.[64]

EPILOGUE

The Geography of Hope

Toward an Ethic of Place and a City of Justice

Ten years returned from Vietnam, gripped by the memories of combat and disabled by bullet wounds, John Beal suffered his third heart attack in almost seven months. It was 1978, Beal was twenty-eight years old, and doctors told him that he had little time to live. In an oft-repeated story, he went down behind his house to the banks of Hamm Creek, a concrete-lined ditch in industrial south Seattle and a tributary of the Duwamish River, and cried. "I looked at this wreck of a stream," he recalled, "filled with refrigerators, old tires, torn garbage bags, broken swings and stinking carpets, and all I wanted to do was clean it up."[1]

Beal started by pulling the trash out of Hamm Creek, then reintroducing native plants and animals—crayfish, beaver, salmon, even injured ospreys and eagles he had rehabilitated—to the raggedy waterway. He dug up and hauled away soil contaminated with motor oil, gasoline, and creosote. He bought a boat, filled the fuel tank with donations to his small nonprofit organization ("heavy on the non-, and non-existent on the profit"), and patrolled the Duwamish River looking for polluters, videotaping and confronting the worst offenders. He invited local schoolchildren to plant cottonwoods and alders along the creek banks to keep the water cool and shady for spawning salmon. Eventually, he persuaded city officials to peel back the pavement and free the creek from its metal-and-concrete culvert to flow, once again, as a stream. The Port of Seattle, King County, even the U.S. Army Corps of Engineers pitched in to help. "It turned from a hobby to a habit to a way of life," he said in a 2002 interview, still alive and now healthy. "I made a deal with God," and in striking his accord to restore the creek, Beal believed he had restored himself.[2]

It is easy to dismiss Beal's story as maudlin. Indeed, many local environmental activists and journalists have portrayed him as a kind of saint. It is a shopworn convention in nature writing, and the subsequent plot is familiar. Long ago, the earth was pure and uncorrupted until human beings despoiled nature's original design. Only by heeding the dire warnings of prophets did we save some of Eden's fragments. Beal's good works thus ennobled him and his cause, just as John Muir, Rachel Carson, David Brower, and other apostles became anointed in the environmental canon. This is not surprising because the force that drives the green fuse of environmentalism is more spiritual than political. It is a secular faith, and its American roots stem from both conventional Judeo-Christian religion and civic culture.[3]

But the creed has taken a beating lately, and some of the harshest blows have come from erstwhile devotees. Political consultants Michael Shellenberger and Ted Nordhaus have thrown the stiffest punch so far. In their August 2004 critique, "The Death of Environmentalism," they lambasted major organizations like the Sierra Club for putting narrowly defined environmental concerns ahead of social and economic justice. This strategy only alienated potential allies while abetting long-standing enemies. A separate group of progressive scholars and activists then threw a left hook with a rejoinder titled "The Soul of Environmentalism." Like Shellenberger and Nordhaus, they attacked the mainstream movement for persistently ignoring racial and economic inequality, but they disagreed that environmentalism was dead. Instead, it was corrupt at its roots and its founders were to blame. John Muir's antipathy toward Indians had tainted the earlier conservation ethic with racism, and the postwar environmental campaigns shamelessly rode on the coattails of the civil rights movement. "Modern environmentalism was, after all, the Elvis of Sixties activism," the authors concluded. Their solution was as straightforward as their critique: return to the genuine grass roots and embrace the environmental justice groups that had long fought the good fight.[4]

Yet neither attacker stepped completely outside the church, nor confronted the challenges of living within history. Shellenberger and Nordhaus stormed the temple, smashed the idols, and called on believers to swap their old verities for new ones. They called it "reframing," using the language of spin doctors and consultants, but it was an old trick long employed by disenchanted religionists. The authors of "The Soul of Environmentalism," however, saw apostasy. Their answer was to dig all the way back to Genesis and excavate the original tenets to smash them. What united the two was an uncritical devotion to pure causes and pure nature.

This allegiance to purity has been a fundamental weakness of environmental politics. Purity is a chimera, and history explains why. As much as environmental fundamentalists of all stripes want to proclaim the virtues of unsullied nature, the world we live in is a messy fusion of the natural and cultural, and Seattle's boosters have capitalized on it relentlessly. In late October 2006, the Seattle Convention and Visitors Bureau rolled out a new slogan to sell the city's charms, "metronatural," a made-up word that can swing as a noun or an adjective. Plastered on banners throughout downtown, emblazoned on top of the Space Needle and on the wings of seaplanes, metronatural is defined as "a world-class metropolis within wild, beautiful surroundings," or "one who respects the environment and lives a balanced lifestyle of urban and natural experiences." Or, simply, "Seattle." The old slogan will endure, the bureau's Web site explains, because "Seattleites and visitors will always think of Seattle as the Emerald City."[5]

Many chided the promotion as vapid or too reminiscent of the term "metrosexual," yet most Seattleites, like most Americans, embrace the confusion of metro and natural—even if they do not puzzle over why. They gather by the thousands every summer and fall to watch salmon push upstream through the Lake Washington Ship Canal. Some try to catch a few of these fish during the occasional short fishing seasons. They visit neighborhood farmers' markets to buy organic fruits and vegetables trucked in from all over the Pacific Northwest, or big chain stores to buy products labeled "Salmon Safe." It is a way of feeling good while doing well. They marvel at cougars and coyotes in their backyards even as they shoo their dogs and cats inside, and every weekend they take to the roads in Subaru Outbacks and hybrid-powered Ford Escapes to ski or hike. Like the salmon that pass through the Ballard Locks, they live in a mixed world, and history has been the cause and consequence of this mixing.

Consider the Duwamish River. The waterway that John Beal's Hamm Creek emptied into was a different Duwamish from the river that the poet Richard Hugo, born and raised in nearby White Center, recalled from his depression-era childhood. Yet even then Hugo remembered the Duwamish as "a curious combination of the industrial and rural," a river of reeds and cattails drifting past sawmills, quarries, and factories, sheltering flocks of "grebes, teals, ducks, pigeons, doves, and other fowl" plus salmon and porgies. As Hugo fished, he also found the carcasses of "bloated dogs" stranded on the mudflats, "gunnysacks of rocks round their swollen necks," the victims of owners that he assumed were too poor to feed them.[6]

Over time, the disregard continued. In Hugo's day, the Duwamish River was home to immigrants and blue-collar workers; by the 1990s, the bordering neighborhoods were even more diverse and destitute. A study by Seattle's Community Coalition for Environmental Justice in 2005 reported that the U.S. Census tracts along the Duwamish corridor had three common characteristics: the lowest household incomes, highest percentage of minority households, and largest number of federally designated toxic Superfund sites and sites under consideration in Seattle. Little had changed since a 1994 study reached the same conclusion; as South Park resident Debbie Carlin observed, "people can't afford to move." "A lot of people live here. That doesn't mean they choose to deal with all the pollution and stuff they dump down here."[7]

Mistreated and deserted, its man-made banks squeezing slack water to the sea, the Duwamish symbolizes the disharmonies of the Emerald City. The river, like the people who live along its banks, remains invisible to those pronouncing judgment on Seattle's future. It is an impure place. Its messy and contradictory state confuses some, resulting in strange pronouncements that only expose our longing for purity. The writer Jonathan Raban, the British expatriate originally so enamored of his new home, quipped in early 2004 that Seattle's "intense proximity to nature" made it an "unsatisfactory" metropolis, since "real cities supplant nature" with "works of epic communal artifice." James Vesely, editorial page editor for the *Seattle Times,* was more measured, calling Seattle "the great in-between, a place that loves nature but rarely is threatened by it, . . . an SUV town with a bike rack" on top. Evergreen and earnest, cozying up to its mountains and forests, Seattle is no Manhattan or London. This garden city seems benign from the lofty heights of Raban's Queen Anne Hill home or Vesely's editorial office on the edges of the Denny Regrade, but to the residents of the Duwamish River valley, wedged between dumps, highways, and a befouled waterway, Seattle is anything but benevolent.[8]

How is it that people look at the same city and see such very different places? The answer lies in history or, more accurately, in how people have chosen to remember the past. The habit of regarding culture and nature as binary categories has shaped how we view cities and their dynamic environments. The result is a kind of intellectual myopia in which "history is experi-

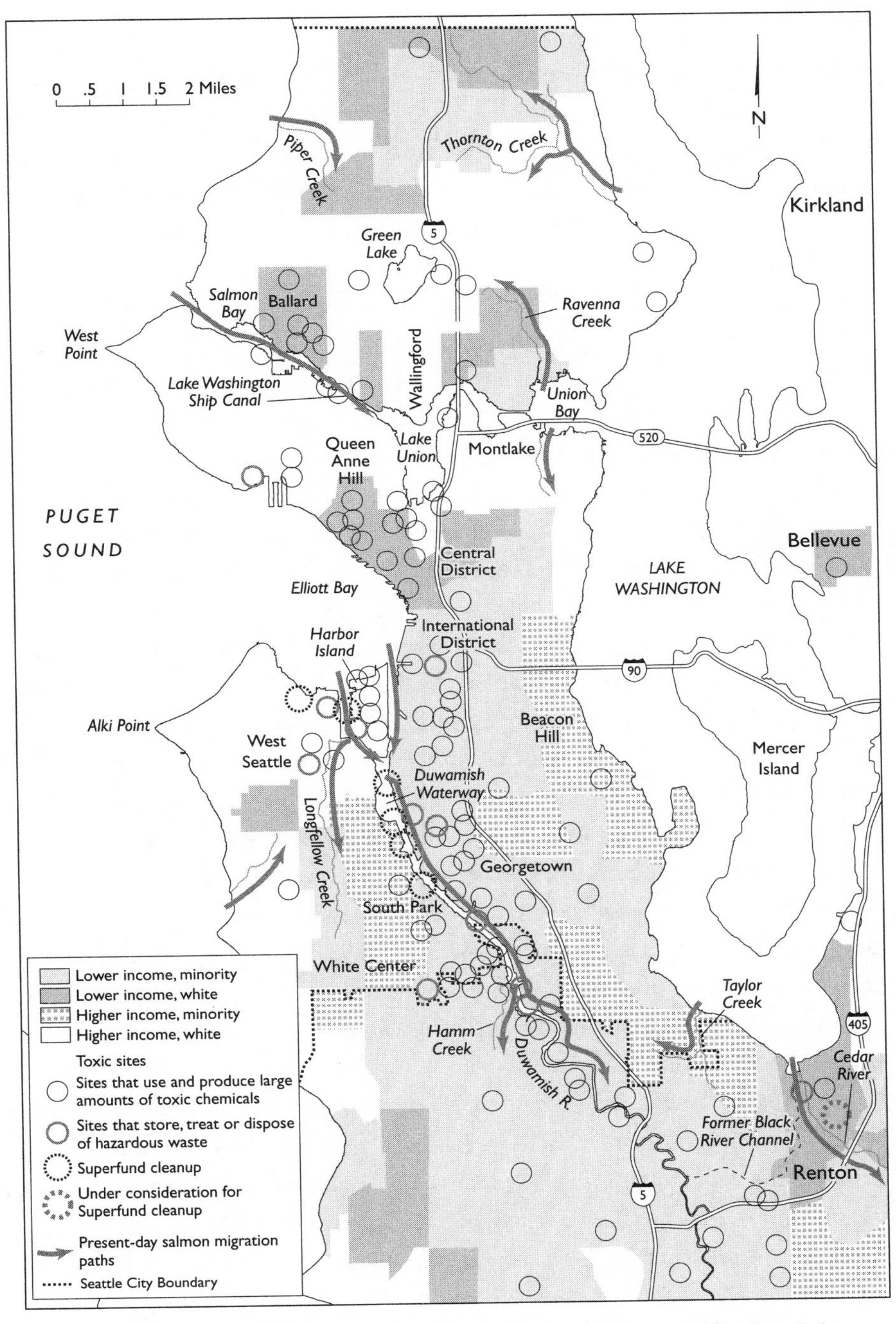

Polluted sites in Seattle, showing low-income and minority neighborhoods by U.S. Census tracts, and major salmon migration routes, c. 2005. Low-income is defined as neighborhoods with household incomes below Seattle's average of $45,736, and minority is defined as neighborhoods with a percentage of minority households above the citywide average of 30 percent. (Both categories are derived from the 2000 U.S. Census; adapted from *Seattle Post-Intelligencer,* August 25, 2005)

enced as nostalgia and nature as regret—as a horizon fast disappearing behind us." In this way, history becomes a collage of tragedies: forests lost to pulp and lumber mills, meadows bulldozed for homes and factories, and rivers dammed up or polluted to ruin. Yet cities and nature can never coexist so long as we constantly lament the passing of each. As the prophet of humane urbanism, Jane Jacobs, said in the conclusion to *The Death and Life of Great American Cities*, "sentimentality about nature denatures everything it touches."[9]

Nostalgia poses the same threat for history. Defined from its Greek roots, nostalgia is a sickness for an ideal past. To think nostalgically is to wish for what might have been or never was. To think historically, however, is to see the past as a different place, the result of complex causes and effects unfolding in particular locations through time. There is no escape. The human and non-human unfold in time and inhere in place, and we live with the outcomes. To ignore our connection to place is to reap the returns of historical ignorance. To face it squarely, though, is first to acknowledge the consequences of human actions, past and present, and then to develop a more expansive ethic of place.

Because the past is integral to place-bound ethics, historians should be part of this process, but it requires inhabiting an intellectually and professionally tricky role. There are good reasons why historians are seldom philosophers. The burdens of research and interpretation make the possibility of creating usable pasts difficult, even dangerous. The past is not a simple toolbox that we can use to fix the problems of the present and prevent troubles in the future. Not all of the past is relevant to us. Much of it can and should remain strange. But if we treat history less as an assemblage of facts than as a practice of mind, then the past becomes inseparable from the values we attach to it. Thus, it is through history and in history that an ethic can evolve to meet the needs of the present and the immediate future.

Seattle's past suggests how such an ethic developed as diverse peoples struggled, sometimes violently, to match their aspirations for community with the challenges of making livable places over time. In this history, there were victors and there were the defeated, but no one was ever wholly virtuous and few were entirely corrupt. Before contact, Indians had a moral accounting system to measure waste and excess, a system learned through working the land to provide material and spiritual sustenance, yet they could and did overexploit their resources. They littered their villages with trash, attracted vermin to their dwellings, and then moved on when conditions became unbearable. The railroad moguls and real estate entrepreneurs who filled the tidelands thought little of the salmon, the Indians, or the working poor living along the waterfront, but their successes made the modern port of Seattle possible. The public-minded engineers, such as Hiram Chittenden or R. H. Thomson, delivered clean water, eradicated disease, and generated commerce even as their ethic of efficiency worsened certain ills. The same was true of John C. Olmsted, the landscape architect, whose treasured greenswards often became battlegrounds and weed patches. Likewise, the conservationists who mobbed the countryside in pursuit of play did rescue many locations from outright destruction. Their desire to defend nature as a consumer entitlement, however, frequently threatened the resulting benefits or restricted them to the worthy few. The rescue of Lake Washington was a promising, even inspiring development that united the region to a degree. But the resulting pollution of the Duwamish River and marginalization of Indians and their fishing rights only underscored the continuity of community narrowly defined.[10]

Nothing better captured these ethical failings than the debates over salmon at the end of the twentieth century. Here was where history mattered most—and was ignored completely. A cacophony of individual interests claimed to speak for the common good by speaking for salmon, all the while ignoring how the fish divided as much as united Seattleites. Turning salmon into a sacred communal symbol masked many old wrongs and revealed some bizarre twists of logic. Many whites began to claim that losing salmon meant losing rights first claimed by the Americans who had settled Puget Sound more than a century before, but middle-class environmentalists and anglers were not Indians. Appalled by such assertions, treaty Indian tribes now maneuvered to protect their share of the fishery, often promoting the construction of still more hatcheries—a self-interested move that contradicted the growing scientific consensus over the dangers of artificial propagation. In this round of recriminations, whites saw themselves as environmentally enlightened and criticized Indians for not being noble savages.[11]

Because salmon have both united and divided people, they present us with an opportunity to think ethically about this place, Seattle. The challenge, though, is to move beyond the habit of trying to dominate nature to the more feasible goal of governing ourselves, and to tame our reflexive impulse to put ideas into neat categories. An evolving ethic of place, like any ethic, is neither a divine commandment nor doctrinaire mandate. It is a product of history and thus it is ever changing. It must be, in other words, a deliberate and enduring dialogue between humans and their environments. The most powerful such statement in the American philosophical tradition is Aldo Leopold's famous "land ethic," added posthumously to his 1949 book, *A Sand County Almanac.* "All ethics so far evolved," wrote Leopold, "rest upon a single premise that the individual is a member of a community of interdependent parts." Leopold's contribution was to enlarge "the boundaries of the community to include soils, waters, plants, and animals, or collectively: the land." Borrowing concepts from the nascent discipline of ecosystem ecology, he spoke of an "ecological conscience" in which "conservation was a state of harmony between men and the land." Seeing the world in ecological terms recast "the role of *Homo sapiens* from conqueror of the land-community to plain member and citizen of it." Leopold identified another challenge as well: our collective responsibility to the non-human world, to the rivers, forests, and salmon to which we can attach so much value. An "ecological conscience," he maintained, required a "conviction of individual responsibility for the health of the land" and its ability for self-renewal. The land was not "merely soil" but a "fountain of energy flowing through a circuit of soils, plants, and animals." For Leopold, repairing the cleavage between the land as commodity and the land as biota was also about an expanded sense of justice.[12]

Since its publication, Leopold's land ethic has been a focal point of debate. In the 1970s, scholars made it a cornerstone of the nascent field of environmental philosophy, but in recent years, postmodern critics have discounted his blindness to nature as a social construction, while some sociobiologists have dismissed the notion of ecological conscience as antiquated since, they believe, all human behavior is genetically predisposed. These criticisms might have some merit but, for all its elegance, the land ethic remains inadequate because, at its core, it is a pastoral vision of nighttime campfires in the woods or a farmer hunting pheasants at dawn. Leopold appealed for ecological interdependence, but he did not speak for social interdependence. His vision of rural tranquillity is not well suited to a diverse metropolitan society riddled with inequalities, antagonistic politics, large-scale technological systems, and multinational corporations. The land ethic is also at odds with contemporary ecology. In Leopold's time, scientists believed in the "balance of nature," that environments were self-regulating systems tending toward a climax of stability, but in recent years, ecologists have reached a new consensus of nature as turbulent, volatile, and in a state of perpetually discordant harmony.[13]

Human diversity and unstable nature test the applicability of Leopold's ethic. So much remains contingent that Leopold alone cannot suffice. In concert with other ideas, however, the land ethic can constitute the nucleus of a more realistic ethic of place. One of Leopold's blind spots, for example, was the problem of inequality. As the history of Seattle demonstrates, an ethic of place produces havoc if it does not confront the question of social and environmental injustice. Bypassing democratic participation and ignoring people's need for empowerment are sure ways to undermine residents' intimate connections to their neighbors and nature. In this sense, a communal ethic must flow from the fundamental principles of justice.

In an ideal situation, each person has an equal right to the most liberties available in a system for all of society, and if inequalities do exist, they work out to the benefit of the worst-off without foreclosing liberties for anyone else. The problem with this view is that justice rests on a presumption of ignorance. But people are not ignorant of their relative social positions. The ideal is ahistorical and blind to the structures of power that fix firmly in particular locations and institutions through time. People remain ferociously loyal to the places they inhabit, and these reasons can become entrenched as ideology or faith. In practice, our places of inclusion—nation, neighborhood, church, house—are also often places of exclusion.[14]

One possible way to avoid this dead end is the concept of cosmopolitanism, which, in very simple terms, extends the obligations to kith and kin to encompass close neighbors and distant peoples alike. The latter group is particularly salient because the local has been the global for some time. We are all dependent on networks that extend far beyond our doorsteps. Although this expanded sense of community complicates the process of citizenship and belonging to place, conversations across cultures can engender greater awareness of why one's own place and people matter. Some values are universal. Some values will remain local. The problem of living in a world of difference is determining which is which.[15]

Appeals to protect Mother Earth or to celebrate cultural diversity also ignore historical processes by which certain individuals and groups have accumulated economic and political power through time. Seen this way, we all seem trapped in global webs of consumption and production that only bolster those corporations and capitalists who feed off our desires for ever more things. It is a neat explanation ideologically, one appealing to many scholars and activists alike, but at a basic level it does not grapple with the complexity of the past. It is easy to pin the label of false consciousness on befuddled Americans alienated from legitimate means of production and, by extension, from legitimate nature. It is far harder to explain why those people came to value nature, to value particular places, through the things they carried in their backpacks or automobiles. Indeed, some authors now claim that our children, obese and hooked on television and video games, suffer from "nature-deficit disorder." Strip away the language of therapy and talk shows, though, and this argument reads suspiciously like the earlier laments of John C. Olmsted or John Muir, and it is freighted with all of the same assumptions about what passes for real nature and who needs it the most.[16]

How and why we consume nature demands a more supple and complex analysis, one that forces us to examine the values we bring to the places we shape and know. Consumerism has never simply been a deception employed by capitalists in the name of class dominance. It is a historical process shaped as much by consumers as by producers. If anything, contests among individuals and groups of unequal economic standing over where and how to consume resources yielded the conservation and environmental movements. Moreover, many activists today, as in the past, are still ideologically averse to living with the resulting historical and political complexities. Indeed, the desire to wall off nature for particular individual or collective prerogatives is not the sole province of the affluent and well connected. Those on the margins have also played the NIMBY card or distorted the past to serve present politics. In the environmental justice movement, for example, activists continue to consume electrical power and plastic products produced by the coal-fired generators and factories that they decry for polluting their bodies and backyards. In simply wanting to displace the waste, those who suffer from injustice can be unjust in their reactions to it.[17]

This is why we need a better ethic of place. Many scholars have called for this, but their ideas either disregard the weight of the past or, like Leopold, prefer the rural and the wild to the urban and the industrial. If an ethic of place is to have any relevance in today's world, it has to begin in that seemingly most unnatural of locales, the city. Nature and cities, the products of human and non-human actions, carry the burdens of our desires. Americans ask few things in their culture to bear so much weight as our environments and cities. We portray them in strikingly similar terms. Both were unsullied and simple before sliding into decline and ruin, and our penchant for instant and painless solutions only fanned our frustrations. By rhapsodizing the past, we have avoided asking questions that could lead to more interesting answers for the present and future. The history of Seattle and other places illustrates why a future ethic of place may thrive if it embraces a more nuanced and honest integration of the human and non-human. We evade that integration at our peril because, as the literary critic Raymond Williams has argued, by alienating "the living processes of which we are a part, we end, though unequally, by alienating ourselves. We need different ideas and different relationships."[18]

If we begin to recognize the degree to which culture and nature have merged in our lives, then we can begin to close the rift between our older, puritanical views of Nature with a capital *N* and Culture with a capital *C*. Cities and gardens are not opposed to wilderness in the order of things. As the environmental philosopher Andrew Light notes, only we "drive conceptual wedges into the world" and injure the environment and ourselves as a result. Cultivating more vibrant connections between groups of people and their environments might, in turn, foster the betterment of both human and natural communities. In essence, this is the principle of civic environmentalism. It emerged in the late 1990s as a collection of ideas focused on giving diverse people a stake in coming together, and it uses an open political process to devise methods to sustain community. Unfortunately, like John Rawls's notions of justice, it also presumes equal power and access—a naive notion in the light of history. Thus if civic environmentalism is to succeed in the future, it will need to account for the imbalances in power and resources among citizens. It is not enough to provide avenues for participation if people cannot or will not put faith in the system, nor is it enough simply to recognize past injustices. A sustainable ethic of place must face the legacies of history, providing redress for wronged communities in some cases, ensuring equity for the future in others, and the principle should apply not only to human communities but to the non-human world as well.[19]

For Seattle, ironically, salmon may help to point the way. Protecting spawning habitat for the fish can also safeguard clean water and open spaces for people, and salmon migration patterns reveal why an ethic of place must aspire to be both local and global. In the course of their lives, salmon may travel from Seattle's shallowest creek to the far reaches of the North Pacific and back again, ferrying with them both the nutrients and pollutants of the lands and waters they traverse. In a sense, a salmon's journey mirrors the circuit of the container ships that crowd the docks along the Duwamish Waterway, arriving with clothes and electronics from East Asia and returning with apples, lumber, and software. In this way, modern environmental and social justice concerns span many scales at once. An attachment to the local is now usually an attachment to the global as well, even if we do not admit the connection. One way to avoid the parochialism of localism is to test any ethic against the ever-changing dynamics of its place. Here, too, salmon can be a guide. Sensitive to changes in temperature and aquatic chemistry, these fish are nevertheless remarkably resilient and adaptable as long as they have clean, cool, and unobstructed waters in which to spawn, seasonally abundant food, and the ability to elude predators. Moreover, if waters are suitable for salmon, they are likely also suitable for people. Salmon may be the best measure for Seattle, but they are not the only measure. What ultimately matters is an ethic that is testable. Without such testing, there is no way to assess past problems, to anticipate future consequences, and to build a common sense of hope tempered by empiricism.

What I am suggesting is that a kind of ethical pragmatism should augment or replace contemporary environmentalism. Here is where the mission of the philosopher and the historian can inform us about what it means to be an ecologically minded citizen. The philosopher Andrew Light, to name just one scholar, has gone to great lengths to remodel pragmatism for the new century. Some philosophers insist on the primacy of anthropocentric values, while others stress the importance of values rooted beyond humans. For Light, an ethic should instead emerge from a discussion of "the largely empirical question of what morally motivates humans to change their attitudes, behaviors, and policy preferences." We do need to evaluate what is true or good and not what merely works, but a blind orthodoxy is usually not an accurate gauge. The more reliable approach would ground our attachment to place in time, so paying attention to history may help steer our thinking and actions.[20]

This is why an ethic of place matters and why historians can help to achieve it. For all of the universal attributes of the human institutions that have built cities—the planning, engineering, social welfare, political systems, financial institutions, cultural impulses, and religious beliefs—the specific manifestations vary dramatically not only from city to city but within each city over time. Parks, utilities, zoning laws, freeways, topography—these and other facets of urban life constantly reshaped where people lived, worked, and played. They also affected environmental relationships within the city—where wildlife thrived, clean water flowed, and pollution concentrated. The structures of the city and its environs imposed limits on inhabitants, and their values inhered in the landscape. Over time, diverse and often antagonistic ambitions and rationales have cohered in particular places and practices. Seen this way, what we today call "environmental injustice" ensnares everyone. Urban landscapes are thus so many ethics of place expressed physically.

The accumulation of events is far too complex and contingent to give us fixed moral lessons. We inevitably interpret the evidence, but the upside is that the stories we tell about the past can provide ethical direction by increasing our burden of awareness. The weight of what has come before thus can dampen any uncritical ardor for progress or lost Edens. Place will continue to divide as much as it unites, and we will never be truly separate from it or each other. Our lives can suddenly change when nature throws our places into chaos. After Hurricanes Katrina and Rita pummeled New Orleans and the Gulf Coast in the late summer of 2005, President George W. Bush visited the region and declared, "the storm didn't discriminate." It was ignorant oratory. The swath of destruction laid bare the patterns of previous inequalities. New Orleans, battered and soggy, may recover in time. What remains unknown is what shape the new city will take. One inescapable fact remains central to New Orleans's future: without a fundamental change in all the residents' ethic of place, this city below sea level will drown again, and those on the margins will suffer the most.[21]

New Orleans has hurricanes and floods, Los Angeles and San Francisco earthquakes; Seattle has its own hazards that could throw it, too, into social and environmental panic. Temblors are common to Puget Sound. The Seattle Fault runs through Bellevue and downtown Seattle. A major displacement, if the epicenter is close to Seattle's waterfront, could liquefy filled tidelands, topple waterfront buildings, and collapse freeways. The remnants of Seattle's missing hills have not gone away, either. Seismologists and oceanographers believe that the raised seafloor of Elliott Bay, created by the spoils from the Denny Hill regrades, could shift during a major earthquake, generating tsunamis that could engulf downtown and Bainbridge Island. And quakes are only one potential threat. The Pacific Northwest is home to the largest number of active volcanoes in the lower forty-eight states. Some sixty miles to the southeast, Mount Rainier remains dormant but not extinct. A large eruption could spew thick ash tens of miles into the atmosphere, melt the ice and snow, and send enormous walls of mud and water, called lahars, rushing toward Tacoma and Puget Sound. The tragedy in New Orleans was unusually dramatic, but the winds of 2005 were a reminder that history has made cities as fragile in their own way as the most imperiled wilderness.[22]

Just as history reveals how natural disasters are misfortunes of our own making, the past should also temper our notions of restoration. The complexity of repairing salmon habitat in urban Puget Sound, for example, may demand projects that could surpass the largest regrade in physical size and potential for disruption. Recent studies suggest that the problem facing the region as a whole is much larger than previously thought. Unchecked urban development is the major reason why, and the estimated cost (upward of $12 billion as of late 2006) to reduce pollution, protect habitat, and clean up Puget Sound is probably too low. Given the scale of this challenge, restoration will not be a matter of picking an arbitrary point in the past, eradicating non-native species, turning creeks back into their beds, and watching nature return. We have created new environments that we can never wholly erase, and we cannot reverse time. Restoration is instead a process of amelioration and moving forward into a more hopeful future. As such, it needs to attend to questions of social justice as much as environmental degradation. At its best, restoration should compel citizens to confront the consequences of the amenities they enjoy today. Someone somewhere will eat tainted salmon—or suffer the consequences of cleaning or restoring one stream at the expense of another. Such projects raise the unavoidably moral question of what makes a sustainable community, and how to remake places we need and cherish without rubbing away histories we would do well to remember.[23]

Americans have taken thinkers like Aldo Leopold too literally, looking to wilderness and open spaces for far too long as the principal embodiments of an ethic of place. Ironically, it was Leopold himself who said "we can be ethical only in relation to something we can see, feel, understand, love, or otherwise have faith in." Middling and wealthy Americans followed this advice selectively, believing that they had to flee the city for country retreats, just as Leopold and Henry David Thoreau had done, to pursue an intimate and balanced relationship with nature. Yet it is in our cities where the majority of Americans, as well as people around the world, now live, and so it is there that we are most likely to find the beginnings of what Wallace Stegner once called "the geography of hope." Stegner was talking about the big open of the American West in his plea to forge "a civilization to match the scenery," but in an era that is increasingly urban by nature, it is more relevant to turn toward the city.[24]

The beginnings of this hope are in our midst. Over the past twenty years Seattle has been a national bellwether for urban environmental politics. That legacy began, as with so many initiatives in this city of neighborhoods, in residents' backyards. Bemoaning the loss of open space, a few concerned citizens founded an organization dedicated to community gardening. Launched in 1978, Seattle Tilth helped set aside many acres of public gardens administered by the city government and tended by local residents. These P-Patches are everywhere from fashionable Belltown to industrial South Park. Other initiatives also suggest hope for the Emerald City, including new government-to-government accords between Seattle and Puget Sound Indian treaty tribes, Seattle's Climate Action Plan, which is one of the most far-reaching commitments to reducing greenhouse gases by an American municipality, and the Shared Salmon Strategy, a collaborative effort to restore habitat all around Puget Sound. The debates over endangered salmon have had a particularly beneficial effect on Seattle's evolving ethic of place. Because of salmon, perhaps more than any other issue, Seattleites are finally talking to one another—even if shouting is more common than conversation.[25]

Still, community gardens, stands against global warming, and meetings to bring back salmon are relatively easy things to embrace because, like so much of modern environmental activism, they focus on places and creatures beautiful and useful. A fuller test of an ethic of place will occur when residents try to love locations and people that seem beyond redemption. One such place is the gloomy Duwamish, a crooked river compelled to flow straight. For some residents, it is not really a river, not even a place deserving of an environmental vision. The Duwamish did not inspire a Henry David Thoreau, Edward Abbey, Barry Lopez, Annie Dillard, or Wallace Stegner. Instead, it got Richard Hugo, a poet prone to depression and alcoholism, a flawed bard for an imperfect river. The symmetry was elegant, and the result has been an allegory for all that is wrong with our broken urban environments. So as perverse as it might seem, the Duwamish is the necessary site to start developing a new ethic of place.

The late John Beal, who passed away in June 2006, may have known this. "When we first found Hamm Creek," he told a reporter, "there wasn't any life in it. There wasn't even a water beetle. There wasn't anything. We had to build an entire ecosystem." It was a bit of an overstatement, but Beal's efforts inspired many around Seattle, including Bernie Hargrave, a project manager for the Army Corps of Engineers, who had started work on the Puget Sound Nearshore Ecosystem Restoration Project. It is a massive undertaking modeled on similar projects under way in Chesapeake Bay, the Florida Everglades, and San Francisco Bay. Filled with salmon and reconnected to the Duwamish, the half-mile creek, according to Hargrave, captured "in a small way what we hope to do in a much larger way around the Sound." Beal, for his part, remained humble. "I never thought I'd get here, because I thought I'd be dead long ago," he said. "It was a place no one else wanted to be and that's why I went there." The new Duwamish will never resemble the old one. It will not be perfect, but it may be enough.[26]

Such is the stuff of our new dreams. Like Dorothy in the story of Oz, we delude ourselves when we search in vain for imaginary shortcuts. When Dorothy, Toto, and her traveling companions reached the Emerald City, they found the Wizard to be a humbug, the city a sham, and their quest for courage, brains, heart, and a ticket back to Kansas within their grasp the entire time. None of this would surprise a child reading the book. After all, L. Frank Baum aspired to write "a modernized fairy tale, in which the wonderment and joy are retained and the heartaches and nightmares are left out." The Emerald City was an illusion, but it mirrors our cities in ways that we may not realize. Oz was a muddle of flying monkeys, traveling salesmen turned sorcerers, tin men and scarecrows come to life, and witches wicked and good. In the real world, history is a similar source of confusion. Nature and culture are inseparable, and sometimes the best of intentions to split the two have produced the worst sort of evils. As creatures of history, we must live with all that it entails. We cannot escape the heartaches of our past, but like the characters in *The Wizard of Oz,* we can, as human beings gifted with courage, compassion, and intellect, accept our limitations and follow a better, if more uncertain, path than the yellow brick road. That journey begins by keeping a clear eye on our past as we move forward into the future together.[27]

Notes

Abbreviations

DSPHC	Don Sherwood Parks History Collection, Seattle Municipal Archives
ES Papers	Eugene Semple Papers, University of Washington Libraries, Special Collections Division
GN-PSF	Great Northern Railway Corporate Records, President's Subject Files, Minnesota Historical Society
JCO Coll.	John C. Olmsted Collection, Frances Loeb Library, Harvard University
JRE Papers	James R. Ellis Papers, University of Washington Libraries, Special Collections Division
KCARM	King County Archives and Records Management Center, Seattle
LID	Local Improvement District
MBSNF	Mount Baker–Snoqualmie National Forest, Supervisor's Office, Record Group 95: United States Forest Service, National Archives and Records Administration, Pacific Northwest Branch
METRO	Municipality of Metropolitan Seattle
MP Files	Margaret Pageler Files, Legislative Department, Seattle Municipal Archives
NL Files	Nick Licata Files, Legislative Department, Seattle Municipal Archives
NP-CESF	Northern Pacific Railway Corporate Records, Chief Engineer's Files, Minnesota Historical Society
OAR	Olmsted Associates Records, Series B, Library of Congress
P-I	*Seattle Post-Intelligencer*
PS Files	Peter Steinbrueck Files, Legislative Department, Seattle Municipal Archives
PSRA	Washington State Archives, Puget Sound Regional Branch, Bellevue
RG77	Record Group 77: United States Army Corps of Engineers, National Archives and Records Administration, Pacific Northwest Branch
RHT Papers	Reginald Heber Thomson Papers, University of Washington Libraries, Special Collections Division
SA	*Seattle Argus*
SCCF	Seattle City Clerk's (Comptroller's) Files, Seattle Municipal Archives
SED	Seattle Engineering Department, Seattle Municipal Archives
SMA	Seattle Municipal Archives
SMH	*Seattle Mail and Herald*
SPL-MB	Seattle Public Library, Main Branch

SS	*Seattle Star*
ST	*Seattle Times*
SWD-HF	Seattle Water Department, Historical Files, Seattle Municipal Archives
SWD-SC	Seattle Water Department, Superintendent's Correspondence, Seattle Municipal Archives
TB Papers	Thomas Burke Papers, University of Washington Libraries, Special Collections Division
UWL-SC	University of Washington Libraries, Special Collections Division
WDNR	Washington State Department of Natural Resources, Washington State Archives, Olympia
WPCC	Washington State Pollution Control Commission, Washington State Department of Ecology, Washington State Archives, Olympia
WSA	Washington State Archives, Olympia
WSDF	Washington State Department of Fisheries (later Fish and Wildlife), Washington State Archives, Olympia
WSDG	Washington State Department of Game (later Fish and Wildlife), Washington State Archives, Olympia

Prologue: The Fish that Might Save Seattle

1. "Endangered and Threatened Species; Threatened Status for Three Chinook Salmon Evolutionarily Significant Units (ESUs) in Washington and Oregon, and Endangered Status for One Chinook Salmon ESU in Washington, Final Rule," *Federal Register,* 64:56 (March 24, 1999): 14308–28. The bull trout, a non-anadromous species, was later added to the list.

2. For quotations, see *New York Times,* April 19, 1998; Lee Hochberg, "Saving Salmon: Back to Market," on *The News Hour with Jim Lehrer,* Public Broadcasting Service, June 23, 1998; *ST,* August 30, 1999; and John M. Findlay, "A Fishy Proposition: Regional Identity in the Pacific Northwest," in *Many Wests: Place, Culture, and Identity,* ed. David M. Wrobel and Michael C. Steiner (Lawrence: University Press of Kansas, 1997), 60. Findlay's critique is, in part, a rejoinder to Timothy Egan's comment, "The Pacific Northwest is simply this: wherever the salmon can get to." See Timothy Egan, *The Good Rain: Across Time and Terrain in the Pacific Northwest* (New York: Knopf, 1990; reprint, New York: Vintage, 1991), 22. For a response to Findlay's argument, see William L. Lang, "Beavers, Firs, Salmon, and Falling Water: Pacific Northwest Regionalism and the Environment," *Oregon Historical Quarterly* 104 (Summer 2003): 151–65. The most comprehensive history of salmon is Joseph E. Taylor III, *Making Salmon: An Environmental History of the Northwest Fisheries Crisis* (Seattle: University of Washington Press, 1999). For salmon as totem, see also Taylor, "Regional Unifier or Cultural Catspaw: The Cultural Geography of Salmon Symbolism in the Pacific Northwest," in *Imagining the Big Open: Nature, Identity, and Play in the New West,* ed. Liza Nicholas, Elaine M. Bapis, and Thomas J. Harvey (Salt Lake City: University of Utah Press, 2003), 3–26.

3. Yi-Fu Tuan, *Space and Place: The Perspective of Experience* (Minneapolis: University of Minnesota Press, 1977), 179. For erasure of place, see Wolfgang Schivelbusch, *The Railway Journey: The Industrialization of Time and Space in the 19th Century* (Berkeley: University of California Press, 1986). For place as history, see Dolores Hayden, *The Power of Place: Urban Landscapes as Public History* (Cambridge: MIT Press, 1995).

4. For creating places, see Paul Carter, *The Road to Botany Bay: An Essay in Spatial History* (London: Faber and Faber, 1987; Chicago: University of Chicago Press); see also Paul Carter and David Malouf, "Spatial History," *Textual Practice* 3 (Summer 1989): 173–83; and Hayden, *The Power of Place,* 2–78. For dominating people, see David Harvey, *Justice, Nature, and the Geography of Difference* (Cambridge, Mass.: Blackwell, 1996); and Henri Lefebvre, *The Production of Space,* trans. Donald Nicholson-Smith (Cambridge, Mass.: Blackwell, 1991). Lefebvre, who uses space and place interchangeably in his work, argues how both emerge from and define historical relationships. In this way, place can serve the interests of those in power or provide opportunities for resistance and change, an insight expressed most forcefully by Lefebvre as well as Harvey, who has refined and tested Lefebvre's original ideas. For other works on space and place that have framed my thinking, see Mark Gottdiener, *The Social Production of Urban Space* (Austin: University of Texas Press, 1985);

Derek Gregory, *Geographical Imaginations* (Cambridge, Mass.: Blackwell, 1994); Neil Smith, *Uneven Development: Nature, Capital, and the Production of Space* (Cambridge, Mass.: Blackwell, 1984, 1991); and Edward W. Soja, *Postmodern Geographies: The Reassertion of Space in Critical Theory* (New York: Verso, 1989), and *Thirdspace: Journeys to Los Angeles and Other Real and Imagined Places* (Cambridge, Mass.: Blackwell, 1996). Smith and Gottdiener are more rigidly materialist in their approach than either Harvey, Gregory, or Soja. One noted geographer who has used these concepts in his research is Don Mitchell, *The Lie of the Land: Migrant Workers and the California Landscape* (Minneapolis: University of Minnesota Press, 1996). Another is Matthew Gandy, *Concrete and Clay: Reworking Nature in New York City* (Cambridge: MIT Press, 2002). However, both scholars, following the lead of many contemporary social and cultural geographers, give relatively scant attention to the dynamic role of the non-human world in structuring social relations through time. Sarah Whatmore, *Hybrid Geographies: Natures, Cultures, Spaces* (Thousand Oaks, Calif., 2002), takes more seriously the contingency of the non-human world as a force that limits human power and knowledge. Four other superb studies by historians that have shaped my ideas of place are Kate Brown, *A Biography of No Place: From Ethnic Borderland to Soviet Heartland* (Cambridge: Harvard University Press, 2004); Jared Farmer, "American Land Marks: A History of Place and Displacement" (Ph.D. diss., Stanford University, 2005); Linda Nash, *Inescapable Ecologies: A History of Environment, Disease, and Knowledge* (Berkeley: University of California Press, 2006); and William Cronon, *Nature's Metropolis: Chicago and the Great West* (New York: W. W. Norton, 1992), a book that inspired my own work in many ways. For a thoughtful synthesis of the idea of place in Western thought, see Edward S. Casey, *The Fate of Place: A Philosophical History* (Berkeley: University of California Press, 1997).

5. Thomas P. Hughes, *Human-Built World: How to Think About Technology and Culture* (Chicago: University of Chicago Press, 2004), 156–61. Hughes derives his idea, in large part, from Richard White, *The Organic Machine: The Remaking of the Columbia River* (New York: Hill and Wang, 1995), which provided me numerous insights into seeing nature, technology, and human culture as intertwined through time and space.

6. Charles F. Wilkinson, *The Eagle Bird: Mapping a New West* (New York: Pantheon, 1992), 137–38. For social justice, see J. Ronald Engel, *Sacred Sands: The Struggle for Community in the Indiana Dunes* (Middletown, Conn.: Wesleyan University Press, 1983).

7. *P-I*, May 11, 1982. For Seattle's promotional tactics, see Walt Crowley, "The Emerald City Hustle," *Seattle Weekly* (June 20, 1984): 33–38.

8. The best scholarly edition of Baum's signature book is L. Frank Baum, *The Annotated Wizard of Oz: The Wonderful Wizard of Oz,* with introduction, notes, and bibliography by Michael Patrick Hearn (New York: W. W. Norton, 2000). For a revisionist take on Oz, see Gregory Maguire, *Wicked: The Life and Times of the Wicked Witch of the West* (New York: Regan, 1995) and *Son of a Witch: A Novel* (New York: Regan, 2005).

9. For nature's agency, see Linda Nash, "The Agency of Nature or the Nature of Agency?" *Environmental History* 10 (January 2005): 67–69, and "The Changing Experience of Nature: Historical Encounters with a Northwest River," *Journal of American History* 86 (March 2000): 1600–1629; and Ted Steinberg, "Down to Earth: Nature, Agency, and Power in History," *American Historical Review* 107 (June 2002): 798–820. Other works that have influenced my understandings of agency and nature include Tim Ingold, *The Perception of the Environment: Essays in Livelihood, Dwelling, and Skill* (New York: Routledge, 2000), which advocates a more complex analysis of how the environment shapes both human actions and intentions through their engagement with the material world; Timothy Mitchell, *Rule of Experts: Egypt, Techno-politics, Modernity* (Berkeley: University of California Press, 2002), which applies a similar approach to illumine the limits the physical environment imposes on human action; Julie Cruikshank, *Do Glaciers Listen?: Local Knowledge, Colonial Encounters, and Social Imagination* (Seattle and Vancouver: University of Washington Press and University of British Columbia Press, 2005), which analyzes how indigenous knowledge of nature arises from and responds to colonialism in time and place; Jon T. Coleman, *Vicious: Wolves and Men in America* (New Haven: Yale University Press, 2005), which applies an evolutionary approach to human-animal relations over time; and Walter Johnson, "On Agency," *Journal of Social History* 37 (Fall 2003): 113–24, a critique of agency within the field of African American history that has implications for how environmental historians might understand the concept.

10. Richard White uses salmon in a similar way to discuss the historical role of space and scale. See "The Nationalization of Nature," *Journal of American History* 86 (December 1999): 976–86.

11. For complex narratives, see William Cronon, "A Place for Stories: Nature, History, and Narrative," *Journal of American History* 78 (March 1992): 1347–76. For ethical obligations, see Tuan, *Space and Place,* 199–203.

12. Anne Whiston Spirn, *The Language of Landscape* (New Haven: Yale University Press, 1998), 192.

Chapter 1. All the Forces of Nature Are on Their Side: The Unraveling of the Mixed World

1. For Native cosmology, see Jay Miller, *Lushootseed Culture and the Shamanic Odyssey: An Anchored Radiance* (Lincoln: University of Nebraska Press, 1999), 49–63.

2. Arthur C. Ballard, *Mythology of Southern Puget Sound* (Seattle: University of Washington Press, 1929), 69–80. Ballard spells the name of his Native informant as "Snuqualmi Charlie," but later accounts list him as "Snuqualmie Charlie," which better reflects the current English spelling of "Snoqualmie," the name of a specific Indian group in Puget Sound. I use the spelling from Ballard's original work. Another clarification: I use the terms "Native" and "Indian" interchangeably when referring to indigenous peoples in aggregate to reflect the historical contingency behind these labels, preferring to use "Indian" in the context of post-contact relations with or comments by non-Indians whenever possible.

3. For problems with dualities, see Melville Jacobs, "A Few Observations on the World View of the Clackamas Chinook Indians," *Journal of American Folklore* 68 (July–September 1955): 283–89. Thanks to anthropologist Bill Seaburg, who helped me to understand the Transformer stories.

4. William James Stark, "The Glacial Geology of the City of Seattle" (master's thesis, University of Washington, 1950); Joe D. Dragovich, Patrick T. Pringle, and Timothy J. Walsh, "Extent and Geometry of the Mid-Holocene Osceola Mudflow in the Puget Lowland: Implications for Holocene Sedimentation and Paleogeography," *Washington Geology* 22 (September 1994): 3–26; and Arthur R. Kruckeberg, *The Natural History of Puget Sound Country* (Seattle: University of Washington Press, 1991), 8–33, 247–84.

5. Kruckeberg, *Natural History,* 36–52, 247–84.

6. Quotation from "Notes for Township 25 North Range 4 East Willamette Meridian" [June–September 1855], 547, King County, Auditor, General Land Office, Record Group 40, series 341, vol. 1, KCARM. For topography and vegetation, see David B. Williams, *The Street-Smart Naturalist: Field Notes from Seattle* (Portland, Ore.: WestWinds, 2005), 43–59.

7. P. E. Church, "Some Precipitation Characteristics of Seattle," *Weatherwise* 27 (December 1974): 244–51; Clifford F. Mass, Mark D. Albright, and Daniel J. Brees, "The Onshore Surge of Marine Air into the Pacific Northwest: A Coastal Region of Complex Terrain," *Monthly Weather Review* 114 (December 1986): 2602–27; and Andrew W. Robertson and Michael Ghil, "Large-Scale Weather Regimes and Local Climate over the Western United States," *Journal of Climate* 12 (June 1999): 1796–1813.

8. George Vancouver, *A Voyage of Discovery to the North Pacific Ocean, and Round the World,* ed. W. Kaye Lamb (London: The Hakluyt Society, 1984), 2:498, 561. For pre-European natural history, see Cathy Whitlock, "Vegetational and Climatic History of the Pacific Northwest During the Last 20,000 Years: Implications for Understanding Present-Day Biodiversity," *The Northwest Environmental Journal* 8 (1992): 5–28.

9. Thomas P. Quinn, *The Behavior and Ecology of Pacific Salmon and Trout* (Seattle: University of Washington Press, 2005), 13–36. The Latinate scientific names for these species are pink or humpback salmon (*O. gorbuscha*), sockeye salmon (*O. nerka*), coho or silver salmon (*O. kisutch*), chum or dog salmon (*O. keta*), chinook or king salmon (*O. tshawytscha*), steelhead trout (*O. mykiss*), and cutthroat trout (*O. clarkii*). Two other species of *Oncorhynchus* are masu (*O. masou*) and amago (*O. rhodurus*) salmon, both found in the western Pacific. Atlantic salmon are *Salmo salar.* Most fisheries biologists believe that anadromy, spawning in freshwater and feeding in the ocean, is a spectrum of behavioral and physiological changes related to specific life-history traits associated

with particular species. Some salmon are always anadramous, and others have non-anadramous populations or individuals. For details, see Quinn, *Behavior and Ecology of Pacific Salmon and Trout,* 5–6; for specific species, see *Pacific Salmon Life Histories,* ed. Cornelius Groot and Leo Margolis (Vancouver: University of British Columbia Press, 1991).

10. For disappearing salmon, see Mary F. Willson and Karl C. Halupka, "Anadromous Fish as Keystone Species in Vertebrate Communities," *Conservation Biology* 9 (June 1995): 489–97. For ecosystems, see James M. Helfeld and Robert J. Naiman, "Effects of Salmon-Derived Nitrogen on Riparian Forest Growth and Implications for Stream Productivity," *Ecology* 82 (September 2001): 2403–9; Scott M. Gende, Thomas P. Quinn, Mary F. Willson, Ron Heintz, and Thomas M. Scott, "Magnitude and Fate of Salmon-Derived Nutrients and Energy in a Coastal Stream Ecosystem," *Journal of Freshwater Ecology* 19 (March 2004): 149–60; Quinn, *Behavior and Ecology of Pacific Salmon and Trout,* 129–42. For El Niños, see Joseph E. Taylor III, "El Niño and the Politics of Blame," *Western Historical Quarterly* 4 (Winter 1998): 437–57.

11. For the whales and the rivers, see Ballard, *Mythology of Southern Puget Sound,* 87–89. For North Wind and Storm Wind, see T. T. Waterman, "The Geographical Names Used by the Indians of the Pacific Coast," *Geographical Review* 12 (April 1922): 194, and Coll Thrush, *Native Seattle: Histories from the Crossing-Over Place* (Seattle: University of Washington Press, 2007), 241–43. Anthropologists use the term "Coast Salish" to describe a subgroup of the Salishan language family as well as the various First Nations or Native American groups occupying present-day southern coastal British Columbia and western Washington. I use "Puget Sound Salish" instead as a more precise term to distinguish those groups living south of the Strait of Georgia in what is now the United States. Whulshootseed is a dialect continuum of Salishan spoken by Puget Sound peoples living roughly south of the King–Snohomish County line. My thanks to Coll Thrush for explaining these particulars to me. There is little consensus on the naming of Native groups in the region. For details, see Wayne M. Suttles, "Introduction," from *Northwest Coast,* ed. Wayne M. Suttles, vol. 7 of *Handbook of North American Indians,* gen. ed. William C. Sturtevant (Washington, D.C.: Smithsonian Institution, 1990), 1–15.

12. Waterman, "Geographical Names," 193; Miller, *Lushootseed Culture,* 15–36; Thrush, *Native Seattle,* 239, 248.

13. Waterman, "Geographical Names," 183, 193–94; Thrush, *Native Seattle,* 229, 238, 241. For fishing and hunting practices, see T. T. Waterman, *Notes on the Ethnology of the Indians of Puget Sound* (New York: Museum of the American Indian/Heye Foundation, 1973), 63–71, 83. I have avoided using contemporary tribal names in the main text thus far because of their historical contingency. For changing Indian identities, see Alexandra Harmon, *Indians in the Making: Ethnic Relations and Indian Identities Around Puget Sound* (Berkeley: University of California Press, 1999), 13–71. For Native place naming, see Keith H. Basso, *Wisdom Sits in Places: Landscape and Language Among the Western Apache* (Albuquerque: University of New Mexico Press, 1996). For details on specific Native place names around Seattle with historical explanations, including their Whulshootseed spellings, see the superb atlas compiled by Coll Thrush, Nile Thompson, and Amir Sheikh in Thrush, *Native Seattle,* 209–55.

14. The historical origins of the term "Muckleshoot" are unclear. See Patricia Slettvet Noel, *Muckleshoot Indian History* (Auburn, Wash.: Auburn School District No. 408, 1980), 5, copy at Muckleshoot Indian Tribe home page, www.muckleshoot.nsn.us/ (accessed November 15, 2006). For the sake of clarity, I have simplified the historical complexity of the various Native bands living in Seattle before contact in my narrative; the people known today as the Duwamish, for example, were once three distinct branches. For details, see Thrush, *Native Seattle,* 23–24.

15. For social intricacies and subsistence patterns, see Miller, *Lushootseed Culture,* 18–19, 79–104; Wayne M. Suttles, *Coast Salish Essays* (Seattle: University of Washington Press/Vancouver, B.C.: Talonbooks, 1987), 209–30. For Indians and property, see William Cronon, *Changes in the Land: Indians, Colonists, and the Ecology of New England* (New York: Hill and Wang, 1983), 54–81, 127–86.

16. For harvesting and hunting, see Miller, *Lushootseed Culture,* 17–21; and Daniel P. Cheney and Thomas Mumford, Jr., *Shellfish and Seaweed Harvests of Puget Sound* (Seattle: University of Washington Sea Grant Program, 1986).

17. Ballard, *Mythology of Southern Puget Sound,* 133–35. Ballard attributes the first story to the

Snoqualmie, as told by Snuqualmi Charlie, the second to the Skokomish. The third story, told by Big John of Green River, is from Arthur C. Ballard, *Some Tales of the Southern Puget Sound Salish* (Seattle: University of Washington Press, 1927), 81. For guiding principles, see Joseph E. Taylor III, *Making Salmon: An Environmental History of the Northwest Fisheries Crisis* (Seattle: University of Washington Press, 1999), 13–38.

18. Miller, *Lushootseed Culture*, 30.

19. Vancouver, *A Voyage of Discovery*, 515, 541–43.

20. Vancouver, *A Voyage of Discovery*, 528. For disease, see Robert Boyd, *The Coming of the Spirit of Pestilence: Introduced Infectious Diseases and Population Decline Among Northwest Coast Indians, 1774–1874* (Vancouver: University of British Columbia Press/Seattle: University of Washington Press, 1999), 30–39, 57–58, 153–60, 262–78; and R. Cole Harris, *The Resettlement of British Columbia: Essays on Colonialism and Geographical Change* (Vancouver: University of British Columbia Press, 1997), 3–30. For cultural continuum, see Suttles, "Cultural Diversity Within the Coast Salish Continuum," in *Ethnicity and Culture*, ed. Reginald Auger et al. (Calgary: University of Calgary Press, 1987), 243–49.

21. For resettlement, see Harris, *Resettlement of British Columbia*, 31–67.

22. For the fur trade, see Richard Somerset Mackie, *Trading Beyond the Mountains: The British Fur Trade on the Pacific, 1793–1843* (Vancouver: University of British Columbia Press, 1997).

23. Harmon, *Indians in the Making*, 1, 3–42. For the idea of the "mixed world," see Richard White, *The Middle Ground: Indians, Empires, and Republics in the Great Lakes Region, 1650–1815* (New York: Cambridge University Press, 1991).

24. First two quotations are from Charles Wilkes, *Diary of Wilkes in the Northwest*, ed. Edmond S. Meany (Seattle: University of Washington Press, 1926), 86–88, 93. Final quotations from Charles Wilkes, *Narrative of the United States Exploring Expedition During the Years 1838, 1839, 1840, 1841, 1842* (Philadelphia: Lea and Blanchard, 1845), 4:332–33, 305, 421.

25. Wilkes, *Diary of Wilkes*, 27, 87. For Koquilton's account, see Pierce County Pioneer Association, "Commemorative Celebration at Sequalitchew Lake" (pamphlet, 1906): 13–14, Tacoma Public Library, as quoted in Harmon, *Indians in the Making*, 45.

26. For Seattle's founding, see Coll-Peter Thrush, "Creation Stories: Rethinking the Founding of Seattle," in *More Voices, New Stories: King County, Washington's First 150 Years*, ed. Mary C. Wright (Seattle: University of Washington Press for the Pacific Northwest Historians Guild, 2002), 34–49.

27. Arthur A. Denny, *Pioneer Days on Puget Sound* (Seattle: C. B. Bagley, 1888), 16–20, 36, 51; Frances Kautz, ed. "Extracts from the Diary of General A. V. Kautz," *Washington Historian* 1 (April 1900): 119; Emily Inez Denny, *Blazing the Way: True Stories, Songs, and Sketches of Puget Sound and Other Pioneers* (Seattle: Rainier Printing, 1909), 70, 74; and A. B. Rabbison, "Growth of Towns," Hubert H. Bancroft Collection, UW Microfilm 155, UWL-SC; also quoted in Harmon, *Indians in the Making*, 60, which identifies the settler as A. B. Rabbeson. For potatoes and Natives, see Suttles, "The Early Diffusion of the Potato Among the Coast Salish," *Southwestern Journal of Anthropology* 7 (1951): 272–88.

28. John R. Finger, "Henry L. Yesler's Seattle Years, 1852–1892" (Ph.D. diss., University of Washington, 1968), 1–17.

29. Finger, "Henry L. Yesler's Seattle Years," 25–26, 33; *Columbian* (Olympia), October 12, 1853; Frederick James Grant, *History of Seattle, Washington* (New York: American Publishing and Engraving, 1891), 82.

30. George Gibbs, "Report on the Indian Tribes of Washington Territory," in *Reports of Explorations and Surveys to Ascertain the Most Practicable Economical Route for a Railroad from the Mississippi River to the Pacific Ocean*, vol. 12, pt. 1 (Washington, D.C.: B. Tucker, Printer, 1855–1860), 432 (hereafter cited as *Pacific Railroad Surveys*); Catherine Blaine to Family (c. January 1854), David E. Blaine to Father, June 21, 1854, *Letters and Papers of Rev. David E. Blaine and His Wife Catherine* (Seattle: Historical Society of the Pacific Northwest Conference of the Methodist Church, 1963), 46, 61.

31. Quotation from Henry L. Yesler, "Henry Yesler and the Founding of Seattle," *Pacific Northwest Quarterly* 42 (October 1951): 274. For the lynchings, see Denny, *Pioneer Days on Puget Sound*, 42–43, 67; and Murray Morgan, *Skid Road: An Informal Portrait of Seattle* (New York: Viking, 1951; Seattle: University of Washington Press, 1982), 35–39. For interracial violence, see Brad Asher, *Be-*

yond the Reservation: Indians, Settlers, and the Law in Washington Territory, 1853–1889 (Norman: University of Oklahoma Press, 1999), 107–53.

32. Charles Prosch, *Reminiscences of Washington Territory: Scenes, Incidents, and Reflections of the Pioneer Period on Puget Sound* (Seattle, 1904), 23–24, 26–27; Denny, *Blazing the Way,* 64–66. For stereotypes, see Robert F. Berkhofer, Jr., *The White Man's Indian: Images of the American Indian from Columbus to the Present* (New York: Vintage, 1978), 153–75.

33. Clarence B. Bagley, *History of Seattle from the Earliest Settlement to the Present Time* (Chicago: S. J. Clarke, 1916), 76. For land plats, see Norbert MacDonald, *Distant Neighbors: A Comparative History of Seattle and Vancouver* (Lincoln: University of Nebraska Press, 1987), 3–7. For social order, see Gregory H. Nobles, "Straight Lines and Stability: Mapping the Political Order of the Anglo-American Frontier," *Journal of American History* 80 (June 1993): 9–35.

34. David and Catherine Blaine, *Memoirs of Puget Sound: Early Seattle, 1853–1856,* ed. Richard A. Seiber (Fairfield, Wash.: Ye Galleon, 1978), 76–77; Sophie Frye Bass, *When Seattle Was a Village* (Seattle: Lowman and Hanford, 1947), 44; and Prosch, *Reminiscences,* 27. For dogfish oil, see Thomas F. Gedosch, "A Note on the Dogfish Oil Industry of Washington Territory," *Pacific Northwest Quarterly* 52 (April 1968): 100–102. For Indians and Seattle's economy, see Thrush, *Native Seattle,* 47–49, 72–74.

35. Thomas Prosch, "A Chronological History of Seattle from 1850 to 1897," typescript (Seattle, 1901), 26, UWL-SC.

36. For settlements, see *Columbian* (Olympia), October 29, November 16, 1853; Harmon, *Indians in the Making,* 13–42, 60–61. For Stevens, see William H. Goetzmann, *Army Exploration in the American West, 1803–1863* (New Haven: Yale University Press, 1959), 262–83; Kent D. Richards, *Isaac I. Stevens: Young Man in a Hurry* (Provo, Utah: Brigham Young University Press, 1979; reprint, Pullman: Washington State University Press, 1993); Sherburn F. Cook, Jr., "The Little Napoleon: The Short and Turbulent Career of Isaac I. Stevens," *Columbia* 14 (Winter 2000–2001): 17–20.

37. George Suckley, "Report Upon the Fishes Collected on the Survey: Report Upon the Salmonidae," in *Pacific Railroad Surveys,* vol. 12, pt. 2, 311; Stevens, "Resources and Geographical Importance of Puget Sound, and Its Relation to the Trade of Asia," and Gibbs, "Report of George Gibbs on a Reconnaissance of the Country Lying Upon Shoalwater Bay and Puget Sound," in *Pacific Railroad Surveys,* vol. 1, 113, 472; and Suckley to John Suckley, January 25, 1854, as quoted in Goetzmann, *Army Exploration in the American West,* 283. For sectional politics, see Richard White, *"It's Your Misfortune and None of My Own": A New History of the American West* (Norman: University of Oklahoma Press, 1991), 125–26.

38. Gibbs, "Indian Tribes of Washington Territory," 424; Harmon, *Indians in the Making,* 78–79.

39. *Puget Sound Courier* (Olympia), November 20, 1855; Finger, "Henry Yesler's Seattle Years," 32–33; and Gibbs, "Indian Tribes of Washington Territory," 422–23.

40. David Blaine to Parents, December 19, 1854, January 24, 1855, *Letters and Papers,* 93, 95. For the treaty process and its later complications, see Harmon, *Indians in the Making,* 72–86, 103–217. The history of Indian reservation policy and federal recognition of tribes in western Washington is exceedingly complex and politically combustible. Many bands that consider themselves original signatories to the various treaties negotiated by Stevens during his tenure as territorial governor remain unrecognized by the federal government. The ongoing and contentious disputes over recognition, which often pit acknowledged treaty tribes against unrecognized tribes, are beyond the scope of this work.

41. Denny, *Pioneer Days on Puget Sound,* 74; Richard White, "The Treaty at Medicine Creek: Indian-White Relations on Upper Puget Sound, 1830–1880" (master's thesis, University of Washington, 1972), 36–42; and Harmon, *Indians in the Making,* 88–90.

42. *Pioneer and Democrat* (Olympia), September 5, 1856; Harmon, *Indians in the Making,* 92–95; and Asher, *Beyond the Reservation,* 3–59. Stevens stayed on as Washington's territorial governor until 1857, served as a congressional delegate until 1861, then joined the Union Army. He was killed at the Battle of Chantilly on September 1, 1862.

43. Hazard Stevens, *The Life of Isaac Ingalls Stevens* (Boston: Houghton, Mifflin, 1901), 1:448; Denny, *Blazing the Way,* 120–21. For Indian laborers, see John Lutz, "Work, Sex, and Death on the Great Thoroughfare: Annual Migrations of 'Canadian Indians' to the American Pacific Northwest," in *Parallel Destinies: Canadian-American Relations West of the Rockies,* ed. John M. Findlay and

Ken S. Coates (Seattle: Center for the Study of the Pacific Northwest in association with University of Washington Press/Montreal: McGill-Queen's University Press, 2002), 80–103; Paige Raibmon, *Authentic Indians: Episodes of Encounter from the Late-Nineteenth-Century Northwest Coast* (Durham: Duke University Press, 2005), 74–97.

44. Charles A. Kinnear, "Arrival of the George Kinnear Family on Puget Sound, and Early Recollections by C. A. Kinnear, One of the Children," n.d., Manuscript Collection, Museum of History and Industry, Seattle; Reverend A. Atwood, *Glimpses of Pioneer Life on Puget Sound* (Seattle: Denny-Correll, 1903), 149; Edward P. Smith, Commissioner, "Information with Historical and Statistical Statements Relative to the Different Tribes," and E. C. Chirouse to Smith, September 21, 1875, in *Report of the Secretary of the Interior,* vol. 1680 (1875), 573, 868–69.

45. *Daily Intelligencer* (Seattle), December 29, 1876. For restricting Indians, see Thrush, *Native Seattle,* 51–55. For anti-miscegenation laws, see Asher, *Beyond the Reservation,* 63–68.

46. "Petition to the Honorable Arthur A. Denny, Delegate to Congress from Washington Territory," n.d., National Archives Roll 909, copy at National Archives and Records Administration, Pacific Northwest Branch, Seattle, and at HistoryLink.org, www.historylink.org/essays/output.cfm?file_id=2955 (accessed November 20, 2005); and *Weekly Intelligencer* (Seattle), July 3, 1875. For Yesler's later years, see Yesler, "Henry Yesler and the Founding of Seattle," 274; Thomas W. Prosch, "Seattle and the Indians of Puget Sound," *Washington Historical Quarterly* 2 (July 1908): 305; and Finger, "Henry L. Yesler's Seattle Years," 114–44.

47. "Seattle Illahee," from Philip J. Thomas, *Songs of the Pacific Northwest,* transcription and notation by Shirley A. Cox (Saanichton, B.C.: Hancock House, 1979), 60–61, as quoted in Thrush, *Native Seattle,* 40–41; Edith Sanderson Redfield, *Seattle Memories* (Boston: Lothrop, Lee, & Shepherd, 1930), 31. For Skid Road, see David M. Buerge, *Seattle in the 1880s* (Seattle: Historical Society of Seattle and King County, 1986).

48. H.H. (Helen Hunt Jackson), "Puget Sound," *Atlantic Monthly* 51 (February 1883): 218–31.

49. *SS,* October 29, 1887; see also Grant, *History of Seattle,* 433–36.

50. Jill Lepore, "Wigwam Words," *American Scholar* 70 (Winter 2001): 98. For the various interpretations, see John M. Rich, *Chief Seattle's Unanswered Challenge* (1932; repr., Fairfield, Wash.: Ye Galleon, 1970); Janice Krenmayr, "'The Earth Is Our Mother': Who Really Said That?" *Seattle Times Sunday Magazine* (January 5, 1975), 4–6; Ted Perry, "Chief Seattle Speaks Again," *Middlebury College Magazine* 63 (Winter 1988–89): 30–38; and David Buerge, "Seattle's King Arthur," *Seattle Weekly* (July 17, 1991), 27–29. For the speech's genealogy, see Albert Furtwangler, *Answering Chief Seattle* (Seattle: University of Washington Press, 1997).

51. Ballard, *Mythology of Southern Puget Sound,* 80.

Chapter 2. The Work Which Nature Had Left Undone: Making Private Property on the Waterfront Commons

1. Welford Beaton, *The City That Made Itself: A Literary and Pictorial Record of the Building of Seattle* (Seattle: Terminal Publishing, 1914), 69.

2. For Seattle's economy, see Roger Sale, *Seattle: Past and Present* (Seattle: University of Washington Press, 1976), 50–53. For the larger context of industrial development in the Far West, see William G. Robbins, *Colony and Empire: The Capitalist Transformation of the American West* (Lawrence: University Press of Kansas, 1994); David Igler, "The Industrial Far West: Region and Nation in the Late Nineteenth Century," *Pacific Historical Review* 69 (May 2000): 159–92.

3. For speculation, see Richard Wade, *The Urban Frontier: Pioneer Life in Early Pittsburgh, Cincinnati, Lexington, Louisville, and St. Louis* (Cambridge: Harvard University Press, 1959), 1–100. For San Francisco, see Gray Brechin, *Imperial San Francisco: Urban Power, Earthly Ruin* (Berkeley: University of California Press, 1999), 1–120.

4. For a provocative study of property and nature, see Theodore Steinberg, *Slide Mountain, or the Folly of Owning Nature* (Berkeley: University of California Press, 1995).

5. Marshall Moore, "Address," December 9, 1867, in Charles M. Gates, *Messages of the Governors of the Territory of Washington to the Legislative Assembly, 1854–1889* (Seattle: University of Washington Press, 1940), 139, 142.

6. For Whitworth's travels, see Diary, January 2–15, April 19–29, May 10–15, May 20–25, June

9–26, 1869, box 1, James Edwin Whitworth Papers, UWL-SC. T. J. McKenney to H. G. Taylor, in *Report of the Secretary of the Interior,* vol. 1366 (1868), 553. For flooding in southern Puget Sound, see Thomas Prosch, "A Chronological History of Seattle from 1850 to 1897," typescript (Seattle, 1901), 185, 187–88, UWL-SC. Major floods occurred in 1867, 1871, 1873, 1876–77, 1888, 1892–94, 1900, 1903, 1906, and 1911. For drainage districts, see "In re Ditch Petition, Lake Washington–Lake Union–Salmon Bay," February 4, 1893, box 1, fol. 7, King County Commissioners, Drainage District Files, PSRA.

7. Robert C. Nesbit, *"He Built Seattle": A Biography of Judge Thomas Burke* (Seattle: University of Washington Press, 1961), 9–56; Kurt Armbruster, *Orphan Road: The Railroad Comes to Seattle, 1853–1911* (Pullman: Washington State University Press, 1999), 49–142.

8. *Rochester Advertiser* (New York), as quoted in *Daily Intelligencer* (Seattle), June 8, 1876; Prosch, "Chronological History," 231.

9. *P-I,* November 6, 1885. For Chinese and dual-labor, see Richard White, *"It's Your Misfortune and None of My Own": A New History of the American West* (Norman: University of Oklahoma Press, 1991), 282–84, 320–22. For anti-Chinese, see Murray Morgan, *Puget's Sound: A Narrative of Early Tacoma and the Southern Sound* (Seattle: University of Washington Press, 1979), 212–52; Art Chin, *Golden Tassels: A History of the Chinese in Washington, 1857–1977* (Seattle: n.p., 1977), 55–56; and Marilyn Tharp, "Story of Coal at Newcastle," *Pacific Northwest Quarterly* 48 (October 1957): 120–26.

10. Burke to Waldo M. Yorke, December 15, 1881, box 37, TB Papers. For legal doctrines, see Bonnie J. McCay, *Oyster Wars and the Public Trust: Property, Law, and Ecology in New Jersey History* (Tucson: University of Arizona Press, 1998), 5–158. For eastern cities, see Ann L. Buttenwieser, *Manhattan, Water-bound: Planning and Developing Manhattan's Waterfront from the Seventeenth Century to the Present* (New York: New York University Press, 1987); Nancy S. Seasholes, *Gaining Ground: A History of Landmaking in Boston* (Cambridge: MIT Press, 2003).

11. *P-I,* August 4, September 26, October 8, 17, 1889.

12. For the Panic of 1873, see Elmus Wicker, *Banking Panics of the Gilded Age* (New York: Cambridge University Press, 2000), 16–33.

13. James Blaine Hedges, *Henry Villard and the Railways of the Northwest* (New Haven: Yale University Press, 1930), 72–77; White, *"It's Your Misfortune,"* 252–57; and Nesbit, *"He Built Seattle,"* 33–56. For waterfront real estate, see *P-I,* November 28, 1880, November 17, 1882.

14. *Railroad Gazette* (New York) 17 (May 29, 1885), 349; *Commercial and Financial Chronicle* (New York) 43 (August 15, 1886), 184; and Gilman to Burke, November 20, 25, 1886, box 6, fol. 1, TB Papers.

15. For real estate, see Burke to David T. Denny, June 15, 1884, Burke to Gilman, April 18, May 31, 1886, boxes 33, 38, TB Papers. For Railroad Avenue, see Fred H. Peterson to Arthur A. Denny and Bailey Gatzert, February 11, 1887; Burke to Crawford, February 13, 1887, box 34, TB Papers; Ordinance 804, *Laws and Ordinances of the City of Seattle* (Seattle, 1887), SMA.

16. *P-I,* January 3, 14, February 19, December 30, 1888, January 10, 1889.

17. Burke to Carrie L. Allen, May 1, 1888, box 20, TB Papers.

18. *Daily Intelligencer* (Seattle), May 15, 1878, July 27, 1879; Prosch, "Chronological History," 259–60. For Indians, see Coll Thrush, *Native Seattle: Histories from the Crossing-Over Place* (Seattle: University of Washington Press, 2007), 63–64.

19. Rudyard Kipling, *From Sea to Sea: Letters of Travel* (New York: Charles Scribner's Sons, 1899), 119–20. For other accounts, see Frederick James Grant, *History of Seattle, Washington* (New York: American Publishing and Engraving, 1891), 212–34; and Clarence B. Bagley, *History of Seattle from the Earliest Settlement to the Present Time* (Chicago: S. J. Clarke, 1916), 419–28.

20. Wilfred J. Airey, "A History of the Constitution and Government of Washington Territory" (Ph.D. diss., University of Washington, 1945), 497–518.

21. *P-I,* August 19, 23, 1889; Jacob Furth to William F. Prosser, April 16, 1890, box 2, Harbor Line Commission, Survey Notes, Correspondence, and Reports, WDNR. For legal doctrines after statehood, see Airey, "History of the Constitution," 518–21; and *Laws, Rules, and Regulations Governing the Appraisement and Sale of Tidelands of the State of Washington* (Olympia, 1893), box 1, Harbor Line Commission Reports, WDNR.

22. For invasion of navigable waters, see *Yesler v. Harbor Line Commissioners,* 146 U.S. 646

(1892); for rights of upland owners, see *Dearborn v. Moran,* 2 Wash. 405 (1891); and *Denny v. Northern Pacific Railway Company,* 19 Wash. 298 (1898).

23. Minutes, July 31 and September 10, 1890, Minutes of the Harbor Line Commission, vols. 25 and 65; and Proceedings of the Board of Tideland Appraisers for King County, July 13, 1894 and January 18, 1895, no. 42, box 2, WDNR.

24. Proceedings of the Board of Tideland Appraisers for King County, September 6, 1894, no. 42, box 2, WDNR.

25. For the Great Northern, see Ralph W. Hidy, Muriel E. Hidy, and Roy V. Scott, with Don L. Hofsommer, *The Great Northern Railway: A History* (Cambridge: Harvard Business School Press, 1988).

26. Nesbit, *"He Built Seattle,"* xvii; Burke to Gilman, June 17, 1890, box 21, fol. 5, TB Papers; and *P-I,* October 30, 1890. For the Northern Pacific, see J. C. Haines to C. J. Smith, May 16, 1891, and C. J. Smith to C. B. Tedcastle, April 7, 1892, boxes 50 and 51, fol. 8 and 12, Oregon Improvement Company Records, UWL-SC; Hill to P. P. Shelby, June 21, 1892, box 2, file 446, GN-PSF; *P-I,* November 12, 1890. For corruption, see Richard White, "Information, Markets, and Corruption: Transcontinental Railroads in the Gilded Age," *Journal of American History* 90 (June 2003): 19–43.

27. W. P. Clough to Hill, February 14, 26, 1890, GN-PSF; R. S. Jones to Hill, November 3, 1892, President's Assistant File, James J. Hill Papers, James J. Hill Reference Library, St. Paul; and Burke to Clough, March 20, April 1, 25, 1890, box 20, TB Papers.

28. For legal battles, see "Agreement Between Columbia and Puget Sound Railroad, Northern Pacific Railroad, and the Seattle and Montana Railroad," September 8, 1894, box 45, fol. 3, TB Papers; *ST,* October 13, 1892; *P-I,* October 16, November 22, 1892, May 18, 1893. For port facilities, see R. H. (Reginald Heber) Thomson, *That Man Thomson,* ed. Grant H. Redford (Seattle: University of Washington Press, 1950), 49–51. For Hill's tactics, see Frank Leonard, "'Wise, Swift, and Sure'?: The Great Northern Entry into Seattle, 1889–1894," *Pacific Northwest Quarterly* 92 (Spring 2001): 81–90. For railroad monopolies, see William Deverell, *Railroad Crossing: Californians and the Railroad, 1850–1910* (Berkeley: University of California Press, 1994), 93–122; and Frank Leonard, *A Thousand Blunders: The Grand Trunk Pacific Railway and Northern British Columbia* (Vancouver: University of British Columbia Press, 1996), 127–64. For an overview of railroad business practices, see Steve W. Usselman, *Regulating Railroad Innovation: Business, Technology, and Politics in America, 1840–1920* (New York: Cambridge University Press, 2002).

29. Thomas W. Symons to Fred Gasch, January 19, 1894, box 1, fol. 7, Drainage District Files, King County Commissioners, RG 011, PSRA.

30. Alan Hynding, *The Public Life of Eugene Semple: Promoter and Politician of the Pacific Northwest* (Seattle: University of Washington Press, 1973), 3–163. For gambling, see John M. Findlay, *People of Chance: Gambling in American Society from Jamestown to Las Vegas* (New York: Oxford University Press, 1986); Ann Fabian, *Card Sharps, Dream Books, and Bucket Shops: Gambling in 19th-Century America* (Ithaca: Cornell University Press, 1990).

31. Eugene Semple, "Proposition for a Ship and Water Power Canal Between Elliott Bay and Lake Washington, near Seattle, Wash," January 3, 1891, box 17, fol. 4, ES Papers. For the South Canal, see Bagley, *History of Seattle,* 380–89; and Alan Hynding, "Eugene Semple's Seattle Canal Scheme," *Pacific Northwest Quarterly* 59 (April 1968): 77–87.

32. For previous canals, see Eugene Ricksecker to Thomas Symons, June 10, 1895, box 51, RG77; Bagley, *History of Seattle,* 371–75; Thomas Canfield, *Northern Pacific Railroad, Partial Report to the Board of Directors of a Portion of a Reconnaissance Made in the Summer of 1869, Between Lake Superior and the Pacific Ocean . . . Accompanied with Notes on Puget Sound by Samuel Wilkeson* (St. Paul, Minn., May 1870), 22–28, UWL-SC; Henry Gorringe, "Survey," box 24, fol. 29, Oregon Improvement Company Records, UWL-SC; Suzanne B. Larson, *"Dig the Ditch!": The History of the Lake Washington Ship Canal* (Boulder: Western Interstate Commission for Higher Education, 1975); and Stacey Patterson, "Interpreting the History of Chinese Labor on the Fremont Cut Through Public Art" (master's thesis, University of Washington, 1998), 9–44.

33. For the 1891 report, see *Annual Report of the Chief of Engineers for 1892* (Washington, D.C.: Government Printing Office, 1892), 2762–93. For flood control, see Jamie W. Moore and Dorothy P. Moore, *The Army Corps of Engineers and the Evolution of Federal Flood Plain Management Policy*

(Boulder: Institute of Behavioral Science, 1989), 1–16; and Samuel P. Hays, *Conservation and the Gospel of Efficiency: The Progressive Conservation Movement, 1890–1920,* rev. ed. (New York: Athenaeum, 1969), 5–26, 91–121, and 199–218. For sense of place, see William Cronon, *Nature's Metropolis: Chicago and the Great West* (New York: W. W. Norton, 1991), 55–340; and Carol Sheriff, *The Artificial River: The Erie Canal and the Paradox of Progress, 1817–1862* (New York: Hill and Wang, 1996), 9–78.

34. Virgil G. Bogue, "Report to Board of Tide Land Appraisers of King County, Washington" (January 10, 1895), 22, no. 42, box 2, WDNR; Record of the Proceedings of the Board of Tideland Appraisers for King County, March 19, 1895, no. 42, box 2, WDNR.

35. Semple to Edgar and Henry Ames, October 16, 1894, box 2, fol. 18, ES Papers. For operations, see *History and Advantages of the Canal and Harbor Improvement Project Now Being Executed by the Seattle and Lake Washington Waterway Company* (Seattle: Lowman and Hanford, 1902), 6–14, 39–40, UWL-SC.

36. Jacob Furth to David E. Durie, July 26, 1894; Seattle Brewing and Malting Company to W. T. Forrest, June 9, 1894, box 17, fol. 4, ES Papers.

37. For court cases, see *Schlopp v. Forrest,* 11 Wash. 640 (1895), and *Mississippi Valley Trust Company and Seattle & Lake Washington Waterway Company v. Hofius* (1898).

38. Gilman to Burke, March 2, 1892, box 6, fol. 1, TB Papers; Gilman to Burke, April 26, 1896, box 6, fol. 2, TB Papers.

39. Greene et al., to Daniel S. Lamont, July 18, 1895, box 10, fol. 7, TB Papers. For the 1895 report, see *Annual Report of the Chief of Engineers for 1896* (Washington, D.C.: Government Printing Office, 1896), 3351–73; and William F. Willingham, *Northwest Passages: A History of the Seattle District, U.S. Army Corps of Engineers, 1896–1920,* based in part on research by Robert E. Ficken (Seattle: U.S. Army Corps of Engineers, 1992), 81. Another proposed and rejected route would have run through Smith's Cove to Salmon Bay and Lake Union, cutting the Great Northern switching yards there in two.

40. *SA,* December 18, 1897. For business and boosters, see Kathryn Morse, *The Nature of Gold: An Environmental History of the Klondike Gold Rush* (Seattle: University of Washington Press, 2003), 166–85.

41. For metaphors of trade, see Ari Kelman, *A River and Its City: The Nature of Landscape in New Orleans* (Berkeley: University of California Press, 2003), 19–86.

42. *History and Advantages of the Canal,* 25–26; Erastus Brainerd, *Lake Union and Lake Washington Waterway* (Seattle, 1902), 52. For testimony, see *Annual Report of the Chief of Engineers for 1903* (Washington, D.C.: Government Printing Office, 1903), 2340–57.

43. Semple to Symons, June 27, 1905, box 4, fol. 15, ES Papers. For financial and legal difficulties, see *ST,* March 24, 1903, May 10, 1905; *P-I,* December 23, 1900, May 1, 1903, May 6, 24, 1904; *SS,* May 5, 1903; and *SMH,* March 15, 1902. For sluicing and dredging, see "Protest of Hall and Paulson Furniture Company," September 18, 1899, "Hydraulicking not Feasible" [c. 1902], box 18, fol. 4, 5, ES Papers. For filling contracts, see "Agreement Between Seattle and Lake Washington Waterway Company and Columbia and Puget Sound Railroad," October 4, 1901, box 1, fol. 71, Great Northern Railway Corporate Records, Pacific Coast Railroad Company, Law Department Files, Minnesota Historical Society.

44. *White River Journal* (Kent), June 25, 1898, January 20, September 22, 1900.

45. T. H. Crosswell to Thomas Cooper, November 30, 1901, C. W. Bunn to Pearson, November 10, 1903, and E. J. Pearson to A. R. Cook, November 11, 1903, box 35, file 1316-1, NP-CESF.

46. For Chittenden's career, see Gordon B. Dodds, *Hiram Martin Chittenden: His Public Career* (Lexington: University Press of Kentucky, 1973). For the 1896 report, see Martin A. Reuss, "Andrew A. Humphreys and the Development of Hydraulic Engineering: Politics and Technology in the Army Corps of Engineers, 1850–1950," *Technology and Culture* 20 (January 1985): 1–33. For Roosevelt, see Richard Slotkin, *Gunfighter Nation: The Myth of the Frontier in Twentieth-Century America* (New York: Athenaeum, 1992; Norman: University of Oklahoma Press, 1998), 29–62.

47. For the Moore plan, see Eugene Semple to James A. Moore, March 23, May 19, 1906, box 4, fol. 17, 18, ES Papers. For Chittenden's maneuvering, see "Report by H. M. Chittenden," December 2, 1907, Chittenden to J. S. Brace, June 11, 1907, and Chittenden, "Statement," n.d., boxes 176, 204, RG77.

For the "Seattle Spirit," see Sharon Boswell and Lorraine McConaghy, *Raise Hell and Sell Newspapers: Alden J. Blethen and the Seattle Times* (Pullman: Washington State University Press, 1996), 195–98.

48. Hiram M. Chittenden et al., *Report of an Investigation by a Board of Engineers of the Means of Controlling Floods in the Duwamish-Puyallup Valleys and their Tributaries in the State of Washington* (Seattle: Lowman and Hanford, 1907), 9–11, 31–32, UWL-SC. For flooding, see *White River Journal* (Kent), November 16, 1906; *SMH,* December 8, 1906; *Argus* (Auburn), November 15, 1906.

49. For quotation, see "Winter Floods," *Pacific Coast Sportsman* 3 (November 1906): 379–80. For the Duwamish Waterway, see John Shorett to M. E. Hay, June 15, 1909, and "Resolution by the Seattle Commercial Club" to Hay, June 1909, box 2G-2-012, Duwamish Waterway File, Governors' Papers, Hay Group, WSA. For the White River diversion, see Wyndham J. Roberts, *Report on Flood Control of White-Stuck and Puyallup Rivers: Inter-County River Improvement, King and Pierce Counties, State of Washington* ([Seattle?]: Inter-County River Improvement Board of Commissioners, 1920), UWL-SC.

50. Joseph Moses, as quoted in Mike Sato, *The Price of Taming a River: The Decline of Puget Sound's Duwamish/Green Waterway* (Seattle: The Mountaineers, 1997), 56–57. For the Black River, see Morda C. Slauson, *Where Was the Black River?* (Renton: Renton Historical Society, 1967), UWL-SC; David M. Buerge, "Requiem for a River," *Seattle Weekly* (October 16–22, 1985): 33–38, 47–49. For changes to Lake Washington, see Eugene Collias and G. R. Seckel, *Lake Washington Ship Canal Data* (Seattle: University of Washington Press, 1954); Michael Chrzastowski, *Historical Changes to Lake Washington and the Route of the Lake Washington Ship Canal, King County, Washington* (Reston, Va.: U.S. Geological Survey, 1983).

51. For the 1914 break, see *Annual Report of the Chief of Engineers for 1914* (Washington, D.C.: Government Printing Office, 1914), 1430; *Annual Report of the Chief of Engineers for 1915* (Washington, D.C.: Government Printing Office, 1915), 3258; and *P-I,* March 14, 1914. For an earlier breach, see *Annual Report of the Chief of Engineers for 1904* (Washington, D.C.: Government Printing Office, 1904), 3612–13.

52. Affidavit of William L. Bilger, April 13, 1910, Case 84: *Bilger et al. v. Washington et al.,* Attorney General Case Files, WSA. For canal construction, see Chittenden to MacKenzie, March 11, 1907, and Chittenden, "Report," December 2, 1907, box 176, RG77; Chittenden to George F. Cotterill, June 30, 1906, box 8, George F. Cotterill Papers, UWL-SC.

53. For quotation, see *Puget Mill Company v. State of Washington,* 93 Wash. 140 (1916). For the two other cases, see *County of King v. Seattle Cedar Lumber Manufacturing Company,* 94 Wash. 84 (1916); *Hewitt-Lea Lumber Company v. County of King,* 113 Wash 431 (1920); for the six additional cases, see *State of Washington ex rel Burke v. Board of Commissioners,* 58 Wash. 511 (1910) and 61 Wash. 684 (1911); *Bilger et al. v. State of Washington,* 63 Wash. 157 (1911); *Blaine v. Hamilton,* 64 Wash. 353 (1911); *Alymore v. Hamilton,* 74 Wash. 433 (1913); *Carlson v. State of Washington,* 234 U.S. 103 (1914); and *Sanderson v. City of Seattle,* 95 Wash. 582 (1917).

54. *Commercial Waterway District No. 1 v. Seattle Factory Sites Company,* 76 Wash. 181 (1913); *Wardell v. Commercial Waterway District No. 1,* 80 Wash. 495 (1915).

55. For river and estuary changes, see G. C. Bortleson, M. J. Chrzastowski, and A. K. Helgerson, *Historical Changes of Shoreline and Wetland at Eleven Major Deltas in the Puget Sound Region, Washington* (Reston, Va.: U.S. Geological Survey, 1980), Sheet 7; and George Blomberg, Charles Simenstad, and Paul Hickey, "Changes in Duwamish River Estuary Habitat over the Past 125 Years," *Proceedings: First Annual Meeting on Puget Sound Research* (Seattle: Puget Sound Water Quality Authority, 1988), 2:437–54.

56. *Duwamish Valley News,* July 27, 1917. For the pastoral description, see *Seattle Telegraph,* August 12, 1894. For Seattle's economy, see Richard C. Berner, *Seattle, 1900–1920: From Boomtown, Urban Turbulence, to Restoration* (Seattle: Charles, 1991), 178–84. For skyscraper views, see David M. Nye, *American Technological Sublime* (Cambridge: MIT Press, 1994), 87–108.

57. Thomson to Shorrett, March 26, 1930, box 6, fol. 1, Acc. 89, RHT Papers. For industrialization, see Lorraine McConaghy, "The Lake Washington Shipyards: For the Duration" (master's thesis, University of Washington, 1987).

58. For salt water, see Victor E. Smith and Thomas G. Thompson, "The Control of Sea Water Flowing into Lake Washington Ship Canal," *Industrial and Engineering Chemistry* 17 (1925): 1084;

Salinity of Lake Washington Ship Canal Engineering Experiment Station Bulletin No. 41 (Seattle: University of Washington Press, 1927). For regional flooding, see Fred F. Henshaw and Glenn L. Parker, *Water Powers of the Cascade Range—Part II: Cowlitz, Nisqually, Puyallup, White, Green, and Cedar Drainage Basins* (Washington, D.C.: Government Printing Office, 1913), 9–24, 95–121. For Duwamish flooding, see *Duwamish Valley News,* December 21, 1917, and January 4, 1918.

59. Andrew Gibson to W. L. Darling, November 25, 1901, box 25, file 1073, NP-CESF; *P-I,* November 2, 1902, April 10, 1904, May 12, 1907; *SMH,* September 12, 1903; and C. B. Bussell, *Tide Lands, Their Story* (Seattle: C. B. Bussell, c. 1903), UWL-SC. For filled tidelands, see Herb Voigt, "Major Reclaimed Tidelands of Seattle" (Seattle, 1952), UWL-SC.

60. L. C. Gilman to Louis W. Hill, May 10, 1907, box 52, file 3938-6, GN-PSF.

61. Padraic Burke, *A History of the Port of Seattle* (Seattle: Port of Seattle, 1976); and Berner, *Seattle, 1900–1920,* 141–52.

62. Hill to Burke, January 16, 1900, box 6, fol. 17, TB Papers. For tunnel and tideland building, see *P-I,* April 5, July 12, 1903, October 27, 1904; Darling to Howard Elliott, April 28, 1908, box 64, file 1971, NP-CESF; L. C. Gilman to G. R. Martin, January 6, 1922, and "Report of Real Estate Committee to Ralph Budd," February 26, 1924, boxes 468, 512, files 10,415 and 11,072, GN-PSF.

63. *ST,* March 7, 1893.

64. *P-I,* April 28, 1899, May 1, 1904.

65. D. C. Govan, "Report of Tulalip Agency," to Commissioner of Indian Affairs, August 28, 1896, in *Report of the Secretary of the Interior,* vol. 3499 (1896), 315. For Indian policy, see Frances Paul Prucha, *The Great Father: The United States Government and the American Indian* (Lincoln: University of Nebraska Press, 1984), 2:657–86; and Alexandra Harmon, *Indians in the Making: Ethnic Relations and Indian Identities Around Puget Sound* (Berkeley: University of California Press, 1999), 131–59.

66. Charles M. Buchanan, "Report of School Superintendent in Charge of Tulalip Agency," to Commissioner of Indian Affairs, October 24, 1901, in *Report of the Secretary of the Interior,* vol. 4290 (1901), 391; and "Report of School Superintendent in Charge of Tulalip Agency," to Commissioner of Indian Affairs, September 10, 1902, in *Report of the Secretary of the Interior,* vol. 4458 (1902), 391.

67. Samuel A. Eliot, *Report Upon the Conditions and Needs of the Indians of the Northwest Coast* (Washington, D.C.: Government Printing Office, 1915), 17, 21.

68. Buchanan, "Rights of the Puget Sound Indians to Game and Fish," reprinted in *Washington Historical Quarterly* 6 (April 1915): 110, 113. For Indians' options, see Thrush, *Native Seattle,* 85–91.

69. *P-I,* January 6, 7, May 11, 1910. For a similar analysis, see Coll Thrush, "City of the Changers: Indigenous People and the Transformation of Seattle's Watersheds," *Pacific Historical Review* 75 (February 2006): 89–111. See also the other articles by Sarah S. Elkind, Matthew Morse Booker, and David S. Torres-Rouff in "Forum: Environmental Inequality and the Urbanization of West Coast Watersheds," *Pacific Historical Review* 75 (February 2006): 53–88, 112–40.

70. For unreliability of property, see Carol M. Rose, *Property and Persuasion: Essays on the History, Theory, and Rhetoric of Ownership* (Boulder: Westview, 1994), 105–62, 267–302.

71. For the commons, see Bonnie J. McCay and James M. Acheson, "Human Ecology of the Commons," and Bonnie J. McCay, "The Culture of the Commoners: Historical Observations of Old and New World Fisheries," in *The Question of the Commons: The Culture and Ecology of Communal Resources,* ed. Bonnie J. McCay and James M. Acheson (Tucson: University of Arizona Press, 1987), 3–34, 195–216.

Chapter 3. The Imagination and Creative Energy of the Engineer: Harnessing Nature's Forces to Urban Progress

1. For this critique, see Elizabeth Blackmar, "Contemplating the Force of Nature," *Radical Historians Newsletter* 70 (May 1994): 4; and Ted Steinberg, "Down to Earth: Nature, Agency, and Power in History," *American Historical Review* 107 (June 2002): 798–820.

2. *P-I,* January 16, 1916.

3. For Thomson's early years, see R. H. Thomson, *That Man Thomson,* ed. Grant H. Redford (Seattle: University of Washington Press, 1950), 13–14; Diaries, 1882, 1885, 1887, box 1, fol. 5, and

Thomson to Mayor and Common Council, March, 20 1885, box 1, book 1, RHT Papers. For California mining, see Andrew C. Isenberg, *Mining California: An Ecological History* (New York: Hill and Wang, 2005).

4. For disease, see *Seattle Department of Health, Annual Reports* (1890–92), Annual Reports, SMA. For Skid Row, see Murray Morgan, *Skid Road: An Informal Portrait of Seattle* (New York: Viking, 1951; Seattle: University of Washington Press, 1982), 116–44.

5. Rose Simmons, "Old Angeline, the Princess of Seattle," *Overland Monthly* 20 (November 1892): 506; and Seattle Department of Health, *Monthly Report of the Health Officer* (June 1891), 1–2, SPL-MB.

6. For Waring and Williams reports, see "Reports on Seattle Water Supply, 1889–1891" [typescript copy], box 3, fol. 21, SWD-HF.

7. Thomson, *That Man Thomson*, 33–34.

8. Thomson to William Carnes, London, England, February 18, 1905, box 1, book 3, RHT Papers. For engineers, see Edwin T. Layton, Jr., *The Revolt of the Engineers: Social Responsibility and the American Engineering Profession* (Baltimore: Johns Hopkins University Press, 1971, 1986), 53–74.

9. For city engineers, see Stanley K. Schultz, *Constructing Urban Culture: American Cities and City Planning, 1800–1920* (Philadelphia: Temple University Press, 1989), 153–205. For international community, see Daniel T. Rodgers, *Atlantic Crossings: Social Politics in a Progressive Age* (Cambridge: Belknap Press of Harvard University Press, 1998), 112–208.

10. *ST*, May 12, 1907.

11. House Resolution 262, *Session Laws for the State of Washington* (Olympia: O. C. White, State Printer, 1893), 189–209.

12. For Seattle's early water system, see John Lamb, *The Seattle Municipal Water Plant: Historical, Descriptive, Statistical* (Seattle: Moulton Printing, 1914), 8–19; Clarence B. Bagley, *History of Seattle from the Earliest Settlement to the Present Time* (Chicago: S. J. Clarke, 1916), 267; Mary McWilliams, *Seattle Water Department History, 1854–1954* (Seattle: Department of Water, 1954), 53–63.

13. Thomson, *That Man Thomson*, 39–42.

14. H. R. Clise, J. Eugene Jordan, et al., "Special Report, Joint Fire and Water Committee," October 7, 1895, in Lamb, *Seattle Municipal Water Plant*, 109. For the court decision, see *Winston v. Spokane*, 12 Wash. 524 (1895).

15. Thomson to Atwell Thompson, October 18, 1907, box 4, book 5, RHT Papers; R. H. Thomson, "The Seattle Regrades" [c. 1930], 4, 12, box 13, fol. 5, RHT Papers; Seattle Department of Engineering, *Annual Report* (1908): 18–19, Annual Reports, SMA. Thomson's comment echoes Henri Lefebvre's depiction of engineers as "doctors of space" and their treatments for "sick" cities. See Lefebvre, *The Production of Space*, trans. Donald Nicholson-Smith (Cambridge, Mass.: Blackwell, 1991), 99. For moral uplift, see Michael Rawson, "The Nature of Water: Reform and the Antebellum Crusade for Municipal Water in Boston," *Environmental History* 9 (July 2004): 411–35.

16. For business and real estate, see Kathryn Morse, *The Nature of Gold: An Environmental History of the Klondike Gold Rush* (Seattle: University of Washington Press, 2003), 166–69, 178–85.

17. For the quotation, see Thomson to C. J. Moore, February 15, 1898, box 1, fol. 1, RHT Papers. For demographics, see Thomson, "The Seattle Regrades," 11; O. A. Piper, "Regrading in Seattle North District" [c. 1910], 1, LID 4818, Letters, fol. 3, SED.

18. Stephen W. Kohl, "Writings of Nagai Kafū: An Early Account of Japanese Life in the Pacific Northwest," *Pacific Northwest Quarterly* 70 (April 1979): 62; Louis P. Zimmerman, "The Seattle Regrade, with Particular Reference to the Jackson Street Section," *Engineering News* 60 (November 12, 1908): 511. For residential patterns, see Quintard Taylor, *The Forging of a Black Community: Seattle's Central District from 1870 Through the Civil Rights Era* (Seattle: University of Washington Press, 1994), 13–45, 106–34; and Janice L. Reiff, "Urbanization and the Social Structure: Seattle, Washington, 1852–1910" (Ph.D. diss., University of Washington, 1981), 145–71. For germ theory, see Martin V. Melosi, *The Sanitary City: Urban Infrastructure in America from Colonial Times to the Present* (Baltimore: Johns Hopkins University Press, 2000), 103–16.

19. Mark A. Matthews, "Roses for Seattle's Children," Sermonettes, 1906, Mark A. Matthews Papers, UWL-SC. For Matthews's work, see Dale E. Soden, *The Reverend Mark Matthews: An Activist in the Progressive Era* (Seattle: University of Washington Press, 2001), 86–103.

20. Samuel P. Hays, "The Politics of Reform in Municipal Government in the Progressive Era," *Pacific Northwest Quarterly* 55 (October 1964): 157–69; Mansel G. Blackford, "Reform Politics in Seattle During the Progressive Era," *Pacific Northwest Quarterly* 59 (October 1968): 177–85; Richard C. Berner, *Seattle, 1900–1920: From Boomtown, Urban Turbulence, to Restoration* (Seattle: Charles, 1991), 109–21.

21. Lee Forrest Pendergrass, "Urban Reform and Voluntary Association: A Case Study of the Seattle Municipal League" (Ph.D. diss., University of Washington, 1972), 29.

22. Thomson to Carnes, February 18, 1905, box 1, book 3, RHT Papers.

23. Wesley Arden Dick, "The Genesis of Seattle City Light" (master's thesis, University of Washington, 1965), 8–33; Gregory Gray Fitzsimons, "The Perils of Public Works Engineering: The Early Development of Utilities in Seattle, Washington, 1890–1912" (master's thesis, University of Washington, 1992), 82–144.

24. For federal lands, see Binger Hermann to City of Seattle, October, 10 1899, box 3, folder 7, SWD-HF.

25. Thomson, "Cedar River Water Supply Design Notes," Diary, 1899, box 1, fol. 6, RHT Papers. For court challenges, see *Moran v. Thomson,* 20 Wash. 525 (1899) on illegal contracts and fraud; *Faulkner v. Seattle,* 19 Wash. 320 (1898), upholding the *Winston* decision; and *Smith v. Seattle,* 25 Wash. 300 (1901), on extending the original waterworks.

26. Thomson to George M. Trowbridge, December 18, 1909, box 3, book 7, RHT Papers; Thomson, "The Seattle Regrades," 11. For Chicago, see Robin Einhorn, *Property Rules: Political Economy in Chicago, 1833–1872* (Chicago: University of Chicago Press, 1991), 61–187. Another useful approach to property is Hendrik Hartog, *Public Property and Private Power: The Corporation of the City of New York in American Law, 1730–1870* (Chapel Hill: University of North Carolina Press, 1983).

27. V. V. Tarbill, "Mountain-Moving in Seattle," *Harvard Business Review* 8 (July 1930): 484; Thomson to Frederick J. Haskin, April 24, 1908, box 4, book 5, RHT Papers. For the 1898 regrade, see "Contract for Local Improvement," March 19, 1898, LID 148, box 5, LID Files, SCCF.

28. Thomson to Goodwin Real Estate Company, May 2, 1908, box 2, book 5, RHT Papers; George H. Emerson, *The Building of a Modern City* (Seattle: Metropolitan Building Company, 1907), UWL-SC.

29. "The Jackson Street Regrade" [c. 1911], LID 1213, Letters, fiche 1, SED; R. M. Overstreet, "Hydraulic Excavation Methods in Seattle," *Engineering Record* 65 (May 4, 1912): 480–83.

30. "The Jackson Street Regrade," 1. For statistics, see Myra L. Phelps, *Public Works in Seattle: A Narrative History of the Engineering Department, 1875–1975* (Seattle: Department of Engineering, 1975), 28, 33; Piper, "Regrading in Seattle North District," 20–21.

31. Overstreet, "Hydraulic Excavation," 481. For shoreline alterations, see George Holmes Moore, "Heavy Regrading by Means of Hydraulic Sluicing at Seattle, Wash.," *Engineering News* 63 (March 31, 1910): 358; and Cook to Darling, July 30, 1906, box 31, file 1234, NP-CESF.

32. "The Jackson Street Regrade," 11–12. For operations, see Overstreet, "Hydraulic Excavation," 480–83. For Boston, see Walter Muir Whitehill and Lawrence W. Kennedy, *Boston: A Topographical History,* 3rd ed. (Cambridge: Belknap Press of Harvard University Press, 2000), 141–73; and Nancy S. Seasholes, *Gaining Ground: A History of Landmaking in Boston* (Cambridge: MIT Press, 2003), 21–104, 135–51.

33. Phillip R. Kellar, "Washing Away a City's Hills," *The World Today* 19 (July 1910): 706–7; Arthur H. Dimock, "Preparing the Groundwork for a City: The Regrading of Seattle, Washington," *Transactions of the American Society of Civil Engineers* 92 (1928): 733–34; Bagley, *History of Seattle,* 22–23, 354.

34. *P-I,* September 18, 1911. For Thomson as Transformer, see Coll Thrush, *Native Seattle: Histories from the Crossing-Over Place* (Seattle: University of Washington Press, 2007), 79–104.

35. George F. Cotterill to City Council, March 17, 1913, file 54815, SCCF. For demographics, see Campbell Gibson, "Population of the 100 Largest Cities and Other Urban Places in the United States, 1790–1990," Population Division Working Paper No. 27 (Washington, D.C.: United States Census Bureau, June 1998), www.census.gov/population/www/documentation/twps0027.html (accessed June 30, 2006).

36. Roger Sale, *Seattle: Past and Present* (Seattle: University of Washington Press, 1976), 110–12; Berner, *Seattle, 1900–1920,* 153–60, 190–91.

37. For watershed expansion, see Lamb, *Seattle Municipal Water Plant,* 286–303; McWilliams, *Seattle Water Department,* 67–69, 154–58. For timber, see Thomas R. Cox, *Mills and Markets: A History of the Pacific Coast Lumber Industry to 1900* (Seattle: University of Washington Press, 1974), 199–296; Bagley, *History of Seattle,* 221–42. For logging and municipal watersheds in the Northeast, California, and Utah, see William G. Robbins, *American Forestry: A History of National, State, and Private Cooperation* (Lincoln: University of Nebraska Press, 1985), 119–20; Sarah S. Elkind, *Bay Cities and Water Politics: The Battle for Resources in Boston and Oakland* (Lawrence: University Press of Kansas, 1998); Ellen Frances Stroud, "The Return of the Trees: Urbanization and Reforestation in the Northeastern United States" (Ph.D. diss., Columbia University, 2001).

38. John R. Freeman, "Report to the State Board of Health of Washington," August 1906, quoted in Lamb, *Seattle Municipal Water Plant,* 186. For sanitation laws, see Lamb, 188–92. For the report, see *Engineering News* 56 (August 30 and December 27, 1906): 238–39, 684–88.

39. Affidavit of J. E. Crichton, July 12, 1911, Cause No. 75069, Superior Court of King County, box 1, Alpha Files: LP-National Forest Planning and Establishment, 1910–1951, MBSNF. For Cedar Lake flooding, see Henry Landes and Milnor Roberts, "A Report on the Proposed Site for the New Cedar River Dam," July 12, 1910, box 1, Alpha Files: LP-National Forest Planning and Establishment, 1910–1951, MBSNF.

40. E. T. Clark, "Report on Cedar River Watershed, Proposed Land Grant to the City of Seattle," February 1, 1910; Burt P. Kirkland to C. J. Beck, November 9, 1910, boxes 1 and 2, fol. 2, MBSNF. For timber surveys, see Clark, "Report," February 1, 1910. For the Forest Service campaign, see F. A. Ames to Stanton Smith, May 31, 1913, Smith to Ames, June 2, 1913, Smith to Mayor, City of Seattle, January 17, 1914, and L. B. Youngs to Smith, November 10, 1914, box 2, fol. 1 and 2, MBSNF.

41. For a watershed ranger, see R. H. Ober to Board of Public Works, March 6, 1912, box 3, fol. 6, SWD-HF. For the report and its rejection, see Ober and Hamilton C. Johnson, "Report on Reforestation of Cedar River Watershed" (February 1913), 6–9, box 4, fol. 4, SWD-HF; Youngs to Finance Committee, City Council, April 16, 1913, and Ober to Finance Committee, May 1, 1913, box 2, fol. 2, MBSNF. For compromise, see Youngs to City Council, February 19, 1915, box 3, fol. 10, SWD-HF. Cooperation by municipal governments with federal or provincial agencies during this period was common; see Robbins, *American Forestry,* 105–21; Louis P. Cain, "Water and Sanitation Services in Vancouver: An Historical Perspective," *BC Studies* 30 (Summer 1976): 27–43; and Rick Harmon, "The Bull Run Watershed: Portland's Enduring Jewel," *Oregon Historical Quarterly* 96 (1995), 242–70. For the Pacific States Lumber Company contracts (hereafter abbreviated PSLC), see Ordinances 37496 and 37645 (1917), SCCF. For an example of one of many complaints against the PSLC, see Youngs to PSLC, April 19, 1923, book HH, SWD-SC.

42. Thomson to E. J. Pierson, January 14, 1909; G. A. Kyle to Thomson, February 2, 1909, box 1, fol. 13, SWD-HF; "Recent Flood Conditions on the Cedar River, Near Seattle, Wash.," *Engineering News* 66 (December 14, 1911): 706–8, 713.

43. "Alder Report" [n.d.]; Stanton S. Smith to Youngs, June 10, 1914, Dimock and J. G. Priestley to Youngs, September 22, 1915, Priestley to L. Murray Grant, October 13, 1926, and Priestley to J. D. Blackwell, October 27, 1924, box 2 and 3, fol. 13, 18, and 23, SWD-HF.

44. For fires, see Allen T. Thompson, "Forest Fire Reports," May 5, 1925, and September 3, 1927, and McWilliams, "Watershed Fires—Research Notes," April 29, 1952, boxes 3, 5, fol. 13, 29, SWD-HF.

45. George F. Russell to C. E. Brakenridge, September 25, 1924, book LL, SWD-SC.

46. *SMH,* July 25, 1903; *Engineering News* 63 (March 31, 1910): 375; Thomson, *That Man Thomson,* 92.

47. For regrading difficulties, see *P-I,* October 15, 1903, December 2, 1904; Piper, "Regrading in Seattle North District," 27; Chemical Engineer to W. D. Barkhuff, October 24, 1929, LID 4848, letters, fol. 1, SED; John Millis to C. J. Erickson, July 19, 1904, box 8, fol. 12, George F. Cotterill Papers, UWL-SC. For Pine Street, see Inspectors' Reports (1904), LID 799, box 26, LID Files, SCCF.

48. James G. Love to Board of Public Works, October 21, 1909, LID 1707, Letters, fol. 2, SED. For conflicts, see Dimock to Streets and Sewers Committee, Seattle City Council, December 27, 1913, LID 1213, Letters, fiche 1, SED; "Remonstrance against improvement of Denny Way et al.," November 10, 1902, LID 701, box 23, LID Files, SCCF; and Thomson to Erickson, July 30, 1908, LID 1310, letters, SED.

49. *P-I,* April 8, 1909. For property and propriety, see Rose, *Property and Persuasion,* 49–70.

50. *P-I*, May 26, 1907.

51. *Knickerbocker Company v. Seattle* 69 Wash. 336 (1912). For the cases discussed here, see *Povine v. Seattle*, 59 Wash. 683 (1910); *Coffer v. Erickson*, 61 Wash. 559 (1911); *Casassa v. Seattle*, 66 Wash. 147 (1911); *Casassa v. Seattle*, 75 Wash. 367 (1913); and *Johanson v. Seattle*, 80 Wash. 527 (1914). For related damage and injury cases, see *In re Fifth Avenue and Fifth Avenue South, Seattle*, 62 Wash. 218 (1911); *Povine v. Seattle*, 70 Wash. 326 (1912); *Williams v. Seattle*, 73 Wash. 377 (1914); *Hinckley v. Seattle*, 74 Wash. 101 (1913); *Dose v. Seattle*, 78 Wash. 571 (1914); *Marks v. Seattle*, 88 Wash. 1915 (1915); and *Rainier Heat and Power Company v. Seattle*, 113 Wash. 95 (1920). For rulings against unfair assessments or in favor of city improvements, see *Lewis v. Seattle*, 28 Wash. 639 (1902); *In re Westlake Avenue*, 40 Wash. 444 (1905); *Smith v. Seattle*, 41 Wash. 60 (1905); *In re Third, Fourth, and Fifth Avenues*, 49 Wash. 109 (1908); *James v. Seattle*, 49 Wash. 347 (1908); *Schuchard v. Seattle*, 51 Wash. 41 (1908); *In re Mercer Street, Seattle*, 55 Wash. 116 (1909); *Manhattan Building Company v. Seattle*, 52 Wash. 226 (1909); *In re Westlake Avenue*, 60 Wash. 549 (1910); *Levy v. Seattle*, 61 Wash. 540 (1911); *In re Fifth Avenue and Fifth Avenue South*, 66 Wash. 327 (1911); *Hapgood v. Seattle*, 69 Wash. 497 (1912); *In re Ninth Avenue*, 79 Wash. 674 (1914); and *Richardson v. Seattle*, 97 Wash. 317 (1917).

52. Piper, "Regrading in Seattle North District," 2–5, 11–15.

53. For spite mounds, see *P-I*, October 2, 1910. For court cases, see *Brinton v. Lewis-Littlefield Company*, 66 Wash. 40 (1911); *Sweeney v. Lewis Construction Company*, 66 Wash. 490 (1912); *Inner-Circle Property Company v. Seattle*, 69 Wash. 508 (1912); *Turner Investment Company v. Seattle*, 70 Wash. 201 (1912); *Scandinavian American Bank v. Washington Hotel and Improvement Company*, 70 Wash. 223 (1912); *In re Twelfth Avenue South*, 74 Wash. 132 (1913); and *Kelly v. Smith*, 101 Wash. 475 (1918).

54. Sandy Moss, interview by Esther Mumford, April 22 and 25, 1975, Acc. BL-KNG 75-2em, Washington State Oral History Project, WSA.

55. J. W. Charlton to Thomson, October 22, 1910, LID 2325, letters, SED; Piper, "Regrading in Seattle North District," 4.

56. *Wong Kee Jun v. Seattle*, 143 Wash. 505 (1927). The so-called Jackson Street regrade cases (as defined by the Court in *Davis v. Seattle*) were: *Jorguson v. Seattle*, 80 Wash. 126 (1914); *Farnandis v. Seattle*, 95 Wash. 587 (1917); *Lochore v. Seattle*, 98 Wash. 265 (1917); *Blomskog, Erickson, and Cotton v. Seattle*, 107 Wash. 471 (1919); *Davis v. Seattle*, 134 Wash. 1 (1925); *Bingaman v. Seattle*, 139 Wash. 68 (1926); *Hamm v. Seattle*, 140 Wash 427 (1926); and *Wong Kee Jun v. Seattle*, 143 Wash. 479 (1927). Only in *Jorguson* did the Court find for the city; the verdict in the *Wong Kee Jun* decision was unanimous. For the Twelfth Avenue Bridge, see "Slide Causes Collapse of Seattle Bridge Approach," *Engineering Record* 75 (March 17, 1917): 444–45; "Restoring a Steel Bridge After Landslide Movement," *Engineering News-Record* 94 (January 15, 1925): 112–13.

57. Dudley Stuart, "Seattle's Coming Retail and Apartment-House District," UW027, Map Collections, UWL-SC.

58. Board of Engineers, "Report on Cedar River Project," May 6, 1912, box 27, Subject Reference Files, SED; Charles Evan Fowler, "Leakage from Cedar Lake Reservoir, Seattle Water-Supply," *Engineering News* 73 (January 21, 1915): 112–15; "Seattle Water-Works Troubles," *Engineering News* 75 (March 30, 1916): 585–88; *Engineering Record* 74 (August 19, 1916): 228–29; *Engineering Record* 79 (July 26, 1917): 155–56.

59. For descriptions of the flood, see *ST*, December 23, 1918, December 24, 1918; *P-I*, December 24, 1918.

60. *ST*, December 27, 1918; *P-I*, December 24, 1918.

61. Timothy Mitchell, *Rule of Experts: Egypt, Techno-Politics, Modernity* (Berkeley: University of California Press, 2002), 34; see also James C. Scott, *Seeing Like a State: How Certain Schemes to Improve the Human Condition Have Failed* (New Haven: Yale University Press, 1998).

62. Thomson, *That Man Thomson*, 127.

Chapter 4. Out of Harmony with the Wild Beauty of the Natural Woods: Artistry Versus Utility in Seattle's Olmsted Parks

1. Thomas Bender, *Toward an Urban Vision: Ideas and Institutions in Nineteenth-Century America* (Lexington: University Press of Kentucky, 1975; Baltimore: Johns Hopkins University

Press, 1982), 187. For another overview of urban planning during this era, see David Schuyler, *The New Urban Landscape: The Redefinition of City Form in Nineteenth-Century America* (Baltimore: Johns Hopkins University Press, 1986).

2. John C. Olmsted (hereafter abbreviated JCO) to Sophia (Fidie) White Olmsted (hereafter abbreviated SWO), May 4, 1903, box 5A, fol. 36, JCO Coll.

3. *P-I,* May 1, 1903. For Olmsted's career, see Norman T. Newton, *Design on the Land: The Development of Landscape Architecture* (Cambridge: Belknap Press of Harvard University Press, 1971), 294–95; and Arleyn Levee, "John Charles Olmsted," in *Pioneers of American Landscape Design,* ed. Charles A. Birnbaum and Robin Karson (New York: McGraw-Hill, 2000), 282–85. For the senior Olmsted's career, see Laura Wood Roper, *FLO: A Biography of Frederick Law Olmsted* (Baltimore: Johns Hopkins University Press, 1973).

4. Amy S. Brown, "Nature in Practice: The Olmsted Firm and the Rise of Landscape Architecture and Planning" (Ph.D. diss., Massachusetts Institute of Technology, 2002), 12–19, 51–104; Susan L. Klaus, "All in the Family: The Olmsted Office and the Business of Landscape Architecture," *Landscape Journal* 16 (Spring 1997): 80–93.

5. John C. Olmsted, "Organization and Management of a City Engineer's Office," *Journal of the Association of Engineering Societies* 13 (October 1894): 594. For new professionals, see Robert H. Wiebe, *Self-Rule: A Cultural History of American Democracy* (Chicago: University of Chicago Press, 1995), 141–43. For Olmsted's professional activities, see Noel Dorsey Vernon, "Toward Defining the Profession: The Development of the Code of Ethics and the Standards of Professional Practice of the American Society of Landscape Architects, 1889–1927," *Landscape Journal* 6 (Spring 1987): 13–20; Bremer Pond, "Fifty Years in Retrospect: A Brief Account of the Origin and Development of the ASLA," *Landscape Architecture* 40 (January 1950): 59–66.

6. *Second Annual Report of the Board of Park Commissioners* (1892), 6, box 1, fol. 3, DSPHC; see also *First Annual Report, 1884–1904* (Seattle: Lowman and Hanford, 1905), 7–18. Schwagerl's remarks were inspired by Frederick Law Olmsted's classic essay from 1870, "Public Parks and the Enlargement of Towns." See Frederick Law Olmsted, "Public Parks and the Enlargement of Towns," in *Civilizing American Cities: Writings on City Landscapes,* ed. S. B. Sutton (New York: Da Capo, 1997), 75. For parks as public space, see Mary P. Ryan, *Civic Wars: Democracy and Public Life in the American City During the Nineteenth Century* (Berkeley: University of California Press, 1997), 203–10, 230–31; and Galen Cranz, *The Politics of Park Design: A History of Urban Parks in America* (Cambridge: MIT Press, 1989).

7. *SA,* September 13, 1900; *P-I,* September 21, 1902; *SMH,* May 24, 1902; *First Annual Report of the Board of Park Commissioners* (1891), 8, box 1, fol. 3, DSPHC; *SA,* November 2, 1902.

8. J. D. Blackwell to P. R. Jones, March 21, 1902; Charles W. Saunders to Frederick Law Olmsted, Jr., December 15, 1902, and January 14, 1903, file 2690, box 130, fol. 1, OAR. For rival cities, see William H. Wilson, *The City Beautiful Movement* (Baltimore: Johns Hopkins University Press, 1989), 149–50; and Janet Anne Northam, "Sport and Urban Boosterism: Seattle, 1890–1910" (master's thesis, University of Washington, 1978), 103–4.

9. JCO, "Seattle Parks," March 19, 1903, file 2690, box 130, fol. 1, OAR; Frederick Law Olmsted, "Preliminary Report to the Commissioners for Laying Out of a Park in Brooklyn, New York: Being a Consideration of Circumstances of Site and Other Conditions Affecting the Design of Public Pleasure Grounds," as quoted in *Civilizing American Cities,* 17; JCO to SWO, May 4, 1903 box 5A, fol. 36, JCO Coll.

10. JCO to SWO, May 4, 1903 box 5A, fol. 36, JCO Coll. For ruined landscapes, see JCO, "Seattle Parks," May 6–June 13, 1903, file 2690, box 130, fol. 1, OAR; JCO to SWO, May 20, 21, 1903, box 5A, fol. 37, JCO Coll. For scenery, see JCO to SWO, May 27, 1903, box 5A, fol. 37, JCO Coll.; JCO, "Seattle Parks," May 1, 5, and 11, 1903, file 2690, box 130, fol. 1, OAR.

11. Seattle Board of Park Commissioners, "Report of Olmsted Brothers," in *First Annual Report, 1884–1904* (1904), 44, Annual Reports, SMA.

12. "Report of Olmsted Brothers," 52–55.

13. For reduced parks system, see "Report of Olmsted Brothers," 48–51. For realtors, see C. P. Dose, "How It Was Done in Chicago and How It Ought to Be Done in Seattle: Address to Seattle Real Estate Association," July 1904, box 130, fol. 2, OAR; Olmsted Brothers to Saunders, December 11, 1903, box 53, fol. 2, DSPHC.

14. Olmsted Brothers to Saunders, December 11, 1903, box 53, fol. 2, DSPHC; *P-I,* March 6, 1904.

15. *ST,* March 7, 1904. For the election, see *SA,* January 9, 1904; *ST,* February 22, 1904; *P-I,* January 17, 1904. For results, see Saunders to JCO, March 11, 1904, file 2690, box 130, fol. 2, OAR; and *Journal of the Proceedings of the City Council of Seattle,* March 11, 1904, SMA. The vote was 3,825 to 3,732. I base my analysis on earlier work by the historians Mansel Blackford and William Wilson. See Blackford, "Sources of Support for Reform Candidates and Issues in Seattle Politics, 1902–1916" (master's thesis, University of Washington, 1967), 10–20; and Wilson, *The City Beautiful Movement,* 156–59.

16. JCO to Saunders, March 13, 1904, file 2690, box 130, fol. 2, OAR.

17. For Thomson quotation, see Thomson to W. T. Hall, October 5, 1910, fol. 9, box 1, RHT Papers. For Olmsted quotations, see JCO to SWO, November 29, 1903, and May 27, 1907, boxes 6A and 15A, fol. 47 and 108, JCO Coll.

18. JCO to SWO, April 27, 1907, October 19, 1906, and June 10, 1910, boxes 13A, 14A, and 22A, fol. 92, 105, and 176, JCO Coll.

19. JCO to SWO, June 9, 1910, box 22A, fol. 176, JCO Coll. For residential hotels, see Paul E. Groth, *Living Downtown: The History of Residential Hotels in the United States* (Berkeley: University of California Press, 1994). For "race suicide," see Gail Bederman, *Manliness and Civilization: A Cultural History of Gender and Race in the United States, 1880–1917* (Chicago: University of Chicago Press, 1995), 170–215; and Linda Gordon, *The Moral Property of Women: A History of Birth Control Politics in America* (Urbana: University of Illinois Press, 2002), 86–104.

20. James A. Dawson, "Seattle Parks," August 1904, file 2690, box 130, fol. 2, OAR; Olmsted Brothers to C. E. Fowler, May 16, 1904, box 53, fol. 3, DSPHC.

21. For Green Lake, see "Report of Olmsted Brothers," 64–65; Myra L. Phelps, *Public Works in Seattle: A Narrative History of the Engineering Department, 1875–1975* (Seattle: Department of Engineering, 1975), 284–85.

22. "Report of Olmsted Brothers," 44, 53. For cars, see J. E. Crichton, "Seattle Parks," *Health Bulletin* 2 (September 1909): 2–4, SPL-MB. For the transition from horses to automobiles, see Joel A. Tarr, "A Note on the Horse as an Urban Power Source," *Journal of Urban History* 25 (March 1999): 434–48; and Clay McShane, *Down the Asphalt Path: The Automobile and the American City* (New York: Columbia University Press, 1995), 1–124.

23. JCO to A. B. Ernst, September 12, 1907, file 2694, box 130, fol. 3, OAR. For the battles between Thomson and Olmsted, see JCO, "Seattle Parks," November 18, 19, 1907, fol. 2690, box 130, fol. 3, OAR; and *P-I,* February 19, 24, 1909.

24. JCO, "Upon the Relation of the City Engineer to Public Parks" [typed draft], January 24, 1894, fol. A10; JCO, "Landscape Gardening" [draft], read before Congress of Horticulture, September 1907, fol. A2; and JCO to SWO, May 30, 1910, box 21A, fol. 174, JCO Coll.

25. Frederick Law Olmsted, Jr., "Introduction," in *City Planning: A Series of Papers Presenting the Essential Elements of a City Plan,* ed. John Nolen (New York: D. Appleton, 1916), 1; and Thomas W. Sears, "The Functions of the Landscape Architect in Connection with the Improvement of a City," *Proceedings of the Engineers' Club of Philadelphia* 28 (April 1911): 147. For the McMillan Plan see Jon A. Peterson, *The Birth of City Planning in the United States, 1840–1917* (Baltimore: Johns Hopkins University Press, 2003), 91–94, 123–38. For merging utility and beauty, see Susan L. Klaus, "Efficiency, Economy, Beauty: The City Planning Reports of Frederick Law Olmsted, Jr., 1905–1915," *Journal of the American Planning Association* 57 (Autumn 1991): 456–70.

26. JCO to Cheasty, June 14, 1909, file 2690, box 130, fol. 4, OAR; JCO to SWO, May 16, 1907, box 15A, fol. 107, JCO Coll.; and JCO to T. P. Revelle, May 17, 1907, file 2690, box 130, fol. 3, OAR.

27. JCO to J. E. Shrewsbury, November 26, 1906, file 2690, box 130, fol. 3, OAR; Olmsted Brothers to Saunders, July 20, 1904, box 53, fol. 4, DSPHC.

28. Mansel G. Blackford, "Civic Groups, Political Action, and City Planning in Seattle, 1892–1915," *Pacific Historical Review* 49 (November 1980): 561; Wilson, *The City Beautiful Movement,* 159–60; *Journal of the Proceedings of the City Council of Seattle,* December 15, 1905, March 9, 1906, SMA.

29. Olmsted Brothers to Frink, January 26, 1908, box 53, fol. 7, DSPHC; *P-I,* April 8, 1909; and JCO to SWO, January 2, 1909, box 19A, fol. 145, JCO Coll. For the 1908 report, see Olmsted Brothers

to Frink, January 27, 1908, file 2690, box 130, fol. 4, OAR. For Way's campaign, see *P-I,* April 29, May 4, 1909.

30. Olmsted Brothers to J. T. Heffernan, October 11, 1910, box 53, fol. 9, DSPHC; Seattle Playground Association, "The Playground Movement in Seattle" (Seattle, 1909), UWL-SC. For playground construction in Seattle before 1930, see Vern Allen Armstrong, "The First Fifty Years of Municipal Recreation Programs in the State of Washington" (master's thesis, University of Washington, 1956), 51–76. For the national playground movement, see Dominic Cavallo, *Muscles and Morals: Organized Playgrounds and Urban Reform, 1880–1920* (Philadelphia: University of Pennsylvania Press, 1981); and Paul S. Boyer, *Urban Masses and Moral Order in America, 1820–1920* (Cambridge: Harvard University Press, 1978), 242–51.

31. Olmsted Associates to Heffernan, October 4, 1910, file 2690, box 130, fol. 5, OAR; Olmsted Brothers to Heffernan, October 11, 1910, box 53, fol. 9, DSPHC. For the election results, see Wilson, *The City Beautiful Movement,* 160; and *Journal of the Proceedings of the City Council of Seattle,* December 29, 1908, March 11, 1910, SMA. The 1908 park bond vote was 6,688 for and 2,359 against, while the 1910 vote was 13,311 for to 7,549 against.

32. *SA,* May 16, 1903.

33. JCO to SWO, May 28, 1910, box 21A, fol. 174, JCO Coll.

34. "Speech at Banquet in Honor of Virgil G. Bogue," November 19, 1910, box 6, fol. 9, RHT Papers. For Olmsted's politicking, see JWO to SWO, December 12, 1909, box 20A, fol. 163, JCO Coll.; Thomson to Olmsted, June 25, 1910, fol. 8, box 3, RHT Papers. For Thomson's recommendation, see Thomson to Municipal Commission, June 16, 1910, box 4, book 8, RHT Papers.

35. Municipal Plans Commission, *Plan of Seattle: Report of the Municipal Plans Commission Submitting Report of Virgil G. Bogue, Engineer* (Seattle: Lowman and Hanford, 1911), 34–47. Seattle's population in 1910 was 237,194. For the Bogue commission, see D. W. McMorris, "The Organization and Work of the Municipal Plans Commission of the City of Seattle," *Proceedings of the Pacific Northwest Society of Engineers* 11 (January 1912): 1–8. For the 1930 Los Angeles plan, see Greg Hise and William Deverell, *Eden by Design: The 1930 Olmsted-Bartholomew Plan for the Los Angeles Region* (Berkeley: University of California Press, 2000).

36. *Municipal League News,* February 3, 1912, and December 9, 1911, UWL-SC. For opposition, see Civic Plans Investigation Committee, "The Bogue Plan Question" (Seattle: R. L. Davis, 1911), UWL-SC; Padraic Burke, "The City Beautiful Movement in Seattle" (master's thesis, University of Washington, 1973), 102–6; Lee Forrest Pendergrass, "Urban Reform and Voluntary Association: A Case Study of the Seattle Municipal League" (Ph.D. diss., University of Washington, 1972), 42–49; Wilson, *The City Beautiful Movement,* 220–27. The final vote was 14,506 for and 24,966 against; see *Journal of the Proceedings of the City Council of Seattle,* March 11, 1912, SMA; also quoted in Mansel G. Blackford, *The Lost Dream: Businessmen and City Planning on the Pacific Coast, 1890–1920* (Columbus: Ohio State University Press, 1993), 123.

37. For union opposition, see *Seattle Union Record,* March 2, 1912. For analysis of the Bogue Plan's defeat, see Wilson, *The City Beautiful Movement,* 227–30; and Blackford, *The Lost Dream,* 121–24.

38. *SA,* December 18, 1909. The exact total was 684 acres. For park acreage, see Wilson, *The City Beautiful Movement,* 167. For park and playground growth, see Garland A. Haas, "Parks and Recreation in the Seattle Metropolitan Area: A Study in Policy Development and Administration" (Ph.D. diss., University of Washington, 1958), 53–56; Armstrong, "The First Fifty Years of Municipal Recreation Programs," 69–85.

39. *Board of Park Commissioners' Annual Report* (1911), Gus Knudson, "Report for May 1925" and "Report for July 1925," and L. Glenn Hall, "Landscape Architect's Report on Seward Park," March 31, 1926, box 43, fol. 5, 11 and 16, DSPHC.

40. For one such warning, see JCO to Cheasty, June 14, 1909, file 2690, box 130, fol. 4, OAR. For initial complaints, see Petition to Board of Park Commissioners, March 21, 1912, box 21, fol. 18, DSPHC. For roads and slides, see Henry Fitch, "WPA Slide Control Drainage Projects, 1935–1967," 3 vols. (Seattle: Department of Engineering, n.d.), SMA.

41. Edward S. Bowman to Frank Edwards, July 5, 1931, and Wheeler to M. O. Sylliaasen, February 15 and May 5, 1934, box 21, fol. 18, DSPHC. For lack of funds, see A. O. Soreng to Park Board, January 2, 1931, and E. R. Hoffman to Soreng, January 19, 1931, DSPHC.

42. A. S. Kerry, September 25, 1930, and "Save the Parks Association" (Seattle, 1930?), box 46, fol. 16, DSPHC. For pro-road arguments, see Greenwood-Phinney, Fremont, and West Green Lake Commercial Clubs and Seattle Building Trades Council, "The Great Park Mystery" (Seattle, 1930?), box 46, fol. 16, DSPHC. For arguments for and against the cross-lake bridge, see Board of Park Commissioners to City Council, December 11, 1928; box 43, fol. 11, DSPHC; "Signatures of Organization Leaders opposed to the Lake Washington Toll Bridge," February 14, 1928, and Mercer Island Community Club to Board of County Commissioners, February 14, 1928, box 1, fol. 7, Lake Washington Toll Bridge Files, KCARM. The controversy over a possible cross-lake bridge at Seward Park continued into the 1930s.

43. *P-I*, July 13, 1908; "Analysis of Silt from Green Lake," April 20, 1915, box 28, fol. 14, DSPHC.

44. For changes to Green Lake, see JCO to Frink, January 22, 1908, Olmsted Brothers to J. W. Thompson, January 21, 1912, and Olmsted Brothers to Otto Roseleaf, December 15, 1917, box 53, fol. 7, 11, and 12, DSPHC. For algal growth, see H. M. Read, Commissioner of Health, to Board of Park Commissioners, c. 1919, box 28, fol. 14, DSPHC.

45. *Second Annual Report of the Board of Park Commissioners* (Seattle, The City: 1892), 4–6, box 1, fol. 3, DSPHC; *P-I*, December 23, 1901; and *Ravenna Park (Big Tree Park)* (Seattle: Pacific, 1909), box 1, fol. 16, DSPHC. For Olmsted's impressions, see JCO, "Seattle Parks," May 5, 11, 1903, file 2690, box 130, fol. 1, OAR. For purchasing Ravenna Park, see *Sixth Annual Report of the Board of Park Commissioners* (1909), 47, box 1, fol. 13, DSPHC.

46. For vandalism, see *P-I*, April 17, 1904. For timber and land values, see L. Missigman to John W. Thompson, November 8, 1910, box 42, fol. 1, DSPHC.

47. *ST*, January 2, 1919. For feeding stations, see *Annual Report of the Board of Park Commissioners, 1906–1907* (Seattle, 1907), 14–15; *The Seattle Wren* 6 (April 1938): 1, 4, and "Bird Feeding in the Arboretum," *The Seattle Wren* 8 (April 1940): 4, box 12, fol. 15, Seattle Audubon Society Records, UWL-SC.

48. W. M. Elliott to Simon Burnett, August 12, 1929, box 45, fol. 22, DSPHC. For crow ecology, see John M. Marzluff and Tony Angell, *In the Company of Crows and Ravens* (New Haven: Yale University Press, 2005), 80–107, 218–51. For wolves, see Jon T. Coleman, *Vicious: Wolves and Men in America* (New Haven: Yale University Press, 2004). For crows in Seattle today, see David B. Williams, *The Street-Smart Naturalist: Field Notes from Seattle* (Portland, Ore.: WestWinds, 2005), 187–200.

49. P. F. Appel to Board of Park Commissioners, September 18, 1916, box 42, fol. 1; P. S. Rowntree to Park Board, September 6, 1915, box 43, fol. 5; and Compton to Board of Park Commissioners, November 24, 1919, box 2, fol. 24, DSPHC.

50. Seattle Audubon Society to Game Commission for King County, January 10, 1924, box 1, fol. 12, Seattle Audubon Society Records, UWL-SC. For nativism and hunting, see Louis S. Warren, *The Hunter's Game: Poachers and Conservationists in Twentieth-Century America* (New Haven: Yale University Press, 1997), 21–70.

51. Edgar S. Hadley to Board of Park Commissioners, November 5, 1929, box 19, fol. 1; and Knudson, "Sanitary Inspections," January 1, 1924, box 3, fol. 9, DSPHC. For Alki Beach shacks, see Scott Calhoun, Corporation Counsel, to Board of Park Commissioners, August 8, 1910, box 19, fol. 2, DSPHC. For squatters in other urban parks, see Roy Rosenzweig and Elizabeth Blackmar, *The Park and the People: A History of Central Park* (Ithaca: Cornell University Press, 1992), 58–91, 291–97; and Terence G. Young, *Building San Francisco's Parks, 1850–1930* (Baltimore: Johns Hopkins University Press, 2004), 33–34, 59–60.

52. William Graefe to George W. Hill, City Council, September 12, 1929 [emphasis in original], and Roland Loeff, Park Maintenance Superintendent, to Board of Park Commissioners, September 6, 1946, box 29, fol. 1 and 2, DSPHC. For algae eradication, see Roy M. Harris to F. F. Powell, December 17, 1937, and "Report on Green Lake Algae Control," March 1938, box 29, fol. 1, and box 30, fol. 7, DSPHC.

53. *Annual Report, Board of Park Commissioners* (1938), box 5, fol. 13, and Donna M. Henderson, Secretary, Lakewood Community Club, to Seattle Park Board, November 14, 1936, box 34, fol. 21, DSPHC. For refuge support, see "History of the Seattle Audubon Society" [Seattle, 1955–56], 9–10, 30, box 1, fol. 1, Seattle Audubon Society Records, UWL-SC. For waterfowl and water pollution in Green Lake, see Read to Board of Park Commissioners, April 27 and May 8, 1920, box 29, fol. 1,

and *Annual Work Report, Parks and Boulevard Division, Seattle Park Departments* (Seattle, 1943), box 5, fol. 17, DSPHC.

54. Grace Cadwallader to John H. Hanley, June 15, 1939 [emphasis in original], box 46, fol. 8, DSPHC.

55. Dawson to Kerry, February 23, 1929, box 53, fol. 13, DSPHC.

56. Rosenzweig and Blackmar, *The Park and the People,* 58–91, 291–97; Young, *Building San Francisco's Parks,* 107–11.

57. For the limits of the Olmsteds' designs, see Anne Whiston Spirn, "Constructing Nature: The Legacy of Frederick Law Olmsted," in *Uncommon Ground: Toward Reinventing Nature,* ed. William Cronon (New York: W. W. Norton, 1995), 91–113; Geoffrey Blodgett, "Frederick Law Olmsted: Landscape Architecture as Conservative Reform," *American Historical Review* 62 (March 1976): 869–89.

58. For crossing lines, see Mark Fiege, "The Weedy West: Mobile Nature, Boundaries, and Common Space in the Montana Landscape," *Western Historical Quarterly* 36 (May 2005): 22–48.

59. For gardens, see Michael Pollan, *Second Nature: A Gardener's Education* (New York: Delta, 1991), 45–64. For restoration ideas at the time, see Ian R. Tyrell, *True Gardens of the Gods: Californian-Australian Environmental Reform, 1860–1930* (Berkeley: University of California Press, 1999); and Marcus Hall, *Earth Repair: A Transatlantic History of Environmental Restoration* (Charlottesville: University of Virginia Press, 2005).

Chapter 5. Above the Weary Cares of Life: The Benefits and High Social Price of Outdoor Leisure

1. "The A.-Y.-P. Exposition," *The World's Work* 18 (August 1909): 11894, 11889. For an overview, see George A. Frykman, "The Alaska-Yukon-Pacific Exposition, 1909," *Pacific Northwest Quarterly* 53 (July 1962): 89–99.

2. "The A.-Y.-P. Exposition," 11890; "The A.-Y.-P. and the Class Struggle," *Seattle Socialist,* May 1, 1909, as quoted in Robert Rydell, *All the World's a Fair: Visions of Empire at American International Expositions, 1876–1916* (Chicago: University of Chicago Press, 1984), 191; and *Miner's Union Bulletin* (Fairbanks, Alaska), reprinted in *Seattle Union Record,* March 20, 1909. For labor troubles, see Rydell, *All the World's a Fair,* 191–93. For landscape and capitalism, Don Mitchell, *The Lie of the Land: Migrant Workers and the California Landscape* (Minneapolis: University of Minnesota Press, 1996).

3. Henri Lefebvre makes a similar point when he explains how "the right to nature entered into social practice thanks to leisure." See "The Right to the City," in *Writing on Cities,* trans. and ed. Eleonore Kofman and Elizabeth Lebas (Cambridge, Mass.: Blackwell, 1996), 157–59.

4. *The City of Seattle, The County of King, The State of Washington* (Seattle: Chamber of Commerce, 1898), 15, UWL-SC. For back to nature, see Peter J. Schmitt, *Back to Nature: The Arcadian Myth in Urban America* (New York: Oxford University Press, 1969).

5. For the park campaign, see Theodore Catton, *National Park, City Playground: Mount Rainier in the Twentieth Century* (Seattle: University of Washington Press, 2006), 3–31. For railroad promotion, see *Wonderland '96* (St. Paul: Northern Pacific Railway, 1896); *Wonderland '97, A Story of the Northwest* (St. Paul: Northern Pacific Railway, 1897); *Wonderland '98* (St. Paul: Northern Pacific Railway, 1898); and *Wonderland '99* (St. Paul: Northern Pacific Railway, 1899), UWL-SC.

6. "An Automobile Trip: Snoqualmie Falls as the Objective Point," *Pacific Coast Sportsman* (September 1903): 24–27; Mrs. E. W. Houghton, "Seattle as a Place of Residence," in *The City of Seattle* (Seattle: Chamber of Commerce, 1900), 8, UWL-SC; and Thomas J. Humes, "Bumping River—A Kittitas County Trout Stream," *Pacific Coast Sportsman* (September 1903): 36–37. For comparisons, see Davis Strauss, "Toward a Consumer Culture: 'Adirondack Murray' and the Wilderness Vacation," *American Quarterly* 39 (Summer 1987): 270–86; Dona Brown, *Inventing New England: Regional Tourism in the Nineteenth Century* (Washington, D.C.: Smithsonian Institution Press, 1995); Richard Judd, *Common Lands, Common People: The Origins of Conservation in Northern New England* (Cambridge: Harvard University Press, 1997), 197–228; and Philip G. Terrie, *Contested Terrain: A New History of Nature and People in the Adirondacks* (Syracuse: Syracuse University Press/The Adirondack Museum, 1997), 44–82.

7. *Seattle and King County: Statement of Facts Showing the Resources, Business Conditions, and Progress of the Chief City and Largest County in the State of Washington* (Seattle: Chamber of Commerce, 1905), 27–28, UWL-SC. For promotional efforts, see "Seattle—A City for Tourists," *Washington Magazine* (July 1906); 386–89; John F. Miller, "The City of Seattle," *The Coast* (September 1909): 129–38; and Janet Northam Russell and Jack W. Berryman, "Parks, Boulevards, and Outdoor Recreation: The Promotion of Seattle as an Ideal Residential City and Summer Resort, 1890–1910," *Journal of the West* 26 (January 1987): 5–17.

8. Carl Abbott, "Regional City and Network City: Portland and Seattle in the Twentieth Century," *Western Historical Quarterly* 23 (August 1992): 293–322; "The A.-Y.-P. Exposition," 11890. For the Olmsteds, see Norman J. Johnson, "The Olmsted Brothers and the Alaska-Yukon-Pacific Exposition: 'Eternal Loveliness,'" *Pacific Northwest Quarterly* 75 (April 1984): 50–61. The Olmsteds had hoped the site would become part their park system, but university officials refused and kept the fairgrounds for the campus.

9. "The A.-Y.-P. Exposition," 11894. University officials razed the Forestry Temple after the fair when the columns succumbed to rot and pests.

10. *ST,* August 8, 1909. For the Pay Streak, see *ST,* October 14, 1906; *P-I,* March 7, 1909. For exhibits of Native peoples, see *P-I,* June 13, July 4, 6, 18, August 15, 1909; *ST,* August 15, 22, September 16, 1909. "Igorot" is the currently preferred spelling for this group of Philippines indigenous people. I use the original spelling for the main text.

11. *Participation in the Alaska-Yukon-Pacific Exposition* (Washington, D.C.: Government Printing Office, 1911), 43, 82; John Harkman, "Address," box 29, fol. 2, Edmond S. Meany Papers, UWL-SC; and "The A.-Y.-P. Exposition," 11890. For Indian education, see Robert A. Trennert, Jr., "Selling Indian Education at World's Fairs and Expositions, 1893–1904," *American Indian Quarterly* 11 (Fall 1987): 203–20. For progress and vanishing Indians, see James Clifford, *The Predicament of Culture: Twentieth Century Ethnography, Literature, and Art* (Cambridge: Harvard University Press, 1988), 212, passim.

12. Janice L. Reiff, "Urbanization and the Social Structure: Seattle, Washington, 1852–1910" (Ph.D. diss., University of Washington, 1981), 126–247.

13. M. W. O'Shea, "Opening Day at the A.-Y.-P.-E.," *Seattle Union Record,* June 5, 1909, as quoted in Rydell, *All the World's a Fair,* 192; *P-I,* July 6, 1909. For the labor boycott, see *P-I,* January 9, 1909. For defenses of the Igorot exhibit, see *ST,* July 7, 8, 9, 12, 18, 1909; *P-I,* July 7, August 1, 1909.

14. For labor and leisure, see Thomas G. Andrews, "'Made by Toile'? Tourism, Labor, and the Construction of the Colorado Landscape, 1858–1917," *Journal of American History* 92 (December 2005): 837–63; Connie Y. Chiang, *Shaping the Shoreline: Fisheries and Tourism on the Monterey Coastline* (Seattle: University of Washington Press, 2008); Benjamin Heber Johnson, "Conservation, Subsistence, and Class at the Birth of Superior National Forest," *Environmental History* 4 (January 1999): 80–99; and Richard White, "'Are You an Environmentalist or Do You Work for a Living?': Work and Nature," in *Uncommon Ground: Toward Reinventing Nature,* ed. William Cronon (New York: W. W. Norton, 1995), 171–85.

15. Caroline Sheldon, "Puget Sound Country and Alaska," *The Chautauquan* 51 (July 1908): 175–77.

16. Thorstein Veblen, *The Theory of the Leisure Class: An Economic Study of Institutions,* rev. ed. (New York: Macmillan, 1899, 1912), 88; "The Spirit of Play," *Welfare* 1 (November 1913): 8–9, UWL-SC. For antimodernism see T. J. Jackson Lears, *No Place of Grace: Antimodernism and the Transformation of American Culture, 1880–1920* (New York: Pantheon, 1981).

17. *Brief History of the Washington State Good Roads Association* (Seattle: Washington State Good Roads Association, 1939), UWL-SC; *SMH,* December 19, 1903, January 13, 1906. For literary commuters, see Schmitt, *Back to Nature,* 20–32. For outdoor sports, see *SMH,* November 16, 1901, and April 16, 1902; *SA,* March 28, 1896, and December 18, 1897; *P-I,* May 11, 1898, and June 1, 1902; and Virginia Miller, "The Development of Leisure Time Activities in Seattle" (master's thesis, University of Washington, 1980), 183–86.

18. Janet Ore, *The Seattle Bungalow: People and Houses, 1900–1940* (Seattle: University of Washington Press, 2007), 20–129 (quotation on 50). For bungalows in other West Coast cities, see James Allen Long, "Greening the City: Environment and City Life in the Far West, 1870–1930" (Ph.D. diss., University of California at Berkeley, 1996), 249–303; and Deryck William Holdsworth,

"House and Home in Vancouver: The Emergence of a West Coast Urban Landscape" (master's thesis, University of British Columbia, 1971), 109–247.

19. *The American Wonderland: The Pacific Northwest* (Chicago Burlington and Quincy RR, Northern Pacific Railway, Great Northern Railway, c. 1920), 19, box 28, Louis W. Hill Papers, James J. Hill Reference Library, St. Paul; Erwin L. Weber, *In the Land of Filtered Sunshine: Why the Pacific Northwest Is Destined to Dominate the Commercial World* (Seattle: Chamber of Commerce, 1924), 2, 4–11, UWL-SC; and *SMH,* August 26, 1905.

20. *P-I,* July 18, 1909. For the Mountaineers' history, see Jim Kjeldsen, *The Mountaineers: A History* (Seattle: The Mountaineers, 1998). For the Mazamas and Sierra Club, see *Mazama History* (Portland, Ore.: The Mazamas, 1932); Michael P. Cohen, *The History of the Sierra Club, 1892–1970* (San Francisco: Sierra Club Press, 1988); and Erik L. Weiselberg, "Ascendancy of the Mazamas: Environment, Identity, and Mountain Climbing in Oregon, 1870 to 1930" (Ph.D. diss., University of Oregon, 1999).

21. Meany, "What It All Means," *Collier's: The National Weekly* 43 (September 18, 1909): 14–15. For the Young Naturalists' Society, see Keith R. Benson, "The Young Naturalists' Society: From Chess to Natural History Collections," *Pacific Northwest Quarterly* 77 (July 1986): 82–93. For Meany's career, see George A. Frykman, *Seattle's Historian and Promoter: The Life of Edmond Stephen Meany* (Pullman: Washington State University Press, 1998), 27–126.

22. *The Mountaineer* 8 (1915): 78, UWL-SC. For examples of social functions, gender roles, and required gear, see *The Mountaineer* 1 (1907): 45–48, and *Mountaineer Bulletin* 2 (June 1912): 2, UWL-SC. For gender and outdoor recreation in early conservation politics, see Susan R. Schrepfer, *Nature's Altars: Mountains, Gender, and American Environmentalism* (Lawrence: University Press of Kansas, 2005).

23. Anna Louise Strong, "Plans Made by Camp Seattle for Camps on Rainier, 1917," box 9, fol. 10, Anna Louise Strong Papers, UWL-SC; see also *Co-Operative Campers Bulletin* (Seattle, 1923–24), UWL-SC. For the Co-operative Campers, see Catton, *National Park, City Playground,* 69–78. For utopian communities, see Charles P. LeWarne, *Utopias on Puget Sound, 1885–1915* (Seattle: University of Washington Press, 1975).

24. Strong to Sydney Strong, c. 1916, box 3, fol. 9a, Anna Louise Strong Papers, UWL-SC; Anna Louise Strong, *I Change Worlds: The Remaking of an American* (New York: Garden City, 1937), 57. For Strong's radicalism, see Tracy B. Strong and Helene Keysser, *Right in Her Soul: The Life of Anna Louise Strong* (New York: Random House, 1983), 64–69.

25. Stephanie Francine Ogle, "Anna Louise Strong: Progressive and Propagandist" (Ph.D. diss., University of Washington, 1981), 112–200; Strong and Keysser, *Right in Her Soul,* 70–83.

26. For private concessions, see Catton, *National Park, City Playground,* 60–89. For cooperative enterprises, see Walt Crowley, *To Serve the Greatest Number, A History of Group Health Cooperative of Seattle* (Seattle: Group Health Cooperative and University of Washington Press, 1995), 3–17; Dana Frank, *Purchasing Power: Consumer Organizing, Gender, and the Seattle Labor Movement, 1919–1929* (New York: Cambridge University Press, 1994), 40–65. For REI, see Lloyd Anderson, "The Beginning of Recreational Equipment Cooperative," October 22, 1976, and "The Cooperative Group," c. 1938, box 1, Lloyd Anderson Papers, UWL-SC; and Harvey Manning, *REI: Fifty Years of Climbing Together, The REI Story* (Seattle: Recreational Equipment Cooperative, 1988), 1–53.

27. Robert S. Lynd and Helen Merrill Lynd, *Middletown: A Study in American Culture* (New York: Harcourt, Brace, 1929), 225–314; Robert Spector, *The Legend of Eddie Bauer* (Lyme, Conn.: Greenwich, 1994), 9–41.

28. Frank B. Cooper to Board of Directors, Seattle School District, April 7, 1915, box 2, Folder: School Gardens, 1915–1919, Superintendent's Files, Seattle School District Archives and Records Center.

29. *Nineteenth Annual Report, Seattle Board of Park Commissioners* (1922): 20, box 3, fol. 7, DSPHC. For programs, see Lou Evans, "A City Teaches Swimming," and Ben Evans, "Seattle's 43 Playfields, 10 Beaches, and 39 Parks Provide Complete Program," *Western City* 7 (May 1931): 26–28; and Armstrong, "First Fifty Years of Municipal Recreation," 49–92.

30. Edmond S. Meany, "The Coming City of Puget Sound," June 24, 1929, box 59, fol. 4, Edmond S. Meany Papers, UWL-SC. For public-private partnerships, see Clark E. Schurman, "Monitor Rock," *The Mountaineer* 31 (December 15, 1938): 20, UWL-SC; and Ben Evans, Address to

Winter Sports Symposium, National Recreation Association Conference, April 13–15, 1938, box 23, fol. 14, Ben Evans Recreation Program Collection, SMA. For two divergent interpretations of the conflicts between automobiles and outdoor recreation at this time, see Paul S. Sutter, *Driven Wild: How the Fight Against Automobiles Launched the Modern Wilderness Movement* (Seattle: University of Washington Press, 2002); and David Louter, *Windshield Wilderness: Cars, Roads, and Nature in Washington's National Parks* (Seattle: University of Washington Press, 2006).

31. Hiram M. Chittenden, "Sentiment Versus Utility in the Treatment of National Scenery," *Pacific Monthly* 23 (January 1910): 37–38.

32. *Pacific Monthly* 18 (July 1907): 114; Paul Whitham, "Seattle Prepares for Great Motor Fleet," *Pacific Motor Boat* 2 (November 1911): 12.

33. "Marine Pageantry Marks Opening of the Famous Lake Washington Canal" *Pacific Motor Boat* 9 (August 1917): 5–6; *Pacific Motor Boat* 10 (September 1917): 49; Daniel L. Pratt, "Seattle to Stage Big International Regatta," *Pacific Motor Boat* 14 (April 1922): 3–4; and *Pacific Motor Boat Handbook Number,* 2nd ed. 23 (April 1931), 81.

34. *P-I,* August 2, 1908.

35. Howard Droker, *Seattle's Unsinkable Houseboats: An Illustrated History* (Seattle: Watermark, 1977), 33–41, 49–62.

36. Barton Warren Evermann and Seth Eugene Meek, "A Report upon Salmon Investigations in the Columbia River and Elsewhere on the Pacific Coast in 1896," *Bulletin of the United States Fish Commission for 1897* (Washington, D.C.: Government Printing Office), 34–47; R. Rathbun, "A Review of the Fisheries in the Contiguous Waters of the State of Washington and British Columbia," *Report of the United States Fish Commission for the Year Ending June 30, 1899* (Washington, D.C.: Government Printing Office, 1890): 251–350. For commercial fishing, see L. H. Darwin to Louis F. Hart, October 1, 1919, box 2J-1-18, Fisheries Commission File, 1919, Governors' Papers, Hart Group, WSA.

37. Jonathan L. Reisland, Commissioner and Game Warden, *Sixteenth and Seventeenth Annual Reports of the State Fish Commissioner and Game Warden, 1905–1906* (Olympia, Wash., 1907), 15–16, WSDF. For further complaints, see the following Annual Reports: *First Annual* (1890), 25–30; *Ninth Annual* (1898), 6; *Tenth/Eleventh Annual* (1899–1900), 16–18; *Twenty-Second/Twenty-Third* (1911–12), 95–104; *Twenty-Fourth/Twenty-Fifth* (1913–15), 27–28, WSDF. For conservation, see Clayton R. Koppes, "Efficiency/Equity/Esthetics: Towards a Reinterpretation of American Conservation," *Environmental Review* 11 (Summer 1987): 127–46.

38. W. T. Isted to Darwin, December 19, 1919, box 2J-1-18, Fisheries Commission File, 1919, Governors' Papers, Hart Group, WSA.

39. *Twenty-Eighth and Twenty-Ninth Annual Reports of the State Fish Commissioner, 1917–1919* (Olympia, Wash., 1919), 10–13, WSA. For hatchery politics, see Joseph E. Taylor III, *Making Salmon: An Environmental History of the Northwest Fisheries Crisis* (Seattle: University of Washington Press, 1999), 166–202. For nativist politics in other western locales, see Arthur F. McEvoy, *The Fisherman's Problem: Ecology and Law in the California Fisheries, 1850–1980* (New York: Cambridge University Press, 1986), 65–119; and Connie Y. Chiang, "'Monterey-by-the-Smell': Odors and Social Conflict on the California Coastline," *Pacific Historical Review* 73 (May 2004): 183–214.

40. First two quotations from "Petitions to Louis F. Hart," January 5, 1920, and Frank L. Jackson to Hart, February 13, 1920, Box 2J-1-18, Fisheries Commission File, 1919, Governors' Papers, Hart Group, WSA. Last three quotations are from "Committee" to Hart, November 28, 1919, box 2J-1-18, Fisheries Commission File, 1919, Governors' Papers, Hart Group, WSA; and "Statement of R. Q. Hall, Hearing on Puget Sound Fishery Matters," June 20, 1921, Box 2, Hearings of State Fisheries Board, vol. H1, Department of Fisheries, Administration Library, Acc. 74-12-1115, WSDF. For closings, see "Before the State Fisheries Board," Order No. 18 (1921), box 2, Hearings of State Fisheries Board, vol. H7, Department of Fisheries, Administration Library, Acc. 74-12-1115, WSDF; and "Test of Duwamish Closing," *Pacific Fisherman* 18 (February 1920): 30.

41. *ST,* July 31, 1928, and August 18, 1929; H. L. Dilaway, "Trout Fishing," *Northwest Sportsman* 1 (April 1932): 2–4, 7. For Ben Paris's store, see *SS,* September 3, 1935. For arrest records, see *Annual Reports of the State Fish Commissioner: Twenty-Eighth and Twenty-Ninth, 1917–1919* (Olympia, Wash., 1919), 162; *Thirtieth and Thirty-First* (1919–21), 202, 293–94; *Thirty-Second and Thirty-Third* (1921–23), 44–45, 104; *Thirty-Fourth and Thirty-Fifth* (1923–25), 63–64, 128; *Thirty-Sixth and Thirty-*

Seventh (1925–27), 123–24, 203, WSDF. For other accounts, see *ST,* October 1, 6, 11, 1928. For banditry, see Karl Jacoby, *Crimes Against Nature: Squatters, Poachers, Thieves, and the Hidden History of American Conservation* (Berkeley: University of California Press, 2001), 2. For the Black Act, see Edward Palmer (E. P.) Thompson, *Whigs and Hunters: The Origin of the Black Act* (London: Allen Lane, 1975); and Douglas Hay, "Poaching and the Game Laws on Cannock Chase," in *Albion's Fatal Tree: Crime and Society in Eighteenth-Century England,* ed. Douglas Hay, Peter Linebaught, John G. Rule, E. P. Thompson, and Cal Winslow (New York: Pantheon, 1975), 189–254. For the South, see Steven Hahn, "Hunting, Fishing, and Foraging: Common Rights and Class Relations in the Postbellum South," *Radical History Review* 26 (October 1982): 36–64; and Stuart A. Marks, *Southern Hunting in Black and White: Nature, History, and Ritual in a Carolina Community* (Princeton: Princeton University Press, 1992). For western North America, see Louis S. Warren, *The Hunter's Game: Poachers and Conservationists in Twentieth-Century America* (New Haven: Yale University Press, 1997); Mark David Spence, *Dispossessing the Wilderness: Indian Removal and the Making of the National Parks* (New York: Oxford University Press, 1999); Douglas C. Harris, *Fish, Law, and Colonialism: The Legal Capture of Salmon in British Columbia* (Toronto: University of Toronto Press, 2001); and Taylor, *Making Salmon,* 166–202.

42. Alexandra Harmon, *Indians in the Making: Ethnic Relations and Indian Identities Around Puget Sound* (Berkeley: University of California Press, 1999), 178–85.

43. The *Duwamish, Lummi, Whidbey Island, Skagit, Upper Skagit, Swinomish, Kikiallus, Snohomish, Snoqualmie, Stillaguamish, Suquamish, Samish, Puyallup, Squaxin, Skokomish, Upper Chehalis, Muckleshoot, Nooksack, Chinook, and San Juan Islands Tribes of Indians, Claimants, vs. the United States of America, Defendant. Consolidated petition. No. F-275.* 2 vols. (Seattle: Argus, c. 1933), 163, 189, 193, 673–74, and 694, copy available at UWL-SC.

44. Almira Bailey, "The Charmed Land," in *Seattle and the Charmed Land* (Seattle: Chamber of Commerce, c. 1930), 2, UWL-SC. For the role of myth and fantasy in promoting the Far West as a tourist destination, see Hal Rothman, *Devil's Bargains: Tourism in the Twentieth-Century American West* (Lawrence: University Press of Kansas, 1998); David M. Wrobel, *Promised Lands: Promotion, Memory, and the Creation of the American West* (Lawrence: University Press of Kansas, 2002); and Bonnie Christensen, *Red Lodge and the Mythic West: Coal Miners to Cowboys* (Lawrence: University Press of Kansas, 2002).

45. For the tragedy of the commoners, see Bonnie J. McCay, "The Culture of the Commoners: Historical Observations of Old and New World Fisheries," in *The Question of the Commons: The Culture and Ecology of Communal Resources,* ed. Bonnie J. McCay and James M. Acheson (Tucson: University of Arizona Press, 1987), 195–216, which builds on the earlier work of E. P. Thomson. McCay's term is a play on the original idea as developed in Garrett Hardin, "The Tragedy of the Commons," *Science* 162 (1968): 1243–48.

46. For consumption and conservation, see Jennifer Price, *Flight Maps: Adventures with Nature in Modern America* (New York: Basic, 1999), 57–110; and Gregory Summers, *Environmentalism in the Fox River Valley, 1850–1950* (Lawrence: University Press of Kansas, 2006).

Chapter 6. Junk-Yard for Human Junk: The Unnatural Ecology of Urban Poverty

1. *Annual Report, Department of Sanitation and Health, 1915* (Seattle, 1915), 94–110; and *Report of the Department of Health and Sanitation, 1912–1914* (Seattle, 1914), 61, Annual Reports, SMA.

2. R. H. Thomson, "The Seattle Regrades" [c. 1930], 12, box 13, fol. 5, RHT Papers.

3. Cotterill, "What Part Has an Engineer in the Development of a Municipality Such as Seattle?" (1928), box 14, fol. 6, George F. Cotterill Papers, UWL-SC.

4. V. V. Tarbill, "Mountain-Moving in Seattle," *Harvard Business Review* 8 (July 1930): 485–86. For creative destruction, see Joseph A. Schumpeter, *Capitalism, Socialism, and Democracy* (New York: Harper & Brothers, 1942), 82–85. For creative destruction in urban planning, see Max Page, *The Creative Destruction of Manhattan* (Chicago: University of Chicago Press, 1999); Jason Gilliand, "The Creative Destruction of Montreal: Street Widenings and Urban (Re) Development in the Nineteenth Century," *Urban History Review/Revue d'histoire urbaine* 31 (October 2001): 37–51.

For Denny Hill, see William Bloch, Jr., "Looking Back, on the First Regrades" (unpublished manuscript, n.d.), Seattle Pamphlet File—Regrading, Pacific Northwest Collections, UWL-SC.

5. *Town Crier* (Seattle), January 4, 1930; *ST,* August 5, 1926. For keystone project, see Blackwell to City Council, January 31, 1924, LID 4818, letters, fol. 1, SED.

6. Joseph I. Granger to Board of Park Commissioners, April 26, 1929, and Cotterill to Board of Park Commissioners and City Council, May 17, 1928, box 24, fol. 8, DSPHC. For preparations, see Tarbill, "Mountain-Moving in Seattle," 485–86. For descriptions of the original park, see JCO to SWO, May 4, 1903, box 5A, fol. 36, JCO Coll.

7. "Unique Regrade Engineering," *Scientific American* 141 (November 1929): 431. For operations, see W. D. Barkhuff, "The Regrading of Denny Hill," *Professional Engineer* 14 (April 1929): 5–7; C. W. Leihy, "Moving a Mountain to Sea," *American City* 42 (June 1930): 153; and "Denny Hill Regrade No. 2, Seattle, Washington," *Western Construction News* 5 (July 25, 1930): 351–54. For the muddy lagoon, see *ST,* August 22, 1929.

8. For lawsuits, see *Nelson v. Seattle,* 180 Wash. 1 (1934); *State ex rel Beecher v. Gilliam,* 146 Wash. 6 (1927); *In re Taylor Avenue Assessment,* 149 Wash. 214 (1928); and *In re Sixth Avenue,* 155 Wash. 459 (1930). For examples of complaints, see Denny Hill Regrade and Improvement Club to Barkhuff, February 6, 1930, and Charles F. Clise to Board of Public Works, June 23, 1930, LID 4818, letters, fol. 1 and 2, SED.

9. "Report of City Planning Commission re 6th Ave. et al. extension and regrading," E. S. Goodwin, President, City Planning Commission to City Council, June 10, 1925, file 100373, SCCF. For neighborhood descriptions, see Irene Burns Miller, *Profanity Hill* (Everett, Wash.: Working Press, 1979).

10. Thomas P. Hughes, "Technological Momentum," in *Does Technology Drive History? The Dilemma of Technological Determinism,* ed. Merritt Roe Smith and Leo Marx (Cambridge: MIT Press, 1994), 101–14. For the proposed regrade, see "Petition of Associated Central Business Properties" to City Council April 25, 1930, file 126258, box 164, SCCF. For various options, see "Report of the City Engineer on Plan for Regrading Yesler Hill," January 16, 1930, file 124810, "Report of the City Engineer on Yesler Hill, Regrade," April 28, 1930, file 126265, boxes 161, 164, SCCF; and "6th Avenue Extension Relief Maps" [c. 1925?], items 3497–3502, Engineering Department Negatives, SMA.

11. *ST,* September 17, 1929; Read Bain, "Seattle: A Harbor Without a Hinterland," *New Republic* 57 (December 19, 1928): 131–34. For later developments, see Alan Jay Razak, "Redeveloping the Redevelopment: The Denny Regrade" (master's thesis, University of Washington, 1981).

12. Carlos Arnaldo Schwantes, *The Pacific Northwest: An Interpretive History,* rev. ed. (Lincoln: University of Nebraska Press, 1989, 1996), 362–67.

13. R. H. Ober and Hamilton C. Johnson, "Report on Reforestation of Cedar River Watershed" (February 1913), 6–9, box 4, fol. 4, SWD-HF; and Richard White, *Land Use, Environment, and Social Change: The Shaping of Island County, Washington,* rev. ed. (Seattle: University of Washington Press, 1980, 1992), 140–41.

14. R. D. McKenzie, "The Ecological Approach to the Study of the Human Community," *American Journal of Sociology* 30 (November 1924): 301.

15. Robert E. Park, "The City: Suggestions for the Investigation of Human Behavior in the Urban Environment," in *The City,* ed. Robert E. Park, Ernest W. Burgess, and Roderick D. McKenzie (Chicago: University of Chicago Press, 1925), 46; McKenzie, "The Ecological Approach," 299. For the Chicago school of sociology, see Dorothy Ross, *The Origins of American Social Science* (New York: Cambridge University Press, 1991), 437–48; and Park Dixon Goist, "City and 'Community': The Urban Theory of Robert Park," *American Quarterly* 23 (Spring 1971): 46–59. For another application of ecological theories, see McKenzie, "Ecological Succession in the Puget Sound Region," in *Roderick D. McKenzie on Human Ecology: Selected Writings,* ed. and intro. Amos H. Hawley (Chicago: University of Chicago Press, 1968), 228–43. For a fascinating analysis of the Chicago school and the putative connections between poverty and spread of weeds in urban environments at the time, see Zachary James Sopher Falck, "Controlling Urban Weeds: People, Plants, and the Ecology of American Cities, 1888–2003" (Ph.D. diss., Carnegie Mellon University, 2004), 56–102.

16. Katherine I. Grant Pankey, "Restrictive Covenants in Seattle: A Study in Race Relations" (B.A. thesis, University of Washington, 1947), 3, UWL-SC; Paul Kitchner Hatt, "A Study of Natural

Areas in the Central Residential District of Seattle" (Ph.D. diss., University of Washington, 1945), 25–26, 104. These neighborhoods included Broadmoor, a private gated community near Lake Washington in Seattle, as well as the upscale Laurelhurst, Queen Anne, and Magnolia districts. For covenants, see the Web site "Segregated Seattle: A Seattle Civil Rights and Labor History Project Special Section," at http://depts.washington.edu/civilr/segregated.htm/ (accessed January 16, 2006).

17. "Mobility of Population in Seattle," *Monthly Labor Review* 22 (February 1926): 277; John Herbert Geoghegan, "The Migratory Worker in Seattle: A Study in Social Disorganization and Exploitation" (master's thesis, University of Washington, 1923), 83, 91. For the Wheatland riot, see Carleton H. Parker, *The Casual Worker and Other Essays,* Cornelia Statton Parker, intro. (New York: Harcourt, Brace, and Howe, 1920).

18. Howard Droker, *Seattle's Unsinkable Houseboats: An Illustrated History* (Seattle: Watermark, 1977), 63–70.

19. Gordon Beebe, "Ecological Aspects of Substandard Housing in Seattle" (master's thesis, University of Washington, 1939), 44–49; *SS,* March 24, 1922; Droker, *Seattle's Unsinkable Houseboats,* 72–75.

20. William Hard, "Giant Negotiations for Giant Power: An Interview with Herbert Hoover," *Survey Graphic* 5 (March 1924): 577, also quoted in Kendrick A. Clements, *Hoover, Conservation, and Consumerism: Engineering the Good Life* (Lawrence: University Press of Kansas, 2000), 78. For the 1927 flood, see John M. Barry, *Rising Tide: The Great Mississippi Flood of 1927 and How It Changed America* (New York: Simon & Schuster, 1997), 361–95.

21. Suellen Hoy, "'Municipal Housekeeping': The Role of Women in Improving Urban Sanitation Practices," in *Pollution and Reform in American Cities,* ed. Martin V. Melosi (Austin: University of Texas Press, 1980), 173–98; *SS,* April 4, 1922. For Landes's career, see Sandra Haarsager, *Bertha Knight Landes of Seattle: Big-City Mayor* (Norman: University of Oklahoma Press, 1994), 3–56, 83–129. For women's politics, see John Putnam, "'A Test of Chiffon Politics': Gender Politics in Seattle, 1897–1917," *Pacific Historical Review* 69 (November 2000): 595–616.

22. Julia Budlong, "What Happened in Seattle," *Nation* 127 (August 29, 1928): 197–98. For Landes's reforms, see Haarsager, *Bertha Knight Landes,* 184–241. For beyond separate spheres, see Maureen A. Flanagan "Gender and Urban Political Reform: The City Club and the Woman's City Club of Chicago in the Progressive Era," *American Historical Review* 95 (October 1990): 1032–50, and "The City Profitable, the City Livable: Environmental Policy, Gender, and Power in Chicago in the 1910s," *Journal of Urban History* 22 (January 1996): 163–90; and Harold L. Platt, "Jane Addams and the Ward Boss Revisited: Class, Politics, and Public Health in Chicago, 1890–1930," *Environmental History* 5 (April 2000): 194–222.

23. Ella B. Rowntree to Mrs. H. F. Alexander, October 15, 1928; Sigrid M. Frisch to Hoffman, Engineer, March 9, 1930, boxes 39, 41, fol. 2, 10; Kerry to City Council and Park Board, July 2, 1937, box 46, fol. 17; and Kerry to Board of Park Commissioners and Community Clubs, September 29, 1928, box 3, fol. 22, DSPHC. For the auto camp, see *Report of the Department of Parks, City of Seattle, 1923–1930* (Seattle, 1930), 47–48, Annual Reports, SMA.

24. Allen R. Potter, "Facilities for Housing, Medical Care, and Other Services for Homeless Men in Seattle, 1929–1930" (Seattle, 1930), 3, 177–78, box 1, Allen R. Potter Papers, UWL-SC; see also Allen R. Potter and Staff, *Occupational Characteristics of Unemployed Persons in Cities of 11,000 or More Persons, State of Washington* (Olympia: Washington Employment Relief Administration, 1935), vii.

25. Frank Edwards, "Mayor's Message to City Council," May 25, 1931, file 131528, box 9, fol. 3, Mayor's Messages, SMA; *Annual Report, Board of Park Commissioners* (Seattle, 1934), 12, Annual Reports, SMA; "Protest of Downtown Local Unemployed Council Against Forced Labor," January 14, 1935, file 145851, SCCF. Unemployment in Seattle rose quickly in the first years of the depression, from 11 percent in April 1930 to 26.5 percent by January 1935. See Potter et al., *Occupational Characteristics of Unemployed Persons,* 2–7. For depression-era politics, see Terry R. Willis, "Unemployed Citizens of Seattle, 1900–1930: Hulet Wells, Seattle Labor, and the Struggle for Economic Security" (Ph.D. diss., University of Washington, 1997); Richard C. Berner, *Seattle, 1921–1940: From Boom to Bust* (Seattle: Charles, 1992), 301–32, 403–15; and William H. Mullins, *The Depression and the Urban West Coast, 1919–1933: Los Angeles, San Francisco, Seattle and Portland* (Bloomington: Indiana University Press, 1991).

26. Don G. Abel, "Letter on W.P.A. Work," January 9, 1937, box 36, fol. 1, SED; Mrs. Alexander F. McEwan to Park Board, April 1, 1936, box 43, fol. 13, DSPHC. For details of specific jobs, see "Progress Report on WPA Projects," June 38, 1938, box 63, fol. 3, DSPHC. For unemployed workers, see E. R. Hoffman to *Municipal News,* November 9, 1931; Seattle Garden Club to Harry M. Westfall March 23, 1936, box 43, fol. 12, DSPHC. For protests, see "Petition to Board of Park Commissioners," October 18, 1934, Josiah Collins to Board of Park Commissioners, March 27, 1936, and Committee to Investigate Seward Park to D. C. Conover, April 9, 1936, box 23, fol. 10, and box 43, fol. 13, DSPHC; and "Petition Against Logging in Seward Park," file 150912, SCCF. For restorative powers, see Neil M. Maher, "A New Deal Body Politic: Landscape, Labor, and the Civilian Conservation Corps," *Environmental History* 7 (July 2002): 435–61; for more on this, see Maher, *Nature's New Deal: Franklin Roosevelt, the Civilian Conservation Corps, and the Roots of the American Environmental Movement* (New York: Oxford University Press, 2007).

27. Board of Park Commissioners to D. W. McMorris, August 27, 1931, box 34, fol. 16, DSPHC; Droker, *Seattle's Unsinkable Houseboats,* 77–81; Myra L. Phelps, *Public Works in Seattle: A Narrative History of the Engineering Department, 1875–1975* (Seattle: Department of Engineering, 1975), 207–9.

28. Gilman to W. P. Kenney, January 15, 1934, box 294, file 7525, GN-PSF. For shack towns, see Beebe, "Ecological Aspects," 40–43, 45; Work Projects Administration, *Report on Land Use Zoning Survey, City of Seattle Projects 667, 4110, 5638* (Seattle, 1938); and Droker, *Seattle's Unsinkable Houseboats,* 66.

29. Donald Francis Roy, "Hooverville: A Study of a Community of Homeless Men in Seattle" (master's thesis, University of Washington, 1935), 1, 97; and Jesse Jackson, "The Story of Hooverville, in Seattle" (Seattle, 1935), 1–3, UWL-SC. For building Hooverville, see *SS,* December 30, 1930; Calvin Schmid, *Social Trends in Seattle* (Seattle: University of Washington, 1944), 286–87; and Willis, "Unemployed Citizens of Seattle," 186–87.

30. "Report of the Sanitation Division, December 31, 1935," from *Annual Report of the Department of Health and Sanitation of the City of Seattle, 1932–1935* (Seattle, 1935), Annual Reports, SMA; "Petition of Downtown Local, National Unemployment Council," May 15, 1935, and Department of Health Response, May 23, 1935, file 147091, SCCF.

31. Roy, "Hooverville," 1, 20–21, 97.

32. Roy, "Hooverville," 77–80; Jackson, "The Story of Hooverville," 3–11; and Schmid, *Social Trends in Seattle,* 286–93. Of the 639 residents in Hooverville in 1934, there were only 7 women. The average age was 45.4 years. Among the non-white residents, there were 120 Filipinos, 29 African-Americans, and 25 Mexicans. Among the 292 foreign-born "white" residents, 207 were from northern Europe and Scandinavia and 85 were from southern and central Europe. Roy classified almost 71 percent as "white" in his survey.

33. Roy, "Hooverville," 21, 29–37. According to Roy, an attack of "feline plague" in 1933 decimated the cat population and sent Hooverville's rat population skyrocketing.

34. "Protest Against Shacks in the Interbay District," April 26, 1937, file 154992, SCCF; Seattle Housing Authority, *Real Property Survey, 1939–1940,* WPA Project 3372, 2 vols. (Seattle: Works Project Administration, 1942); Housing Authority of the City of Seattle to City Council and "Real Property Survey," March 5, 1941, and Shack Elimination Committee to City Council, April 14, 1941, file 168237, SCCF.

35. *Seattle Housing Authority Annual Report* (Seattle, 1940), Annual Reports, SMA; *ST,* April 10, 1941; Richard C. Berner, *Seattle, 1921–1940: From Boom to Bust* (Seattle: Charles, 1992), 183–87.

36. Roy, "Hooverville," 75–76.

37. *ST,* December 27, 1942; Schmid, *Social Trends in Seattle,* 136–37.

38. Quotation from sample covenant at "Segregated Seattle" (accessed January 16, 2006). For a retrospective of Innis Arden, see *ST,* June 3, 2005.

Chapter 7. Death for a Tired Old River: Ecological Restoration and Environmental Inequity in Postwar Seattle

1. "Duwamish Duck Marsh of Century Ago Has Developed into Seattle's 'Golden Shore,'" *Seattle Business* 33 (September 15, 1949): 2; *The Duwamish Diary, 1849–1949* (Seattle: Cleveland High School, 1949, 1996), viii, 116.

2. *ST,* October 26, 1959; Minutes, December 17, 1959, METRO Council, Minutes, box 2, PSRA. For Miller's further response, see *ST,* November 2, 1959.

3. Samuel P. Hays, *Beauty, Health, and Permanence: Environmental Politics in the United States, 1955–1985* (New York: Cambridge University Press, 1987), 71–98; Andrew Hurley, *Environmental Inequalities: Class, Race, and Industrial Pollution in Gary, Indiana, 1945–1980* (Chapel Hill: University of North Carolina Press, 1995); and Adam Rome, *The Bulldozer in the Countryside: Suburban Sprawl and the Rise of American Environmentalism* (New York: Cambridge University Press, 2001). For a new study of one metropolitan region and the interplay between place and environmental politics, see Richard Walker, *The Country in the City: The Greening of the San Francisco Bay Area* (Seattle: University of Washington Press, 2007).

4. "Northgate Beginnings: Jim Douglas Remembers the First Year," People's History, HistoryLink.org, www.historylink.org/essays/output.cfm?file_id=2289 (accessed January 20, 2006). For promotional gimmicks, see *ST,* February 22, 1948, February 1, 1950, May 7, 23, 1950, March 18, August 17, 1965, and March 21, 1968; and "Shopping Center: 1949 Model," *Business Week* (July 23, 1949): 47–50. For Bellevue, see Lucile McDonald, *Bellevue: Its First 100 Years* (Bellevue, Wash.: Bellevue Historical Society, 2000), 136–42. For suburbs and shopping malls, see Lizabeth Cohen, *A Consumer's Republic: The Politics of Mass Consumption in Postwar America* (New York: Knopf, 2003), 112–65, 257–89. For a representative housing advertisement, see *P-I,* August 24, 1953.

5. For postwar growth, see *ST,* March 20, 1949, November 20, 1956; Leagues of Women Voters of King County, *The Municipality of Metropolitan Seattle,* 2nd ed. (Seattle: The Leagues, 1963), 2; and Lorraine McConaghy, "No Ordinary Place: Three Postwar Suburbs and Their Critics" (Ph.D. diss., University of Washington, 1993), 258–63. For an overview of Seattle politics during the 1940s, see Richard C. Berner, *Seattle Transformed: World War II to Cold War* (Seattle: Charles, 1999). For Levittown, see "Up from the Potato Fields," *Time* 56 (July 3, 1950): 67–68, 72; and Rome, *The Bulldozer in the Countryside,* 15–43. For military-metropolitan-industrial complex, see Richard S. Kirkendall, "The Boeing Company and the Military-Metropolitan-Industrial Complex, 1945–1953," *Pacific Northwest Quarterly* 85 (October 1994): 137–49. For California industry and cities, see Roger W. Lotchin, *Fortress California, 1910–1961: From Warfare to Welfare* (New York: Oxford University Press, 1992); and Greg Hise, *Magnetic Los Angeles: Planning the Twentieth-Century Metropolis* (Baltimore: Johns Hopkins University Press, 1997), 186–215.

6. For Seafair, see Sharon Boswell and Lorraine McConaghy, *One Hundred Years of a Newspaper and Its Region: Reprinted from the Seattle Times Centennial Project, 1996* (Seattle: Seattle Times Company, 1997), 36. For the Golden Potlatch, see *P-I,* July 16–18, 23, 1911. For fishing derbies and clambakes, see *P-I,* August 24, 1953. For Mumford and growth, see Lewis Mumford, *The Culture of Cities* (New York: Harcourt, Brace, 1938), 300–493.

7. Material in this and the preceding paragraph is from James R. Ellis, interview by John F. Henry, March 13 and July 1, 1985, vols. 1 and 2, box 9, METRO Library, PSRA.

8. League of Women Voters of Seattle-Bellevue-Renton-Highline, *Municipality of Metropolitan Seattle* (Seattle: The League, 1961), 4, UWL-SC; *Municipal News,* November 28, 1953, UWL-SC; and Robert Lee Peabody, "Financing a Proposed Metropolitan Government for Greater Seattle" (master's thesis, University of Washington, 1956), 4. The League of Women Voters report put the number at 138, the *Municipal News* at 180.

9. Municipal League of Seattle and King County, *Seattle—The Shape We're In!* (Seattle: The League, 1955), 2, 5, and *Seattle—The Shape . . . Of Things to Come!* (Seattle: The League, 1956), 4, 7, box 1, JRE Papers. For Toronto, see James T. Lemon, *Liberal Dreams and Nature's Limits: Great Cities of North America Since 1600* (New York: Oxford University Press, 1996), 253–67. For principled terms, see Joel A. Tarr, *The Search for the Ultimate Sink: Urban Pollution in Historical Perspective* (Akron: University of Akron Press, 1996), 309–22.

10. For septic tanks, see *Lake Washington and the Seattle Metropolitan Area,* Information Series 9 (Olympia: Washington State Pollution Commission, 1955), WPCC; Rome, *The Bulldozer in the Countryside,* 87–118. For Duwamish pollution, see "Stream and Creek Survey Reports, Duwamish River and tributaries," July 22, 1930, Stream Survey—Duwamish (Green) River, Department of Wildlife, Research Projects and Studies, 1940–85, box 2, Acc. 90-9-1255, WSDG. For one early protest, see Commissioner, King County Game Commission, to Charles A. Pollock, January 24, 1928, Pollution-Pulp, Administration Series, box 706, WSDG.

11. For sewers, see Paul Dorpat and Genevieve McCoy, *Building Washington: A History of Washington State Public Works* (Seattle: Tartu, 1998), 313–16, 322; Myra L. Phelps, *Public Works in Seattle: A Narrative History of the Engineering Department, 1875–1975* (Seattle: Department of Engineering, 1975), 186–203; and Tarr, *Search for the Ultimate Sink*, 131–58. For regional pollution, M. S. Campbell, *Sources and Extent of Lake Washington Pollution*, Pollution Series, No. 29 (Olympia: Washington State Pollution Commission, 1943); Richard F. Foster, *Sources of Pollution in Lake Washington Canal and Lake Union*, Series Bulletin No. 28 (Olympia: Washington State Pollution Commission, 1943), WPCC. For Alki Beach, see Park Engineer to Dr. E. T. Hanley, June 5, 1929; and Edgar S. Hadley to Board of Park Commissioners, November 5, 1929, box 19, fol. 1, DSPHC.

12. Abel Wolman, *City of Seattle: Report on Sewage Disposal* (Baltimore, 1948), UWL-SC. For shoreline changes, see Chester A. Hockett, "Urbanization and Shoreline Development of Lake Washington" (master's thesis, University of Washington, 1976), 82–135.

13. For regional solutions, see E. F. Eldridge and Wallace Bergerson, *The Seattle Sewage Treatment Problem, with Comments on the Wolman Report* (Olympia: Washington State Pollution Commission, 1948), 12, WPCC. For West Seattle, see Memorandum 537, Pollution of Longfellow Creek, King County, March 22, 1948; Memorandum 539, Conference Regarding Pollution Conditions in White Center Area, King County, March 24, 1948, Numbered Memoranda, box 2, WPCC. For beach closures, see "Sanitation Report on Water Quality, Seattle Beaches," 1956, box 9, fol. 9, DSHPC.

14. Eldridge to Arthur B. Langlie, May 12, 1949, Activity and Progress Reports 1969, Correspondence 1946–1966, box 17, WPCC.

15. For Edmondson's career, see Nelson G. Hairston, Jr., "In Memoriam: W. Thomas Edmondson: A Distinguished Limnologist and an Intelligent Man of Nobility, Generosity, Kindness and Honesty," *Archiv für Hydrobiologie* 148 (April 2000): 3–8.

16. W. T. Edmondson, *The Uses of Ecology: Lake Washington and Beyond* (Seattle: University of Washington Press, 1991), 13. For earlier visibility, see Victor B. Scheffer and Rex J. Robinson, "A Limnological Study of Lake Washington," *Ecological Monographs* 9 (January 1939): 94–143.

17. *ST*, July 11, 1955; Edmondson to Ellis, February 13, 1957, box 1, fol. 5, JRE Papers. For initial findings, see W. T. Edmondson, G. C. Anderson, and Donald R. Peterson, "Artificial Eutrophication of Lake Washington," *Limnology and Oceanography* 1 (January 1956): 47–53; R. O. Sylvester, W. T. Edmondson, and R. H. Bogan, "A New Critical Phase of the Lake Washington Problem," *Trend in Engineering* 8 (April 1956): 8–14; and Washington State Pollution Control Commission, *Lake Washington and the Seattle Metropolitan Area* (Olympia: The Commission, 1955).

18. Carl Abbott, *The Metropolitan Frontier: Cities in the Modern West* (Tucson: University of Arizona Press, 1993), 39. For ecology and environmental politics, see Stephen Bocking, *Ecologists and Environmental Politics: A History of Contemporary Ecology* (New Haven: Yale University Press, 1997). For local politics, see Thomas J. Sugrue, "All Politics Is Local: The Persistence of Localism in Twentieth-Century America," in *The Democratic Experiment: New Directions in American Political History*, ed. Meg Jacobs, William J. Novak, and Julian E. Zelizer (Princeton: Princeton University Press, 2003), 301–26. For funding and cooperation, see Edmondson, *The Uses of Ecology*, 14, 21. For the fusion of the state and academia, which Brian Balogh calls the "prominiistrative state," see *Chain Reaction: Expert Debate and Public Participation in American Commercial Nuclear Power, 1945–1975* (New York: Cambridge University Press, 1991), 1–94.

19. Edmondson, *The Uses of Ecology*, 286; Minutes, Governor's Metropolitan Problems Advisory Committee, January 1956, box 1, JRE Papers; Ellis Interview, vol. 2, PSRA.

20. Daniel Jack Chasan, *Speaker of the House: The Political Career and Times of John L. O'Brien* (Seattle: University of Washington Press, 1990), 92–94; Madeline Lemere, "Introduction: Metro Action Committee Scrapbook, May–September 1958," box 6, fol. 20, METRO Library, PSRA; "'Metro' Government Is Debated at Washington Institute Sessions," *Western City* 33 (August 1957): 57.

21. Nicholas A. Maffeo, "Speech Against 'Metro,'" Seattle: Metropolitan District, 1958, Vertical Subject Files, UWL-SC.

22. Edmondson, *The Uses of Ecology*, 24, 26, 28–29; *Bellevue American*, March 5, 1958; *Highline Times*, March 6, 1958; *Washington Sentinel*, April 14, 1958; *ST*, August 9, 1958. For Edmondson's warning, see Edmondson to Metro Action Committee, January 13, 1958, box 1, fol. 5, JRE Papers.

23. For conservative politics, see Lisa McGirr, *Suburban Warriors: The Origins of the New American Right* (Princeton: Princeton University Press, 2001).

24. For the Canwell Committee, see Melvin Rader, *False Witness* (Seattle: University of Washington Press, 1969, 1977); and Michael Reese, "The Cold War in Washington State: A Curriculum Project Developed by the Center for the Study of the Pacific Northwest, University of Washington" www.washington.edu/uwired/outreach/cspn/curcan/main.html (accessed April 12, 2005). For Lake Washington suburbs, see McConaghy, "No Ordinary Place," 326–61.

25. *Boeing News* (March 6, 1958), 1, box 1, fol. 5, JRE Papers; *P-I,* February 18, 1958; *ST,* March 9, 1958; see also "Eastsiders for Metro" [draft statement] (c. Feb. 1958), vol. 19, Scrapbooks, METRO Library, PSRA. For the election results, "Official Canvass Book," March 11, 1958, King County Records and Elections Department, KCARM; *P-I,* March 11, 1958.

26. Richard H. Riddell to John F. Henry, March 14, 1958, box 1, fol. 2, John F. Henry Papers, UWL-SC; Henry, "What Happens to East Side Sewers if Metro Fails?: An Open Letter to Residents of the Eastside by John F. Henry," February 11, 1958, box 2, fol. 6, JRE Papers. For Bellevue's early efforts, see Bellevue Sewer District, *Preliminary Report, East Side Trunk Sewer Project* (March 19, 1956), and Henry, "General Information Concerning Bellevue Sewer District," September 18, 1953, box 1, fol. 2, John F. Henry Papers, UWL-SC. For campaign tactics, see Ellis Interview, vol. 2; Edmondson, *The Uses of Ecology,* 31–32.

27. For the revamped Metro plan, see Henry to Mayor and Council of the City of Bellevue, April 3, 1958, box 1, fol. 2, John F. Henry Papers, UWL-SC; *Bellevue American,* April 10, 1958; *P-I,* April 15, 1958; and Ellis Interview, vol. 2. For the engineering report and quotation, see Brown and Caldwell, *Metropolitan Seattle Sewerage and Drainage Survey: A Report for the City of Seattle, King County, and the State of Washington on the Collection, Treatment, and Disposal of Sewage and the Collection and Disposal of Storm Water in the Metropolitan Seattle Area* (San Francisco and Seattle: Brown and Caldwell, 1958), 1, UWL-SC.

28. *ST,* July 4, 1958; "The 'Metro Monster,'" (c. 1958) [emphasis in original], box 6, METRO Library Collection, PSRA. For beach closures, see Ruth Ittner, "Formation of the Municipality of Metropolitan Seattle," July 10, 1959, Public Relations Files, METRO Library Collection, PSRA; *ST,* March 6, July 29, August 4, 24, and September 2, 1958.

29. For the advertisement, see *P-I,* September 8, 1958; *ST,* September 8, 1958. For marches, see *West Seattle Herald,* August 28, 1958; *ST,* August 27, September 8, 1958; *P-I,* September 8, 1958. For pro-Metro editorials, see *Tacoma News-Tribune,* August 2, 1958; *ST,* May 19, September 8, 1958; and *P-I,* August 14, September 4, 1958. For anti-Metro editorials, see *Eastsider* (Bellevue), August 6, 27, 1958. For the staged photograph, see Boswell and McConaghy, *One Hundred Years of a Newspaper,* 34. For election returns, see *P-I,* September 9, 1958; "Official Canvass Book," September 9, 1958, King County Records and Elections Department, KCARM. For women and postwar environmentalism, see Adam Rome, "'Give Earth a Chance': The Environmental Movement and the Sixties," *Journal of American History* 90 (September 2003): 525–54.

30. *Municipality of Metropolitan Seattle v. City of Seattle, et al.* 57 Wn.2nd 452 (1960). For initial progress, see Roy W. Morse, "Metro Takes Action on the Sewage Problem," *American City* 74 (September 1959): 111–12.

31. *Renton Record-Chronicle,* August 16, 1961, July 18, 1962; Robert P. Hillis to Fred E. Lange, December 30, 1965, General Correspondence, 1958–75, box 1, METRO Library, PSRA. Ellis, a moderate Republican, took the charges of being a Communist fellow traveler hard. "There is often little distinction between organizations of the extreme right and of the extreme left," he confided to a friend. The only hope, in his mind, was "for citizens to bend every effort to make government work *at the local level.* This is the fundamental purpose of Metro." He concluded that opposition to Metro came "primarily from persons of extreme viewpoint," an opinion that does not entirely mesh with the historical record. See Ellis to William Goodloe, May 15, 1959, box 2, JRE Papers (emphasis in original).

32. *P-I,* February 19, 1963; *Bellevue American,* February 21, 1963. For the Century 21 fair, see John M. Findlay, *Magic Lands: Western Cityscapes and American Culture After 1940* (Berkeley: University of California Press, 1991), 228–39.

33. *P-I,* October 5, 1953; John Fischer, "Seattle's Modern-Day Vigilantes," *Harper's Magazine* (May 1969): 14–17; John T. Lehman, "Control of Eutrophication in Lake Washington," in *Ecological*

Knowledge and Problem Solving: Concepts and Case Studies, Committee on the Applications of Ecological Theory to Environmental Problems, Commission on Life Sciences, National Research Council (Washington, D.C.: National Academy of Sciences Press, 1986), 302. For the bet, see Hairston, "W. Thomas Edmondson," 6. For other accounts of the Metro "miracle," see Daniel Jack Chasan, "The Seattle Area Wouldn't Allow the Death of Its Lake," *Smithsonian* 2 (July 1971): 6–13; Roberto Brambilla and Gianni Longo, *What Makes Cities Livable?: Learning from Seattle,* ed. Ingrid Bengis (New York: Institute for Environmental Action, 1979); and J. R. McNeill, *Something New Under the Sun: An Environmental History of the Twentieth-Century World* (New York: W. W. Norton, 2000), 136–37. For Lake Erie, see William McGucken, *Lake Erie Rehabilitated: Controlling Cultural Eutrophication, 1960s–1990s* (Akron: University of Akron Press, 2000).

34. Ken McLeod to Roy Harris, Director, WPCC, May 9, 1964, Protests and Complaints of Pollution, 1963–66, box 48, WPCC. For fish kills, see *P-I,* March 15, 17, 23, April 4, 1960; *ST,* March 17, April 11, 1960; Alfred T. Neale to Gil Holland, February 29, 1960, and Neale to Allen, June 11, 1962, Central Files, box 1010-24, WSDF. For sewer complaints and Metro's response, see *P-I,* March 15, 17, and 25, 1960.

35. Milo Moore and Don Clarke, *Report on the Fisheries Resource of the Green-Duwamish River* (Olympia: Washington State Department of Fisheries and Washington State Department of Game, 1947).

36. B. M. Brennan to M. W. Bryner, March 1, 1939, Brennan to Thomas Voyce, September 26, 1941, Duwamish River–Green River File, 1930–47, box 744, WSDF. For the Cedar River, see Director of Fisheries and Director of Game (unsigned) to H. D. Towler, February 27, 1936, Brennan to W. C. Morse, September 1, 1939, and Morse to Brennan, September 12, 1939, Cedar River Watershed Collection, Cedar River Watershed Educational Center, Landsburg, Seattle Public Utilities.

37. Alexandra Harmon, *Indians in the Making: Ethnic Relations and Indian Identities Around Puget Sound* (Berkeley: University of California Press, 1999), 190–244 (quotation on 244); *P-I,* April 20, 1964. For Indian fishing, see Moore to Indian Agency Superintendents, Tribal Councils, Wholesale Fish Dealers, and Fisheries Inspectors, October 7, 1946, and Robert E. Robison "Indian Salmon Fisheries," September 15, 1962, Administrative Correspondence, 1945–49, Indian Affairs, box 1, WSDF.

38. Edward M. Mains, "The Relationship of Washington's Unregulated Indian Fisheries to the State's Salmon Management Program" (Address to Quarterly Meeting of the Washington State Sportsmen's Council, Port Angeles, September 15, 1962), and Wesley A. Hunter, Chief, Special Services Division, WSDG, "Indian Hunting and Fishing Privileges," May 17, 1961, Administrative Correspondence, 1945–49, Indian Affairs, box 1, WSDF.

39. Jack Schwabland to Albert D. Rosellini, September 24, 1963, Muckleshoot, 1942–63, box 22, WSDF.

40. Mains to Larry Conniff, Assistant Attorney General, September 13, 1963, Muckleshoot, 1942–63, box 22, WSDF; and Transcripts, *Washington v. James Starr, Louie Starr, and Leonard Wayne,* King County Superior Court Cause 37072, Case 23435, Attorney General Case Files, WSA. For Indian fishing and watershed changes around Puget Sound, see "The Indian Fisheries of Washington State," May 11, 1961, Robert Robison, "The Indian Fisheries," 1959, Administrative Correspondence, 1958–60, box 1, WSDF; see also Robert Robison, "The Muckleshoot Indian Fishery on the White River," October 1952, Muckleshoot, 1942–63, box 22, WSDF.

41. Minutes, Special Meeting of the Water Quality Monitoring Review Board, September 24, 1963, Water Quality Monitoring Review Board Quarterly Reports, 1963–72, acc. 2740, METRO Library Collection, PSRA.

42. C. Carey Donworth to WSDF, October 3, 1963, Pollution-Sewage-METRO File, 1959–67, box 1010-27, WSDF; and Neale to Miller, March 8, 1960, Pollution File, box 1010-24, WSDF. Neale's letter cited evidence dating back to 1954.

43. Bertha McJoe and Olive Hungary to Rosellini, February 27, 1964, Muckleshoot, 1942–63, box 22, WSDF. For the court decision, see *Washington v. McCoy,* 387 Pacific Reporter 2nd 942 (1963); Harmon, *Indians in the Making,* 229.

44. Richard Hugo, selections from "Duwamish Head" and "The Towns We Know and Leave Behind, the Rivers We Carry with Us," from *Making Certain It Goes On: The Collected Poems of Richard Hugo* (New York: W. W. Norton, 1984), 65, 68, 343. For earlier versions, see "Drift Fishing"

(draft) and "The Towns We Know [and Leave Behind], the Rivers [We Carry with Us]" (draft), box 4, fol. 11 and 49, Richard Hugo Papers, UWL-SC. For Hugo's childhood, see Hugo, *The Real West Marginal Way: A Poet's Autobiography,* ed. Ripley S. Hugo, Lois M. Welch, and James Welch (New York: W. W. Norton, 1986), 3–18.

45. For environmental pasts and citizenship, see Christopher Sellers, "Body, Place, and the State: The Makings of an 'Environmentalist' Imaginary in the Post–World War II U.S.," *Radical History Review* 74 (Spring 1999): 31–64; and Matthew S. Booker, "'Metro is *not only* sewers!': Environment and Politics in Postwar Metropolitan Seattle" (seminar paper, Stanford University, Department of History, 1999, copy in author's possession).

46. For anti-growth politics in the Northeast, see Peter Siskind, "Suburban Growth and Its Discontents: The Logic and Limits of Reform on the Postwar Northeastern Corridor," in *The New Suburban History,* ed. Kevin M. Kruse and Thomas J. Sugrue (Chicago: University of Chicago Press, 2006), 161–82.

Chapter 8. Masses of Self-Centered People: Salmon and the Limits of Ecotopia in Emerald City

1. Ernest Callenbach, *Ecotopia: The Notebooks and Reports of William Weston* (Berkeley: Banyan Tree, 1975; reprint, New York: Bantam, 1990), 10, 31, 47, 50.

2. Callenbach, *Ecotopia,* 39, 51, 64, 181.

3. Callenbach, *Ecotopia,* 107–10.

4. Miller Freeman to Henry Ramwell, American Packing Company, December 18, 1934, box 4, fol. 44, Miller Freeman Papers, UWL-SC. For halibut and sockeye fisheries, see Jozo Tomasevich, *International Agreements on Conservation of Marine Resources, with Special Attention to the North Pacific* (Stanford, Calif.: Food Research Institute, 1943), 125–215, 219–65; Joseph E. Taylor III, "The Historical Roots of the Canadian-American Salmon Wars," in *Parallel Destinies: Canadian-American Relations West of the Rockies,* ed. John M. Findlay and Ken S. Coates (Seattle: Center for the Study of the Pacific Northwest, 2002), 155–80; and John Thistle, "'As Free of Fish as a Billiard Ball Is of Hair': Dealing with Depletion in the Pacific Halibut Fishery, 1899–1924," *BC Studies* 142/143 (Summer/Autumn 2004): 105–25.

5. *Great Northern Daily News* (Seattle), January 15, 1921, Miller Freeman Papers, fol. 7, box 12, UWL-SC. For Freeman's biography, see David A. Neiwert, *Strawberry Days: How Internment Destroyed a Japanese American Community* (New York: Palgrave Macmillan, 2005), 22–23; and Robert F. Karolevitz, *Kemper Freeman, Sr., and the Bellevue Story* (Mission Hill, S.D.: Homestead, 1984), 3–66. The former book is a critical study; the latter little more than simple hagiography.

6. Miller Freeman, *The Memoirs of Miller Freeman, 1875–1955* (Seattle? 1956), 107–8, UWL-SC. For Freeman's anti-Japanese activities, see Neiwert, *Strawberry Days,* 48–69, 91–94, 111–40. There is no direct evidence that Freeman launched his tirades against the Japanese with the intention of taking their land, but he certainly benefited from his campaign.

7. *ST,* May 15, 1962; Rillmond Schear, "Seattle—1984," *Seattle Magazine* 2 (May 1965): 26–27; Schear, "The Great Big Boom Over in Bellevue," *Seattle Magazine* 2 (July 1965): 28, 47. For the Freemans' postwar real estate activities, see Neiwert, *Strawberry Days,* 205–17, 224–28; and Karolevitz, *Kemper Freeman,* 74–121. Population statistics come from the U.S. Bureau of the Census, *U.S. Census of Population, 1960,* vol. 1, *Characteristics of the Population, Part 49: Washington* (Washington, D.C.: Government Printing Office, 1963), 10, 17–18. The Standard Metropolitan Statistical Area for Seattle in 1960 encompassed both King and Snohomish Counties. For the bridge, see Payton Smith, *Rosellini: Immigrants' Son and Progressive Governor* (Seattle: University of Washington Press, 1997), 160–74. For Northgate, see Meredith L. Clausen, "Northgate Regional Shopping Center: Paradigm from the Provinces," *Journal of the Society of Architectural Historians* 48 (May 1984): 144–61. For *Seattle Magazine*'s politics, see Roger Sale, *Seattle: Past to Present* (Seattle: University of Washington Press, 1976), 212–15.

8. Ruth Wolf, "Freeways Under Fire," *Seattle Magazine* 6 (February 1969): 32; *ST,* June 1, 1961; *Afro-American Journal* (Seattle), May 28, 1970. For the first Forward Thrust, see Patrick Douglas, "Building a City: The Ins and Outs of Forward Thrust," *Seattle Magazine* 5 (January 1968): 29–32; and Walt Crowley, *Routes: An Interpretive History of Public Transportation in Metropolitan Seattle*

(Seattle: Crowley Associates, 1993), 43–62. For Interstate 5 construction, see Paul Dorpat and Genevieve McCoy, *Building Washington: A History of Washington State Public Works* (Seattle: Tartu, 1998), 62–103. For an excellent survey of downtown revitalization efforts, see Soyhun Park Lee, "From Redevelopment to Preservation: Downtown Planning in Postwar Seattle" (Ph.D. diss., University of Washington, 2001). For freeway protests, see Jeffrey Craig Sanders, "Inventing Ecotopia: Nature, Culture, and Urbanism, 1960–2000" (Ph.D. diss., University of New Mexico, 2005), 29–77; Margaret Cary Tunks, *Seattle Citizens Against Freeways, 1968–1980: Fighting Fiercely and Winning Sometimes* (Marina Del Rey, Calif.: By the author, 1996), UWL-SC; William Issel, "'Land Values, Human Values, and the Preservation of the City's Treasured Appearance': Environmentalism, Politics, and the San Francisco Freeway Revolt," *Pacific Historical Review* 68 (November 1999): 611–46; and Ari Kelman, *A River and Its City: The Nature of Landscape in New Orleans* (Berkeley: University of California Press, 2003), 197–221. For Emmett Watson, see *ST,* May 12, 2001; and *P-I,* May 12, 2001.

9. For the second Forward Thrust and billboard reference, see Crowley, *Routes,* 54–55. None of the bond initiatives met the 60 percent threshold, and the transit bonds received only 46 percent approval. For Boeing's crash, see "Appalachia in Seattle?" *Newsweek* 76 (August 17, 1970): 56–57; and T. M. Sell, *Boeing and the Politics of Growth in the Northwest* (Seattle: University of Washington Press), 26–27. For the 1972 freeway vote, see *ST,* February 9, 1972.

10. "Report on the Condition of the Spawning Tributaries of the Lake Washington Watershed," 1948, box 0061-23, Lake Washington–Cedar River, 1929–54, Stream Improvement and Hydraulics, WSDF; and Robert J. Schoettler and John A. Biggs to Charles J. Bartholet, August 19, 1953, box 0061-6, Duwamish River–Green River File, 1928–62, Stream Improvement and Hydraulics, WSDF. For a deft summary of regional habitat loss, see Alvin Anderson to Egil L. Peterson, August 19, 1949, box 1010-103, WSDF.

11. Lars Langloe, *Report on Development of Industrial Sites in the Duwamish–Green River Valley* (Seattle: City Planning Commission, 1946), 16–18, UWL-SC. For opposition to the dam, see Kenneth McLeod to Corps of Engineers, April 4, 1946, box 5, fol. 14, Kenneth McLeod Papers, UWL-SC. For salmon and dredging, see George C. Starlund and Biggs to King County Engineer, September 22, 1964, box 0061-23, Lake Washington–Cedar River, 1962–64, Stream Improvement and Hydraulics, WSDF.

12. "Statement of the Washington State Department of Fisheries Concerning Modification of the Lake Washington Ship Canal," June 1967, box 1010-103, Stream Improvement–Lakes–Washington–Ship Canal, 1955–1971, WSDF.

13. Thor C. Tollefson to Fred H. Weber, May 18, 1971; William H. Rodgers, Jr., to Tollefson, August 8, 1969, box 1010-103, Stream Improvement–Lakes–Washington–Thornton Creek, 1949–1971, WSDF; and Ed Sierer, "A Bold Proposal to Save Our Salmon," *Seattle Magazine* 2 (August 1965): 10.

14. *ST,* November 22, 1970. For sockeye plantings, see *Twenty-Eighth and Twenty-Ninth Annual Reports of the State Fish Commissioner, 1917–1919* (Olympia, Wash., 1919), 89, WSA; L. A. Royal and Allen Seymour, "Building New Salmon Runs," *Progressive Fish Culturist* 52 (1940): 1–7; Andrew Paul Hendry, "Sockeye Salmon (*Oncorhynchus nerka*) in Lake Washington: An Investigation of Ancestral Origins, Population Differentiation and Local Adaptation" (Ph.D. diss., University of Washington, 1995), 1–13, 135–39. The 1917 and 1935 or 1937 plantings were from the Baker Lake hatchery in northwest Washington; subsequent plantings in 1942–45 were from Baker Lake and Cultus Lake in British Columbia. The anadromous sockeye planted in Lake Washington were evolutionarily distinct from indigenous kokanee or landlocked sockeye observed by earlier fisheries scientists. The latter population of fish, if they were indeed kokanee, was likely trapped in the lake after the last major glacial period.

15. Edmondson, *The Uses of Ecology,* 38–43; Stephanie E. Hampton, Pia Romare, and David E. Seiler, "Environmentally Controlled Daphnia Spring Increase with Implications for Sockeye Salmon Fry in Lake Washington, USA," *Journal of Plankton Research* 28 (April 2006): 399–406. For dredging, see Starlund and Biggs to Winters, King County Engineer, September 22, 1964, box 0061-23, Lake Washington–Cedar River, 1962–64, Stream Improvement and Hydraulics, WSDF.

16. *ST,* November 22, 1970.

17. Valerie Bridges, as quoted in Charles Wilkinson, *Messages from Frank's Landing: A Story of*

Salmon, Treaties, and the Indian Way (Seattle: University of Washington Press, 2000), 43. For Indian fish fights, see *Uncommon Controversy: Fishing Rights of the Muckleshoot, Puyallup, and Nisqually Indians,* A Report Prepared for the American Friends Service Committee (Seattle: University of Washington Press, 1970); and Fay G. Cohen, *Treaties on Trial: The Continuing Controversy over Northwest Indian Fishing Rights,* A Report Prepared for the American Friends Service Committee, with contributions by Joan La France and Vivian L. Bowden (Seattle: University of Washington, 1986).

18. *U.S. v. Washington,* 384 F. Supp. 312 (1974), quotation at 407. For the trial and aftermath, see Wilkinson, *Messages from Frank's Landing,* 51–62; and Alexandra Harmon, *Indians in the Making: Ethnic Relations and Indian Identities Around Puget Sound* (Berkeley: University of California Press, 1999), 230–44.

19. For Wolf's work, see Susan Starbuck, *Hazel Wolf: Fighting the Establishment* (Seattle: University of Washington Press, 2002), 206–13, 310–13.

20. Hazel A. Wolf to "Friends" of the Western Federation of Outdoor Clubs, March 1, 1979, as quoted in Cohen, *Treaties on Trial,* 140–41; *P-I,* July 2, 1979. For pro-Indian editorials, see *ST,* July 3, 1979; *P-I,* July 5, 1979. For pro-white viewpoints, see *ST,* July 2, 1979; Cohen, *Treaties on Trial,* 116–17.

21. Interview by Ken Woo with George Boldt, KIRO-TV (Seattle), July 3, 1979, as quoted in Cohen, *Treaties on Trial,* 141; *ST,* July 2, 1979; Wilkinson, *Messages from Frank's Landing,* 62.

22. Quotation from a 1991 *P-I* editorial cartoon by David Horsey in Wilkinson, *Messages from Frank's Landing,* 92.

23. Hanley to L. Murray Grant, May 17, 1928, box 2, fol. 18, SWD-HF; *ST,* August 20, 1928, April 27, 1930. For the newspaper series, plus editorials, see *ST,* April 20, 22, 24, 27, 29, May 4, 8, 11, 14, 18, 23, 25; June 16, 1930. For reforestation, see Winkenwerder to Russell, September 27, 1923, box 46, Cedar River Subject Series, College of Forest Resources Records, UWL-SC; Hugo Winkenwerder and Allen E. Thompson, *Report on the Reforestation of the Cedar River Watershed* (Seattle: Water Department, 1924), UWL-SC. For reforestation efforts during the 1920s, see "Forestry Activities on the Cedar River Watershed during 1926 and 1927," March 6, 1928, box 46, Cedar River Subject Series, College of Forest Resources Records, UWL-SC. For lumber camps and towns, see Hanley to Winkenwerder, March 8, 1927, box 46, College of Forest Resources Records, UWL-SC; and Linda Carlson, *Company Towns of the Pacific Northwest* (Seattle: University of Washington Press, 2003), 44–47, 65–66, 165–66, 174, 213. For litigation, see Memo Decision, cause 8162, Kittitas County Superior Court, Civil Case Files, box 122, Central Regional Branch, Ellensburg, WSA; and *Seattle v. Pacific States Lumber Company,* 166 Wash. 517 (1932). For the timber economy, see Robert E. Ficken, *The Forested Land: A History of Lumbering in Western Washington* (Seattle: University of Washington Press/Durham, N.C.: Forest History Society, 1987), 182–224.

24. *104 Reporter,* October 21, October 28, 1943; James Scavotto to Community Clubs in Seattle, August 5, 1943 [copy], box 2, fol. 1, MBSNF; and Carl E. Green, Bror L. Grondal, and Abel Wolman, *Report on the Water Supply and the Cedar River Watershed* (Seattle: Cedar River Watershed Commission, February 1944), 58, UWL-SC. On the timber-for-land deal, see U. M. Dickey to W. E. Holt, October 31, 1941, box 1, file 199-1/2, Northern Pacific Railway Corporate Records, Land Department, Minnesota Historical Society. For sustained-yield management, see "Logging Agreement Between City of Seattle, Northern Pacific Railway Company, Soundview Pulp Company, Weyerhaeuser Timber Company, and Cascade Timber Company," Ordinance 74105, SCCF. For missed quotas, see Allen E. Thompson, "Trees and Water: A Dual Crop" (speech, Western Forestry and Conservation Association Annual Meeting, November 28–30, 1951), box 1, fol. 42, SWD-HF.

25. Diana Gale to Margaret Pageler, April 3, 1996, box 83, fol. 1, MP Files.

26. Notes on HCP Briefing to City Council, March 12, 1997, box 75, fol. 2, MP Files. For human activity, see Lynne M. Getz, *Cedar River Watershed Cultural Resource Study* (Seattle: City Water Department, 1987), UWL-SC.

27. For forest structure, see *Cedar River Watershed Habitat Conservation Plan for the Proposed Issuance of a Permit to Allow Incidental Take of Threatened and Endangered Species* (Seattle: April 2000), 3.3-1-3.5, SMA. For an earlier assessment, see L. A. Nelson, "Report on Cruise and Appraisal," July 21, 1913, Alpha Files, box 1, MBSNF.

28. For the old-growth controversy, see William Dietrich, *The Final Forest: The Battle for the*

Last Great Trees of the Pacific Northwest (New York: Penguin, 1992); James D. Proctor, "Whose Forests?: The Contested Moral Terrain of Ancient Forests," in *Uncommon Ground: Toward Reinventing Nature,* ed. William Cronon (New York: W. W. Norton, 1995), 269–97; and Bruce Braun, *The Intemperate Rainforest: Nature, Culture, and Power on Canada's West Coast* (Minneapolis: University of Minnesota Press, 2002).

29. Charles C. Raines to Stan Moses, May 6, 1997 (copy), box 75, fol. 1, MP Files.

30. Moses to David Mann, April 4, 1997 (copy), box 75, fol. 4, MP Files. For Indians' environmentalism, see *ST,* August 20, 1982, January 25, 1989; *P-I,* October 2, 1988; and Coll Thrush, "City of the Changers: Indigenous People and the Transformation of Seattle's Watersheds," *Pacific Historical Review* 75 (February 2006): 112–17.

31. Virginia Cross to Bridgett Chandler, November 1, 1996; Moses to Mann, April 4, 1997, box 75, fol. 1, MP Files.

32. Frank Urabeck to Seattle City Council, July 7, 1999, box 30, fol. 5, NL Files; John C. Evensen to Jan Drago, March 1, 1998, box 27, fol. 6, Jan Drago Files, SMA; Mark Scalzo to Schell, February 1, 1999, box 90, fol. 5, MP Files.

33. Jo Ostgarden, "Fish, Trees, and Clean Water," *Cascadia Times* (July–August 1998): 3–4; *ST,* April 20, June 23, 1998; "Save Your Watershed for One Latté a Year" (c. 1997), box 3, fol. 11, PS Files; and Schell to Pageler, February 10, 1998, box 13, fol. 4, NL Files.

34. Paul Schell, "Cascadia: The North Pacific West" (Paper presented at the meeting of the North American Institute, Seattle, October, 1992), 4, as quoted in John M. Findlay, "A Fishy Proposition: Regional Identity in the Pacific Northwest," in *Many Wests: Place, Culture, and Identity,* ed. David M. Wrobel and Michael C. Steiner (Lawrence: University Press of Kansas, 1997), 60; see also Paul Schell and John Hamer, "Cascadia: The New Binationalism of Western Canada and the U.S. Pacific Northwest," in *Identities in North America: The Search for Community,* ed. Robert L. Blake and John D. Wirth (Stanford: Stanford University Press, 1995), 140–56. For Cascadia's history, see Janet Johnson, "Cascadia: It's as Much a State of Mind as a Geographic Place," *Seattle University News* (Spring 1994): 21–23. For bioregionalism, see Dan Flores, "Place: An Argument for Bioregional History," *Environmental History Review* 18 (Winter 1994): 1–18; and William L. Lang, "Beavers, Firs, Salmon, and Falling Water: Pacific Northwest Regionalism and the Environment," *Oregon Historical Quarterly* 104 (Summer 2003): 151–65. For a pointed critique of bioregionalism, focused on the American West, see Joseph E. Taylor III, "The Many Lives of the New West," *Western Historical Quarterly* 35 (Summer 2004): 141–60. In a 2005 interview, Ernest Callenbach defended the idea of bioregionalism, explaining how the region encompassed in his imaginary nation of Ecotopia had consistent climate, similar biota, and "a certain geographical unity." "And my contention," Callenbach explained, "as well as that of a lot of professional geographers, is that in the long run the characteristics of your bioregion help to determine what you might call your regional character." Left unstated in his interview was how this argument, like those made by other advocates of bioregionalism, played the trump card of an idealized nature against the complexity of both human and natural history, a position that most professional historians and geographers would roundly reject. Callenbach also did not explain how in his novel *Ecotopia* those living the closest to a pure nature were the whitest as well. See *Seattle Weekly,* March 23, 2005. In many ways, the border between Washington and British Columbia was becoming firmer even before September 11, 2001. See Carl Abbott, "That Long Western Border: Canada, the United States, and a Century of Economic Change," in *Parallel Destinies,* ed. Findlay and Coates, 203–17.

35. John Daniels, Jr., to Schell, April 20, 1999, box 90, fol. 5, MP Files; Susan Bolton et al. to Schell, February 18, 1999, box 3, fol. 11, PS Files; Diana Spencer to Seattle City Council, June 15, 1999, box 30, fol. 5, NL Files. For media endorsements, see *P-I,* December 15, 1998; *ST,* December 18, 1998; *Seattle Weekly,* December 9, 1998; and *The Stranger* (Seattle), December 31, 1998.

36. *ST,* March 14, 1999. For American Rivers, Pageler and Schell to Rebecca Wodder, April 9, 1999, box 3, fol. 13, PS Files. For utility districts, Ron Speer to Schell, July 21 and April 22, 1998, box 90, fol. 5, MP Files.

37. Rick Polich to Pageler, February 16, 1998, box 13, fol. 7, NL Files; Maxine Keesling to Editor, *P-I* (copy), May 25, 1998, box 90, fol. 5, MP Files.

38. For the various changes to the habitat conservation plan, see Resolution 29977 (July 12, 1999) and Resolution 30091 (December 6, 1999), SCCF.

39. Robert F. Roth to Pageler, June 2, 1999, box 105, fol. 15, Legislative Department, Central Staff Analysts' Working Papers, SMA; *New York Times,* April 19, 1998.

40. *New York Times,* September 26, 1996. For REI's flagship store, see *P-I,* September 6, 1996; *ST,* August 19, 1996. For overseas manufacturing, see *ST,* March 24, 1996, May 14, 2000; and Andy Ryan, "Who Owns REI?" *Seattle Weekly* (June 18, 2003): 15–17.

41. *New York Times,* August 18, 2000; Jonathan Raban, "The New Last Frontier," *Harper's Magazine* 287 (August 1993): 48; *P-I,* May 11, 1982; *ST,* May 21, 2000. For other stories about Seattle's rise, see Amy Tucker and Stephen Tanzer, "All of the Best and None of the Worst of Seattle," *Forbes* 157 (May 8, 1995), 141–50; Jerry Alder, "Seattle Reigns," *Newsweek* 127 (May 20, 1996), 48–53; and Geoffrey Precourt and Anne Faircloth, "Best Cities: Where the Living Is Easy," *Fortune* 134 (November 11, 1996), 126–34.

42. Emily Baillargeon Russin, "Seattle Now: A Letter," in *Reading Seattle: The City in Prose,* ed. Peter Donahue and John Trombold (Seattle: University of Washington Press, 2004), 250; Raban, *Hunting Mister Heartbreak: A Discovery of America* (New York: Edward Burlingame, 1991), 299–300. For consumerism, see Andrew Kirk, "Appropriating Technology: *The Whole Earth Catalog* and Counterculture Environmental Politics," *Environmental History* 6 (July 2001): 374–94. For redevelopment, see Sohyun Park Lee, "From Redevelopment to Preservation: Downtown Planning in Post-war Seattle" (Ph.D. diss., University of Washington, 2001). For activism, see Jeffrey Craig Sanders, "Building an 'Urban Homestead': Survival, Self-Sufficiency, and Nature in Seattle, 1970–1980" (paper presented at "The Place of Nature in the City in Twentieth-Century Europe and North America," German Historical Institute, Washington, D.C., December 3, 2005, copy in author's possession).

43. For the Commons, see *Seattle Commons/South Lake Union Plan: Final Environmental Impact Statement.* 4 vols. (Seattle: City of Seattle, Office of Management and Planning, 1995); *ST,* May 16, 1996. For Paul Allen, see *ST,* February 4, 2005; *P-I,* February 4, 2005.

44. Fred Moody, *Seattle and the Demons of Ambition: From Boom to Bust in the Number One City of the Future* (New York: St. Martin's, 2004), 11; Seattle City Council, *Report of the WTO Accountability Review Committee* (Seattle, September 14, 2000), 3; "Answering WTO's Big Questions," *Seattle Weekly* (August 3, 2000): 20. Moody's book is a wonderful, if highly personal, account of Seattle's meteoric rise during the 1980s and 1990s.

45. Kristine Wong, interview by Monica Ghosh, July 28, 2000, WTO History Project, http://depts.washington.edu/wtohist/ (accessed May 2, 2005). For incinerator protests, see *P-I,* September 26, 1998.

46. For South Park, see *P-I,* October 3, 1994; *ST,* February 24, 1994; Ian Ith, "The Road Back: From Seattle's Superfund Sewer to Haven Once More," *Pacific Northwest: The Seattle Times Magazine* (October 2, 2004): 22–29; and *ST,* September 18, 2005. For persistent pollutants, see Gary W. Isaac, Glen D. Farris, and Charles V. Gibbs, *Special Duwamish River Studies,* Water Quality Series No. 1 (Seattle: Municipality of Metropolitan Seattle, February 1964); J. F. Santos and J. D. Stoner, *Physical, Chemical, and Biological Aspects of the Duwamish River Estuary, King County, Washington, 1963–67,* Geological Survey Water-Supply Paper 1873-C (Washington, D.C.: U.S. Government Printing Office, 1972). For contaminated fish, see *P-I,* July 30, 2000, June 24, 2002, January 15, 2004, and October 27, 2006; see also "Final Report, State Board of Health Priority: Environmental Justice" (Olympia: Committee on Environmental Justice of the Washington State Board of Health, June 13, 2001).

47. Quotation from *Puget Sound Business Journal* (September 25–October 1, 1998): 33. For details listed here, see *P-I,* April 21, 2000; *ST,* July 25, 1999, August 9, 1999, May 28, 2000.

48. Joseph E. Taylor III, *Making Salmon: An Environmental History of the Northwest Fisheries Crisis* (Seattle: University of Washington Press, 1999), 3–12; *ST,* August 29, 2000.

49. *P-I,* January 21, April 13, 16, November 28, 1987, July 28, 1993, November 23, 1989, September 1, 1993, February 27, 1995; *ST,* December 22, 1994; and William Dietrich, "Conspicuous Consumers," *Pacific Northwest: The Seattle Times Magazine* (August 5, 2001): 9–17.

50. *ST,* November 30, 1999; and Joseph E. Taylor III, "El Niño and the Politics of Blame," *Western Historical Quarterly* 4 (Winter 1998): 437–57. For salmon aquaculture, see S. Einum and I. A. Fleming, "Genetic Divergence and Interactions in the Wild Among Native, Farmed, and Hybrid

Atlantic Salmon," *Journal of Fish Biology* 50 (1997): 634–51; and "Interactions of Atlantic Salmon in the Pacific Northwest Environment," *Fisheries Research* 62 (June 2003): 235–347.

51. *P-I,* March 14, 1999; *ST,* August 17, 1999; *P-I,* March 14, 1999; Patrick Mazza and Eben Fodor, "Sprawl: Threatened Salmon, Staggering Costs," *Seattle Free Press* (June 14–27, 2000): 19; see also *New York Times,* March 17, 1999.

52. James Fallows, "Saving Salmon, or Seattle?" *Atlantic Monthly* 286 (October 2000); 20–26; see also Taylor, *Making Salmon,* 237–57.

53. *U.S. v. Washington,* 873 F. Supp. 1422 (1994); *U.S. v. Washington,* 898 F. Supp. 1453 (1995); *U.S. v. Washington,* 909 F. Supp. 787 (1995); and 157 F.3d 630 (9th Cir. 1998). For other accounts of legal developments, see *P-I,* December 21, 1994, February 1, March 4, May 9, 1995, January 29, 1998; *ST,* February 15, 1998, April 5, 1999.

54. O. Yale Lewis III, "Treaty Fishing Rights: A Habitat Right as Part of the Trinity of Rights Implied by the Fishing Clause of the Stevens Treaties," *American Indian Law Review* 27 (2002–3): 311; *P-I,* January 21, 2001. For the relevant cases, see *U.S. v. Washington,* 443 U.S. 658 (1979); *U.S. v. Washington,* 506 F. Supp. 187 (1980); and *U.S. v. Washington,* 759 F. 2d. 1353 (9th Cir. 1985). See also Martin H. Belsky, "Indian Fishing Rights: A Lost Opportunity for Ecosystem Management," *Journal of Land Use and Environmental Law* 12 (Fall 1996): 45–62.

55. *P-I,* April 10, 2006; for details on the accord, see *P-I,* March 28, April 22, and November 4, 2006.

56. *P-I,* October 1, 2003, March 30, 2004.

57. Quotation from "Chinook in the City: Restoring and Protecting Salmon Habitat in Seattle" (Seattle, 2004), SMA. This was a pamphlet version of the larger plan, *Seattle's Blueprint for Habitat Protection and Restoration* (Seattle: City of Seattle, December 2003), SMA.

58. *P-I,* March 6, 2006; *Seattle Weekly,* July 19, 2006. In the latest twist, many homeowners and businesses claimed that storm drain screens designed to keep sediment and debris out of salmon streams worsened flooding during disastrous winter storms in late 2006. See *P-I,* December 29, 2006.

59. *New York Times,* March 5, 2006; see also *ST,* April 26, 2005, June 11, November 6, 17, 2006; *P-I,* October 31, 2006.

60. *P-I,* April 6, 2005. For road building, see *Puget Sound Business Journal,* July 9, 1999; *King County Journal,* April 16, 2004.

61. *P-I,* April 9, 2003, April 6, 2006, December 25, 2006; and Frank Colich, "Trace Metals Concentrations in Storm Water Runoff from the Evergreen Point Floating Bridge in the Seattle Washington Area" (Poster presented at the 2003 Puget Sound/Georgia Basin Research Conference, copy in author's possession). Many researchers have found that the pharmaceuticals and other organic compounds mentioned here can disrupt the endocrine system, which produces and regulates hormones vital to the metabolism, growth, and development of humans and other animals. For a thought-provoking overview of this science and its implications for environmental historians, see Nancy Langston, "Gender Transformed: Endocrine Disruptors in the Environment," in *Seeing Nature Through Gender,* ed. Virginia Scharff (Lawrence: University Press of Kansas, 2003): 129–66.

62. D. H. Baldwin, J. F. Sandahl, J. S. Labenia, and N. L. Scholz, "Sublethal Effects of Copper on Coho Salmon: Impacts on Non-overlapping Receptor Populations in the Peripheral Olfactory Nervous System," *Environmental Toxicology and Chemistry* 22 (2003): 2266–74; J. P. Incardona, T. K. Collier, and N. L. Scholz, "Defects in Cardiac Function Precede Morphological Abnormalities in Fish Embryos Exposed to Polycyclic Aromatic Hydrocarbons," *Toxicology and Applied Pharmacology* 196 (2004): 191–205; and J. F. Sandahl, D. H. Baldwin, J. J. Jenkins, and N. L. Scholz, "Comparative Thresholds for Acetylcholinesterase Inhibition and Behavioral Impairment in Coho Salmon Exposed to Chlorpyrifos," *Environmental Toxicology and Chemistry* 24 (2005): 136–45.

63. Climate Impacts Group, University of Washington, *Uncertain Future: Climate Change and Its Effects on Puget Sound* (Olympia, Wash.: Puget Sound Action Team, October 2005), www.psat.wa.gov/Publications/climate_change2005/climate_home.htm (accessed January 15, 2006); *P-I,* February 6, 2003; James Battin, Matthew W. Wiley, Mary H. Ruckelshaus, Richard N. Palmer, Elizabeth Korb, Krista K. Bartz, and Hiroo Imaki, "Projected Impacts of Climate Change on Salmon Habitat Restoration," *Proceedings of the National Academy of Sciences of the United States of America*

104 (April 17, 2007): 6720–25. For restoration science, see *Restoration of Puget Sound Rivers,* ed. David R. Montgomery, Susan Bolton, Derek B. Booth, and Leslie Wall (Seattle: Center for Water and Watershed Studies in association with University of Washington Press, 2003).

64. Callenbach, *Ecotopia,* 157, 165.

Epilogue. The Geography of Hope: Toward an Ethic of Place and a City of Justice

1. Francesca Lyman, "Restoring Nature, Restoring Yourself," *YES! A Journal of Positive Futures* 25 (Spring 2003): 36–37.

2. *P-I,* November 22, 2002. For Beal's work, see also Ingrid Lobet, "Salmon's Best Friend," on *Living on Earth,* National Public Radio, February 4, 2000; *P-I,* April 2, 2002; and Ian Ith, "The Road Back: From Seattle's Superfund Sewer to Haven Once More," *Pacific Northwest: The Seattle Times Magazine* (October 2, 2004): 22–29.

3. For environmentalism as religion, see Thomas Dunlap, *Faith in Nature: Environmentalism as Religious Quest* (Seattle: University of Washington Press, 2004); and Mark Stoll, *Protestantism, Capitalism, and Nature in America* (Albuquerque: University of New Mexico Press, 1997). For appraisals of nature writing, see Lawrence Buell, *Writing for an Endangered World: Literature, Culture, and Environment in the U.S. and Beyond* (Cambridge: Belknap Press of Harvard University Press, 2001), and *The Future of Environmental Criticism: Environmental Crisis and Literary Imagination* (Malden, Mass.: Blackwell, 2005); Michael P. Cohen, "Blues in the Green: Ecocriticism Under Critique," *Environmental History* 9 (January 2004): 9–36; Jennifer Price, "Thirteen Ways of Seeing Nature in L.A.," in *Land of Sunshine: An Environmental History of Metropolitan Los Angeles,* ed. William Deverell and Greg Hise (Pittsburgh: University of Pittsburgh Press, 2005), 220–44; and Richard White, "The Natures of Nature Writing," *Raritan* 22 (Fall 2002): 145–61.

4. Michael Shellenberger and Ted Nordhaus, *The Death of Environmentalism: Global Warming Politics in a Post-Environmental World* (August 2004), copy at www.thebreakthrough.org/ (accessed December 10, 2006); Michel Gelobter et al., *The Soul of Environmentalism: Rediscovering Transformational Politics in the 21st Century* (August 2004), 10, copy at www.rprogress.org/soul/ (accessed December 10, 2006). For two assessments by historians, see Aaron Sachs, *The Humboldt Current: Nineteenth-Century Exploration and the Roots of Modern Environmentalism* (New York: Viking, 2006), 338–53; and Richard White, "On Rediscovering Our Inner Empiricist," *Raritan* 26 (Fall 2006): 62–78.

5. For a critique of purity, see Richard White, "The Problem with Purity," Tanner Lecture on Human Values, delivered at the University of California at Davis, May 10, 1999, available at www.tannerlectures.utah.edu/ (accessed November 20, 2006). For "metronatural" Seattle, see *P-I,* October 21, 2006; *ST,* October 21, 2006; and the Seattle Convention and Visitors Bureau Web site at www.metronatural.com/ (accessed December 24, 2006).

6. Richard Hugo, *The Real West Marginal Way: A Poet's Autobiography,* ed. Ripley S. Hugo, Lois M. Welch, and James Welch (New York: W. W. Norton, 1986), 9, 14.

7. *P-I,* August 25, 2005. According to the report, Seattle's citywide income average was $45,736, and the average percentage of minorities in given neighborhoods was 30 percent. For the 1994 comparison, see *P-I,* October 3, 1994; *ST,* February 24, 1994.

8. *ST,* April 4, 2004.

9. Henri Lefebvre, *The Production of Space,* trans. Donald Nicholson-Smith (Cambridge, Mass.: Blackwell, 1991), 51; Jane Jacobs, *The Death and Life of Great American Cities,* rev. ed., Modern Library (New York: Random House, 1993), 583.

10. For Indians and resource exploitation, see Shepard Krech III, *The Ecological Indian: Myth and History* (New York: W. W. Norton, 1999).

11. John M. Findlay, "A Fishy Proposition: Regional Identity in the Pacific Northwest," in *Many Wests: Place, Culture, and Identity,* ed. David M. Wrobel and Michael C. Steiner (Lawrence: University Press of Kansas, 1997), 40.

12. Aldo Leopold, *A Sand County Almanac, with Essays from Round River* (New York: Oxford University Press, 1966; reprint, New York: Ballantine, 1970), 239–40, 243, 253–58, 63; Joseph E. Tay-

lor III, *Making Salmon: An Environmental History of the Northwest Fisheries Crisis* (Seattle: University of Washington Press, 1999), 237–57.

13. For critiques of Leopold's ideas, see J. Baird Callicott, *Beyond the Land Ethic: More Essays in Environmental Philosophy* (Albany: State University of New York Press, 1999). For an overview of changes within ecology, see Daniel Botkin, *Discordant Harmonies: A New Ecology for the Twenty-first Century* (New York: Oxford University Press, 1990).

14. John Rawls, *A Theory of Justice,* rev. ed. (Cambridge: Belknap Press of Harvard University Press, 1999), and *Justice as Fairness: A Restatement,* ed. Erin Kelly (Cambridge: Harvard University Press, 2001). What I phrase as a "presumption of innocence" Rawls terms, famously, as the "veil of ignorance."

15. Kwame Anthony Appiah, *Cosmopolitanism: Ethics in a World of Strangers* (New York: W. W. Norton, 2006). For the local being global, see Appiah, *The Ethics of Identity* (Princeton: Princeton University Press, 2005), 213–72.

16. David Harvey makes this point powerfully in *Spaces of Hope* (Berkeley: University of California Press, 2000), 111–13, and *Justice, Nature, and the Geography of Difference,* 176–204, 329–438. For "nature-deficit disorder," see Richard Louv, *Last Child in the Woods: Saving Our Children from Nature-Deficit Disorder* (Chapel Hill, N.C.: Algonquin Books of Chapel Hill, 2005).

17. For politics and class, see Karl Jacoby, *Crimes Against Nature: Squatters, Poachers, Thieves, and the Hidden History of American Conservation* (Berkeley: University of California Press, 2001); Benjamin Heber Johnson, "Conservation, Subsistence, and Class at the Birth of Superior National Forest," *Environmental History* 4 (January 1999): 80–99; Richard Judd, *Common Lands, Common People: The Origins of Conservation in Northern New England* (Cambridge: Harvard University Press, 1997); and Jennifer Price, *Flight Maps: Adventures with Nature in Modern America* (New York: Basic, 1999). For consumption, see Susan G. Davis, *Spectacular Nature: Corporate Culture and the Sea World Experience* (Berkeley: University of California Press, 1997); James Morton Turner, "From Woodcraft to 'Leave No Trace': Wilderness, Consumerism, and Environmentalism in Twentieth-Century America," *Environmental History* 7 (July 2002): 462–84; Bruce Stadfeld, "Electric Space: Social and Natural Transformation in British Columbia's Hydroelectric Industry to World War II" (Ph.D. diss., University of Manitoba, 2002); Matthew Klingle, "Spaces of Consumption in Environmental History," *History and Theory* 42 (December 2003): 94–110; and Ramachandra Guha, *How Much Should a Person Consume?: Environmentalism in India and the United States* (Berkeley: University of California Press, 2006). For environmental justice, see Martin V. Melosi, "Environmental Justice, Political Agenda Setting, and the Myths of History," *Journal of Policy History* 12 (2000): 43–71.

18. Raymond Williams, *Problems in Materialism and Culture: Selected Essays* (London: Verso, 1980), 84–85. For another take on what could be construed as ethic of place, see Robert Gottlieb, *Environmentalism Unbound: Exploring New Pathways for Change* (Cambridge: MIT Press, 2001), 273–87.

19. Andrew Light, "Elegy for a Garden: Thoughts on an Urban Environmental Ethic," *Philosophical Writings* 14 (Summer 2000): 47. For civic environmentalism, see DeWitt John, *Civic Environmentalism: Alternatives to Regulation in States and Communities* (Washington, D.C.: Congressional Quarterly Press, 1993).

20. Light, "Contemporary Environmental Ethics: From Metaethics to Public Philosophy," *Metaphilosophy* 33 (July 2002): 449; see also "Urban Ecological Citizenship," *Journal of Social Philosophy* 34 (Spring 2003): 44–63. For pragmatism and environmental politics, see Ben A. Minteer, *The Landscape of Reform: Civic Pragmatism and Environmental Thought in America* (Cambridge: MIT Press, 2006), 153–98. For one geographer's analysis of the ethics of place, see Robert D. Sack, *Homo Geographicus: A Framework for Action, Awareness, and Moral Concern* (Baltimore: Johns Hopkins University Press, 1997), and *A Geographical Guide to the Real and the Good* (New York: Routledge, 2003). Many scholars have vehemently critiqued using nature to justify social relations or ethical systems, a point made best by Anthony Ross, *The Chicago Gangster Theory of Life: Nature's Debt to Society* (New York: Verso, 1994). But as I have argued here, an attention to history works against reifying nature as something that transcends human history. On the relevance of historians and their ethical responsibilities, while I do not agree with many of his positions, I find much to

admire in Thomas L. Haskell, *Objectivity Is Not Neutrality: Explanatory Schemes in History* (Baltimore: Johns Hopkins University Press, 1998). For another set of provocative essays on the subject, see Richard Wightman Fox and Robert B. Westbrook, eds., *In the Face of the Facts: Moral Inquiry in American Scholarship* (New York and Washington D.C.: Cambridge University Press and Woodrow Wilson Center Press, 1998).

21. For Bush's statement, see Ted Steinberg, "A Natural Disaster, and a Human Tragedy," *Chronicle of Higher Education* (September 23, 2005): B12. For the environmental history of New Orleans preceding the 2005 storms, see Ari Kelman, *A River and Its City: The Nature of Landscape in New Orleans* (Berkeley: University of California Press, 2003); and Craig E. Colten, *An Unnatural Metropolis: Wresting New Orleans from Nature* (Baton Rouge: Louisiana State University Press, 2005).

22. For regrades and tsunamis, see Robert D. Loeffler, Mark L. Holmes, and Richard E. Sylwester, "In Search of the Denny Regrade: Fate of a Large Spoil Bank in Elliott Bay, Puget Sound," and Richard E. Sylwester and Mark L. Holmes, "Marine Geophysical Evidence of a Recent Submarine Slope Failure in Puget Sound Washington," in *Oceans '89: Proceedings from the Seattle Meeting, September 18–21, 1989* (Washington, D.C.: Institute of Electrical and Electronic Engineers), 84–89, 1524–29; and *P-I,* January 22, 2001. For volcanic eruptions, see R. P. Hoblitt, J. S. Walder, C. L. Driedger, K. M. Scott, P. T. Pringle, and J. W. Vallance, *Volcano Hazards from Mount Rainier, Washington,* Open-File Report 98-428 (Washington, D.C.: U.S. Geological Survey, 1998). For the environmental fragility of cities and the social conflicts it entails, see Mike Davis, *Ecology of Fear: Los Angeles and the Imagination of Disaster* (New York: Metropolitan, 1998).

23. For restoration, see William R. Jordan III, *The Sunflower Forest: Ecological Restoration and the New Communion with Nature* (Berkeley: University of California Press, 2003), 28–53; see also Andrew Light, "Restoring Ecological Citizenship," in *Democracy and the Claims of Nature,* ed. Ben A. Minteer and Bob Pepperman Taylor (Lanham, Md.: Rowman & Littlefield, 2002), 153–72. For restoring Puget Sound, see *P-I,* October 23, 2006. For criticisms of the initial plan offered by the state government by several prominent scientists, see *P-I,* November 3, 2006. For an excellent history that explores the complexities of restoration over time in the Malheur Basin in southeastern Oregon, see Nancy Langston, *Where Land and Water Meet: A Western Landscape Transformed* (Seattle: University of Washington Press, 2003).

24. Leopold, *A Sand County Almanac,* 251; Wallace Stegner, *The Sound of Mountain Water* (Garden City, N.Y.: Doubleday, 1969; reprint, Lincoln: University of Nebraska Press, 1985), 153. For selective environmentalism, see William R. Jordan, III, "Ten Thousand Thoreaus," *Ecological Restoration* 18 (Winter 2000): 215; and William Cronon, "The Riddle of the Apostle Islands: How Do You Manage a Wilderness Full of Human Stories," *Orion Magazine* (May-June 2003), 36–42.

25. For gardens and neighborhoods, see Jeffrey Craig Sanders, "Inventing Ecotopia: Nature, Culture, and Urbanism, 1960–2000" (Ph.D. diss., University of New Mexico, 2005), 171–229, 387–93. For the city-tribal government accord, see *P-I,* July 21, 2004. For climate policy, see Mayor's Green Ribbon Commission on Climate Protection, "Seattle: A Climate of Change: Meeting the Kyoto Challenge," March 2006, www.ci.seattle.wa.us/climate/ (accessed June 25, 2006). For the Shared Salmon Strategy, see the Web site at www.sharedsalmonstrategy.org/ (accessed November 15, 2006). For a recent documentary on these and other changes to the region's political culture, see *Edens Lost and Found: Seattle, The Future Is Now,* DVD, directed by Harry Wiland (Oley, Pa.: Bullfrog Films, 2007).

26. *P-I,* April 2, November 22, 2002. For a retrospective of Beal's work, see *P-I,* June 26, 2006.

27. L. Frank Baum, *The Annotated Wizard of Oz: The Wonderful Wizard of Oz* (New York: W. W. Norton, 2000), 4.

Index

Page references in italics refer to maps or illustrations